THE BODY OF THE COMBATANT IN THE ANCIENT MEDITERRANEAN

Also available from Bloomsbury

AFTER THE CRISIS: REMEMBRANCE, RE-ANCHORING AND RECOVERY IN ANCIENT GREECE AND ROME
edited by Jacqueline Klooster and Inger N. I. Kuin

MILITARY DEPARTURES, HOMECOMINGS AND DEATH IN CLASSICAL ATHENS
by Owen Rees

THE TREATMENT OF THE WAR DEAD IN ARCHAIC ATHENS: AN ANCESTRAL CUSTOM
by Cezary Kucewicz

TRAGIC BODIES: EDGES OF THE HUMAN IN GREEK DRAMA
by Nancy Worman

THE BODY OF THE COMBATANT IN THE ANCIENT MEDITERRANEAN

Edited by Hannah-Marie Chidwick

BLOOMSBURY ACADEMIC
LONDON • NEW YORK • OXFORD • NEW DELHI • SYDNEY

BLOOMSBURY ACADEMIC
Bloomsbury Publishing Plc
50 Bedford Square, London, WC1B 3DP, UK
1385 Broadway, New York, NY 10018, USA
29 Earlsfort Terrace, Dublin 2, Ireland

BLOOMSBURY, BLOOMSBURY ACADEMIC and the Diana logo are trademarks of Bloomsbury Publishing Plc

First published in Great Britain 2024

A catalogue record for this book is available from the British Library.

Library of Congress Cataloging-in-Publication Data
Names: Chidwick, Hannah-Marie, editor.
Title: The body of the combatant in the ancient Mediterranean / edited by Hannah-Marie Chidwick.
Description: London ; New York : Bloomsbury Academic, 2024. |
Includes bibliographical references and index.
Identifiers: LCCN 2023058347 (print) | LCCN 2023058348 (ebook) |
ISBN 9781350240858 (hardback) | ISBN 9781350240865 (paperback) |
ISBN 9781350240872 (pdf) | ISBN 9781350240889 (ebook)
Subjects: LCSH: Military history, Ancient | Soldiers–Mediterranean Region–History–To 1500. |
Mediterranean Region–History, Military.
Classification: LCC U29 .B64 2024 (print) | LCC U29 (ebook) |
DDC 355/.020937—dc23/eng/20240222
LC record available at https://lccn.loc.gov/2023058347
LC ebook record available at https://lccn.loc.gov/2023058348

ISBN: HB: 978-1-3502-4085-8
ePDF: 978-1-3502-4087-2
eBook: 978-1-3502-4088-9

Typeset by RefineCatch Limited, Bungay, Suffolk

To find out more about our authors and books visit www.bloomsbury.com
and sign up for our newsletters.

CONTENTS

List of Illustrations vii
List of Contributors ix
List of Abbreviations xii
Acknowledgements xiii

Introduction: Writing the Body into Ancient Warfare
Hannah-Marie Chidwick 1

Part I Brutality in the Field

1 **Fighting Fit: Visualizing the Body of the Combatant** *Helen Snell* 17

2 **Battlefield Decapitation and Mutilation in the Ancient Greek Imagination** *Fiona McHardy* 29

3 **Bodies of the Military Community: A Perspective from Roman Britain** *Rebecca Redfern and John Pearce* 45

Part II Embodying Soldierly Identity

4 **Clothing the Shami Prince: The Dress of Parthian Horsemen**
Lloyd Llewellyn-Jones 77

5 **Roman *Milites* and Defeated '*Barbari*': Soldierly Bodies, Dress and Identities in First-Century Tilurium, Dalmatia** *Ewan Shanks Coopey and Danijel Džino* 99

6 **Embodying Military Femininity and Virgil's Camilla**
Hannah-Marie Chidwick 119

Part III The Post-Combat Body

7 **Rethinking Ancient Greek War Monuments: Bodies, Violence and Nature** *Lucia Nováková* 139

8 **Astyanax's Fate and Second-Phase Warfare in the *Iliad***
Hannah Sorscher 159

9 **Repairing and Reconstructing the Body of the Combatant: The Commission, Design and Manufacture of Assistive Technology for Wounded Soldiers and Veterans** *Jane Draycott* 175

Conclusion: Beyond the Body of the Combatant *Annemarie Ambühl* 189

References 197
Index 231

ILLUSTRATIONS

Figures

1.1, 1.2 RED DOTS 28
1.3, 1.4, 1.5, 1.6 FIGHTING FIT TRIPTYCH 28
1.7 FIGHTING FIT TRIPTYCH 28
1.8 ILLUSTRIOUS 28
1.9 *HMS Broke* Damage sustained in the Battle of Jutland 28
1.10 Meet the Marines (First Blood) 28
1.11 Liferaft Goggles 28
3.1 The stele of Flavius Fuscinus and Flavius Romanus 50
3.2 The stele of Titus Valerius Pudens 52
3.3 The stele of Marcus Favonius Facilis 55
3.4 Age of death on soldiers' epitaphs 55
3.5 The rounding of age of death on soldiers' epitaphs 56
3.6 The rounding of years of service on soldiers' epitaphs 57
3.7 Injury recidivist demography 62
3.8 Population affiliations by age group 63
4.1a–b Bronze statue from Shami, known as the 'Shami Prince' 78
4.2 The Shami Prince, drawing detailing the combinations of garments 81
4.3 Textile fragment of felt, cotton and silk 82
4.4 The probable shape and cut of the tunic worn by the Shami Prince 83
4.5 Possible cut and construction of the crossover fur-trimmed jacket worn by the Shami Prince 84
4.6 Rear view of the Shami Prince 85
4.7 Detail showing distinctive bulky fabric at the rear of breeches 86
4.8a–b Possible shape of the Shami Prince's breeches 87
4.9 *Vraka*, or full-seated trousers, traditionally worn by men on Cyprus 88
4.10 The construction of the Prince's silk chaps 89
4.11 Framework of variables related to Parthian mounted combat 90
4.12 Body schema, peripersonal space and extrapersonal space 92
4.13 The body schema of the Parthian warrior and his horse 93
4.14 Stretch and 'give' needed in modern trousers and jeans 94
4.15 Wrinkling and 'ride-up' in contemporary garments like shirts 94
4.16, 4.17 Open underarm seams or seam inserts allow for complete flexibility of the arms 95
4.18 Reconstruction of the Shami Prince in battle 96

5.1 Map of Dalmatia, the military fortress of Tilurium and the provincial capital Salona 100
5.2a–e Small metal finds from Tilurium 104
5.3 *Aucissa fibula* AMS H 4372 from Gardun (Tilurium) 106
5.4 Illustration of the 'military friezes' from two stelae at Tilurium 110
5.5 Two seated and bound individuals below a Roman battlefield 'trophy' or 'scarecrow' 111
7.1 Bronze statue of a warrior reflecting a portrait tradition of Alexander 140
7.2 Siege scenes in Greek art 143
7.3 The Corinthian helmet 145
7.4 Wounded Amazon 149
7.5 Black-figured amphora with Ajax carrying the dead body of Achilles 154

Tables

3.1 Roman military personnel excavated from across Europe 46
3.2 Number of injury recidivists in the study by location and site name 59
3.3 Mobility data by location, site name, individual, population affiliation and origin 64
3.4 Injury recidivists with myositis ossificans by age group 65
3.5 Distribution of sharp-force weapon injuries by timing (ante- or perimortem) 66
3.6 Distribution of dental trauma by age group 69
3.7 Distribution of skull fractures by bone and side 69
3.8 Distribution of post-cranial fractures by bone/body area and side 70

CONTRIBUTORS

Annemarie Ambühl is Associate Professor in Classical Philology at Mainz University, Germany. She has published widely on representations of war and civil war in Greek and Latin literature, with a focus on epic, tragedy and historiography. Among her publications are a monograph on Lucan's civil war epic (*Krieg und Bürgerkrieg bei Lucan und in der griechischen Literatur: Studien zur Rezeption der attischen Tragödie und der hellenistischen Dichtung im Bellum civile*, 2015), a special issue of the journal *thersites* (2016) on the senses in war, and several articles on aspects of war narratives such as landscape, gender and the body.

Hannah-Marie Chidwick is Senior Lecturer in Classics and Ancient History at the University of Bristol, UK. She specializes in exploring different methodological approaches to narratives of war and violence. Her interdisciplinary work draws from war studies to read Roman history and literature, and to inform outreach projects that enrich the way ancient warfare is studied in higher and further education. Alongside developing her PhD thesis on Lucan's *Civil War* for publication, Hannah-Marie recently contributed to a special edition of *Ramus*, 'Deterritorializing Classics: Deleuze, Guattari and Antiquity' (2021), and to Bloomsbury's volume *Women in Classical Video Games*, edited by Jane Draycott and Kate Cook (2022).

Ewan Shanks Coopey is a PhD candidate and sessional academic in the Department of History and Archaeology, Macquarie University, Sydney, Australia. His research interests include the archaeology of community and identity in the Roman world, particularly the south-eastern European provinces, and open archaeology and epigraphy. His doctoral project examines the social dynamics of military communities in Dalmatia through archaeological and epigraphic material. He works at the Macquarie University Centre for Applied History and is passionate about public engagement.

Jane Draycott is Lecturer in Ancient History at the University of Glasgow, UK. Her research investigates the history and archaeology of medicine, impairment and disability in the ancient world. Her publications include the monographs *Approaches to Healing in Roman Egypt* (2012), *Roman Domestic Medical Practice in Central Italy from the Middle Republic to the Early Empire* (2019) and *Prosthetics and Assistive Technology in Ancient Greece and Rome* (2022), and the edited volumes *Prostheses in Antiquity* (2019) and *Bodies of Evidence: Ancient Anatomical Votives Past, Present and Future* (2017). Her current research project, which has been funded by the Wellcome Trust and the Royal Society of Edinburgh, focuses on prostheses and assistive technology in the ancient world.

Danijel Džino is Senior Lecturer in the Department of History and Archaeology, Macquarie University, Sydney, Australia. His research centres on the eastern Adriatic and its hinterland, from the Iron Ages through to the medieval period. His most recent books are *From Justinian to Branimir: The Making of the Middle Ages in Dalmatia* (2021), *Liburnians and Illyrian Lembs: Iron Age Ships of the Eastern Adriatic* (with L. Boršić and I. Radić Rossi, 2021) and the co-edited volume *Dissidence and Persecution in Byzantium: From Constantine to Michael Psellos* (with R. W. Strickler, 2021).

Lloyd Llewellyn-Jones holds the Chair in Ancient History at Cardiff University and is the Director of the Ancient Iran Program for the British Institute of Persian Studies, UK. He is a specialist in the histories and cultures of ancient Iran, particularly the Achaemenid period (559–331 BCE), the Near East and the ancient Mediterranean, and has published widely on ancient history. Recent monographs include *Designs on the Past: How Hollywood Created the Ancient World* (2018) and *The Persians: The Age of the Great Kings* (2022). He has spent extensive time in Iran, where his books have received Farsi translations.

Fiona McHardy is Professor of Classics at the University of Roehampton, UK. She is author of *Revenge in Athenian Culture* (2008) and has co-edited five volumes: *Women's Influence on Classical Civilization* (with Eireann Marshall, 2004), *Lost Dramas of Classical Athens* (with James Robson and David Harvey, 2005), *From Abortion to Pederasty* (with Nancy Sorkin Rabinowitz, 2014), *Revenge and Gender in Classical, Medieval and Renaissance Literature* (with Lesel Dawson, 2018) and *Diversity and the Study of Antiquity in Higher Education* (with Daniel Libatique, 2023). She is currently writing a book on gender violence in ancient Greece and is editor of *A Cultural History of Violence in Antiquity* (2022).

Lucia Nováková currently works at the Department of Classical Archeology at Trnava University, Slovakia. She researches the archaeology of archaic and classical Greece, and focuses on Hellenistic art and architecture, burial rites in Anatolia, epigraphy and the use of written sources. Lucia participated in archaeological excavations in Turkey (Selçuk University), Syria (Free University Berlin), Italy (Humboldt University Berlin, Albert-Ludwigs University Freiburg) and Montenegro (University of Warsaw). She is the author of several monographs, including *Tombs and Burial Customs in the Hellenistic Karia* (2016) and *Greeks Who Dwelt beyond the Sea: People, Places, Monuments* (2019), and numerous journal articles.

John Pearce is Senior Lecturer in Archaeology at King's College London, UK, where he has taught and researched Roman archaeology since 2003. His principal research focus lies on the society and culture of Rome's north-west provinces. From his doctoral thesis (Durham University) onwards, he has published widely on funerary rituals and commemoration in this region. Since his employment as a postdoctoral researcher on the Vindolanda tablets at the Centre for the Study of Ancient Documents, Oxford (1999–2003), he has continued to work on inscriptions, including those from frontier settings, and the archaeology of writing in Roman Britain and beyond.

Rebecca Redfern is Senior Curator of Archaeology at the Museum of London, which she joined in 2008, having previously worked at the British Museum and Museum of London Archaeology. Rebecca currently serves as President of the British Association for Biological Anthropology and Osteoarchaeology, and is Visiting Research Fellow at Newcastle University, UK. She is a bioarchaeologist, whose research interests focus on trauma and the Roman world, and recent publications are concerned with enslavement, population affiliation and health inequalities. Rebecca has over seventy publications, including her book *Interpreting Violence in Past Lives: Injury and Trauma in Bioarchaeology* (2016).

Helen Snell has been Artist in Residence in the Department of Sociology, Philosophy and Anthropology at the University of Exeter, UK, since 2020 as part of the national research project, 'Inequality, Identity and the Media in Brexit-Covid-19-Britain', funded by the ESRC as part of UK Research and Innovation's rapid response initiative to Covid-19. From 2017 to 2019, she was Artist in Residence at the Captain Cook Memorial Museum Whitby supported by The Box, Plymouth; and from 2014 to 2017, she was inaugural Artist in Residence at the National Museum of the Royal Navy, Portsmouth, UK.

Hannah Sorscher is Visiting Assistant Professor of Classics at Colby College, ME, USA, and was recently the 2022/3 Assistant Professor at the Intercollegiate Center for Classical Studies in Rome, Italy. Her research interests focus on Latin poetry, especially Roman comedy; women and the family in Graeco-Roman literature, and women and warfare in antiquity. She received her PhD in Classics in 2021 from the University of North Carolina at Chapel Hill, USA, with a dissertation entitled '*Pro filia, pro uxore*: Young Women in the Conventional and Unconventional Families of Roman Comedy'.

ABBREVIATIONS

AJA	*American Journal of Archaeology*
AJP	*American Journal of Philology*
AJPA	*American Journal of Physical Anthropology*
AMS	Archaeological Museum in Split
ANRW	Temporini and Haase (1972–)
BCH	*Bulletin de Correspondance Hellénique*
CJ	*Classical Journal*
CQ	*Classical Quarterly*
FMS	Archaeological Collection of the Franciscan Monastery in Sinj
HSCP	*Harvard Studies in Classical Philology*
ILJug	*Inscriptiones Latinae quae in Iugoslavia inter annos MCMII et MCMXL repertae et editae sunt* (*Situla* 25), J. Šašel and A. Šašel (eds), Vol. 3 (1986)
JHS	*Journal of Hellenic Studies*
JRA	*Journal of Roman Archaeology*
JRGZM	*Jahrbuch des Römisch-germanischen Zentralmuseums, Mainz*
JRMES	*Journal of Roman Military Equipment Studies*
JRS	*Journal of Roman Studies*
MCK	Museum of Cetinska Krajina
MTK	Trilj Regional Museum
RFFZd	*Radovi Filozofskog Fakulteta u Zadru*
RIB I	Collingwood and Wright (1995)
RIB III	Tomlin, R. (2009)
Tab. Vindol.	Bowman and Thomas (1994)
TAPA	*Transactions of the American Philological Association*
Tilurium Cat.	Radman-Livaja (2010)
TRAC/TRAJ	*Theoretical Roman Archaeology Conference* (1991–2017), *Journal* (2018–)
VAHD/VAPD	*Vjesnik za arheologiju i historiju dalmatinsku (1920–2004, 2013–), Vjesnik za arheologiju i povijest dalmatinsku (2005–12)*

ACKNOWLEDGEMENTS

I am grateful to this volume's contributors, whose inspiring work has helped to shape the conversation concerning the nature of ancient warfare as embodied.

The topic for this book was explored at a colloquium, 'The Body of the Combatant in the Classical World', held at the University of Bristol in 2018. I am grateful to Bristol's Institute for Greece, Rome and the Classical Tradition for making the event possible, and to Ellen O'Gorman for her invaluable input and guidance. Thanks to the speakers and chairs: Ben Angell, Rodrigo Gomes, Juliette Harrison, Marian W. Makins, Elizabeth Mitchell, Elaine Sanderson, Effimia Stavropoulou, Sara Tirrito and Korneel Van Lommel.

Thank you to the tireless editorial team at Bloomsbury: Georgina Leighton, Lily Mac Mahon, Zoe Osman and Alice Wright; also, to the anonymous peer reviewers for their comments and suggestions. My gratitude also goes to Sophie Beardsworth and the publishing team. I would like to thank Jess Farr-Cox for her scrupulous editing and indexing.

In addition, many thanks are owed to my colleagues at Bristol for their support during this volume's development, particularly Lyndsay Coo, Esther Eidinow, Patrick Finglass, Benjamin Folit-Weinberg, Shelley Hales, Laura Jansen, Amy King, Paul Martin, Edwin Shaw, Shaun Wallace and, especially, Vanda Zajko.

INTRODUCTION
WRITING THE BODY INTO ANCIENT WARFARE

Hannah-Marie Chidwick

War is about bodies.[1] The mechanics of warfare concern preparing, motivating, mobilizing bodies to *make war*, a phenomenon that infected and was affected by every aspect of daily life in the Mediterranean between the fifth century BCE and the fourth century CE.[2] The human body and its essential role in war-making regularly feature in the extant literature of the ancient Mediterranean, as well as in epigraphy, art and archaeological evidence that has survived millennia where its fleshly makers have not. These texts and artefacts exhibit patent recognition of the intimate connection between war and the actions of the human body, as demonstrated in this passage from the Roman politician and philosopher, Cicero: 'our army [*exercitus*], you see firstly how they earn their name; then, the great toil of the march: carrying more than half a month's provisions and anything else they might need, carrying the camp walls themselves' (*Tusc.* 2.37.4-7).[3]

The Latin synonym *exercitus*, meaning 'a trained, disciplined group of men', like *agmen* (something driven, in motion), neatly conveys the life of the soldier as fundamentally physically demanding and regulatory (see also Pl. *Symp.* 219e-220a). The perception of military life as 'pushing to the limit' one's body, evident in Cicero's meditation, persists in the Western imagination today (Higate et al. 2019: 10). The classical languages in fact provide ample evidence for a conceptual link between the body and war-making underlying the discourse of everyday life in antiquity: Greek and Latin synonyms for 'army' like *exercitus* or στόλος (equipment or journey) imply physical activity as vital, while body parts act as metonyms: *manus* (hand, troop), *caput* ('head, leader') and *caligati* ('boots'), the nickname for Roman soldiers (Suet. *Aug.* 25.3.5; Quint. *Decl. Maior.* 3.16.2) that resonates with the modern epithet, 'boots on the ground'.

Furthermore, the human body and its actions permeate ancient military vocabulary (particularly Roman) and military metaphors, suggesting a connection between war and bodily matter more integral than the routine physical hardship evoked by Cicero. *Manus* supplied the 'manipular' structure of the mid-Republican army, suggesting both that maniples grouped together many 'hands' and that the group itself was 'under the thumb' of the commander. The Latin *dextra* signifies not only the right hand, but the sword wielded by it, a duality that reflects how the *dextra* was designed to complement bodily movements, to ensure its capacity to injure another body in close combat – stabbing and slashing at the torso and face (Veg. *Mil.* 1.12; Frontin. *Str.* 4.7.32; Caes. *BCiv.* 3.99; Tac. *Ann.* 2.21, *Agr.* 36). As the chapters in this volume broach in different ways, the frequent conflation of bodily terminology with strategic, tactical and technical language generates and sustains a recoding of the soldierly body as a(nother) *technology of war* – one to be deployed, exercised and repaired when necessary.

Beyond fitness and regulation, source materials clearly reveal that ancient Mediterranean cultures recognized the centrality of bodily *wounds* in the pursuit of violence and victory.[4] In the opening lines of Homer's epic *Iliad*, the war between the Trojans and Achaeans is described as having 'sent forth to Hades many valiant souls of heroes, and made them themselves spoil for dogs and every bird' (1.4-6, trans. Murray). Later, the Roman historian Sallust condemned in his *Histories* the constant bodily injuring and 'slaughter' of the *plebs* for the sake of a slim political elite (3.48.27-8). Then there is the trend in early imperial Latin epic of laying out, in stomach-churning detail, the horror and gore of the battlefield experience, evoking a pop culture sensationalism familiar to twenty-first-century audiences (McClellan 2019; Connolly 2016). It is no wonder that the body is ever-present in ancient texts concerning war, if sometimes concealed by metaphor.

Building on the long fascination with the body in antiquity,[5] this volume seeks to showcase and explore the variant roles played by the body in literature, art and archaeology concerning war, to ask what these materials tell us about the perceptions and functions of the *combatant*'s body and, by extension, war itself. In texts, the body frequently appears figuratively: for instance, in Julius Caesar's *Commentaries* on the 40s BCE civil wars, where a council of war features this exclamation from the general, Curio: 'Should not, like the wounds of the body, the troubles of the army be concealed so as not to raise the hopes of our adversaries?' (*BCiv.* 2.31.6-7). Curio here employs the metaphorical conflation of the army with a body metaphor to make his point (see Dinter 2012), a metaphor that has made its way into modern discourse by means of the semantic similarity between the military 'corps' and the Latin *corpus*.

Yet, there are other logics and meanings at play: firstly, Curio's suggestion indicates a conceptual link not only between the army and *a body*, but the individual bodies within its divisions. We can, of course, read this as a political simile in which the wounding and internal haemorrhaging of the body work as an allegory for institutional strife.[6] More literally, however, this statement reflects how the wounded body acted as a key visual cue to enemies of the army's weakness, as the health and strength of an individual combatant signalled that of the unit and the state. The spectacle of a robust host of bodies had the opposite effect, as the imperial historian Tacitus describes, wherein just *seeing* the whole army, bodily present on open ground, had a significant psychological impact on enemy troops (*Agr.* 35; see also Joseph. *BJ.* 1.143; Frontin. *Str.* 4.7.20; Gilliver 2007: 126).

For this reason, as evoked by the synonym *exercitus*, physical preparations were paramount to cultivate a body fit for combat (see Caesar ensuring his legionaries are well fed, *BCiv.* 3.49). The fourth-century CE Vegetius' *Epitome of Military Science* (the only intact Roman manual for warfare) takes care to describe the most desirable features of a potential recruit – height, build, age, occupation, place of origin (1.2-7) – emphasizing that physical appearance could be an essential indicator of military competence.[7] It is partly for this reason that Helen Snell, in the opening chapter of this volume, chooses to parallel her vivid renderings of the twentieth-century body in war with Vegetius' instructions for selecting and training military personnel. Created during her residency at the National Museum of the Royal Navy in Portsmouth, Snell's artwork provides an intersection point for the other chapters in this volume: her images show the modern soldier as recruit, as wounded victim, as techno-human hybrid. Read in parallel with Vegetius' writing, these images convey starkly transhistorical features concerning the body's fundamental role in state security and aggressive imperialism.

This is a book about combat in *war*, not the gladiatorial arena, street brawls, or mob violence. The term 'combatant', with its Latin *con-* prefix, implies a body or collected bodies involved in a conflict situation: 'fighting with', 'a person engaged in fighting', specifically 'fighting during a war'. The word carries an additional synecdochic meaning, 'a nation at war with another', which equates the fighting individual with the nation state.[8]

Upon enlistment, the soldier's body becomes no longer his own, no longer civilian. To effectively withstand and embody threat, the soldierly body had to be subject to certain disciplinary powers – control and regulation, punishment for wrongdoing, as (in)famously described by the second-century BCE historian Polybius (6.19-42), who recounts corporal punishments such as beatings to discourage disobedience (6.37).[9] This kind of physical abuse was more typically inflicted on slaves or gladiators, almost never on civilian bodies (Richlin 2005) – just one aspect of military phenomena that directly contradicted civilian norms and codes. The introduction of tattoos in the Roman army made the body physically, permanently, branded as military (Caston and Weineck 2016: 15; Watson 1983: 50). Still in military institutions today, a soldier's basic training *un*makes civilian bodies, erases muscle memory and *re*makes bodies into 'biopolitical subjects par excellence, the objects and instruments of sovereign violence and disciplinary power,' as military anthropologist Kenneth MacLeish (2012: 6) describes.[10]

Although the weapons and ways of warfare have evolved dramatically in the last two millennia, it is hardly universalist or globalizing to suppose that the human body responds to the military experience, including physical threat, in similar ways.[11] For historians, the question becomes not *was* the human body a central component in war-making and war reporting in the ancient Mediterranean, but in what ways? If evidence for this phenomenon is so readily available in ancient sources, then scholarship about ancient warfare should confront the biophysical ramifications of war being *about bodies*, above and beyond the strategic, political, economic and scientific studies that dominate research into state-sanctioned violence in antiquity. Ever since the publication of John Keegan's *The Face of Battle* (1976, inspiring Philip Sabin's response, 'The Face of Roman Battle' in 2000), there has been rapidly growing interest amongst scholars to rethink ancient warfare at the micro- as well as the macro-political level.[12] There is a wealth of excellent research concerning the psychological impact of 'the emotionally heightened experience of war' in the ancient Mediterranean (Meineck 2016: 187).[13] This book enters the conversation by examining the *physical* embodiment of internal responses to war, with several chapters unearthing what Joy Connolly calls the 'sensible experience' of war (2016: 277).[14] Annemarie Ambühl recently argued for an 'aesthetic' approach to the reading of war narratives (2019: 120), one that focuses on the sensory reception of conflict evident in ancient texts. Ambühl writes, '[f]or most of us, war (fortunately) is a mediated phenomenon. This holds true for antiquity as well' (ibid., 119). Civilians, ancient and modern, are reliant on literary and pictorial representations of battle, like war memorials, all heavily shaded with variant agenda, and these agenda critically affect our understanding of the bodies involved.

An 'aesthetic' approach is one way of highlighting the *human* in a pitifully small canon of extant source materials concerning ancient defence and peace-keeping. While sources indicate that war has always been concerned with the somatic, fighting is often represented or discussed either as romanticized heroics, or by using abstract terminology and statistics.[15] One of the more pointed stumbling blocks is that we only really have literary descriptions of war – whether strategic, political or poetic – written by elite members of society, who were more likely to be military leaders or observers than on the front lines (Melchior 2011: 211). Armed conflict was as popular a subject for ancient writers as it is today,[16] yet ancient literature gives negligible space to the front-line *miles gregarius*, an epithet readily applied to the rank and file 'herd' soldier in Roman historiography (Cic. *Planc.* 72.13; Sall. *Cat.* 37.6.1, *Jug.* 93.2.3; Val. Max. 5.3. ext.1). Even in historiography and poetry, literary spaces suited to relaying the combatant's plight, cohorts are predominantly denoted by collective terms and mass formations, juxtaposed with the outlandish exploits of individual generals or semi-divine 'heroes' (Lendon 2017: 45). Aside from some notable exceptions (discussed below), bodily vulnerability is often camouflaged by quantitative data or euphemisms. For instance, while the conflicts between Rome and Carthage (264–146 BCE), recounted in Polybius' *Histories*, left a deep imprint on the Roman imagination for their sociopolitical and economic costs, both Polybius and the Augustan historian Livy concentrate on the pretexts and prosecution of the wars, with soldierly bodies often narrativized as pocketable statistics. Following a skirmish in 216 BCE against Hannibal's forces, Livy writes, 'near 1,700 [Carthaginians] were cut down, with no more than a hundred Romans and allies killed' (22.41).[17] There is little reminder of the corporeality of loss, little that encourages reflection on the way war is embodied.

These kinds of narratives render the military body a dehumanized conduit for political power. Military manuals, like Vegetius', predominantly advocate an approach to waging war that displaces the human with the tactical and the strategic, evoking the modern derivation of 'strategy' from the Greek, στρατιά (army) and στρατιώτης (soldier). The 'philosophical' diagrams of Asclepiodotus' *Tactics* (see also Aelian, *On the Military Arrangements of the Greeks*; Apollodorus of Damascus, *Siegecraft*) provide a specific example of how these texts strip bodies of their individuality and agency, rendering them deployable instruments of war (Ward 2017). Likewise, Roman war manuals such as Frontinus' *Stratagems* provide ample technical and engineering instruction, or cite renowned historic or mythic *exempla* of military success or failure, but give little guidance for the practical application of weapons in the field (Gilliver 2007: 125) and pay virtually no heed to their somatic aftereffects. Notable exceptions include the writing tablets discovered at Vindolanda (Bowman 1994), which offer rare insight into the daily activities and thoughts of legionaries stationed at Hadrian's Wall, including lists of men 'out of action' due to ill health (tablet 154; on daily life in the Greek armies, see Lee 2019).

In the ancient Mediterranean, many aspects of army life provided a template for good personal conduct (see Pl. *Rep.* 374c-375c, 525b; Val. Max. 3.2; Langlands 2018: 113–14), especially in Sparta or Republican Rome, where military service was, for elite men, an inescapable milestone. When ancient or modern writing about warfare *does* focus on the soldierly body, discussion has often concentrated on the training regimes and bodily *disciplina* central to a well-oiled war machine (Phang 2008; Peddie 1994; Hanson 1993). These studies certainly engage with ideas and constructions of the body as war-maker, but the soldier is still discussed through a strategic lens, with 'emotion' or 'trauma' replaced by the concept of maintaining 'morale' (Levithan 2019; Phang 2008: 51–2; see Frontin. *Str.* 2.8, 10-11). 'Despite it being the manifest heartsickening reality of war,' writes military sociologist, Kevin McSorley, 'many of the ways in which war is thought and talked about frequently fail to hold steady to this central fact of bodily injuring' (2014: 109). In response to the recent burgeoning interest in *all aspects* of experiencing war and warring in the ancient world (Brice 2019; Cosmopoulos 2007; Fabre-Serris and Keith 2015), this volume puts the role of the body as *fighter*, as wounder and wounded, centre stage in order to reflect on the corporeal realities of 'making war' in the past.

War as 'Politics Incarnate'

Drawing on modern theory and new research to focus on the *fighting* quality of the soldierly body can lead us to fresh conclusions about old conflicts. Caroline Holmqvist has argued that 'to take seriously research into the human in war [is] to ask, fundamentally, what it means to be a human being living the condition of war' (2013: 551). The justification for this edited volume resides in the need for wider recognition that we as historians should also be asking what it meant 'to be a human being living the condition of war' in the distant past, in order to reflect on the (in many ways, unchanged) war-making practices pervading societies today.

To address the issues at the heart of this volume, it can therefore be productive to integrate approaches gaining ground in modern war studies, politics, sociology and international relations, where the war-making body receives considerable critical attention. Elaine Scarry told us in *The Body in Pain* that 'the main purpose and outcome of war is injuring' (1985: 63) and such a pronouncement does not seem to lack traction nearly four decades later, no matter the specific scholarly subsect.[18] Studying war should mean considering the anatomy of state security, starting with its base matter, human bodies (see Foucault 1991: 135–41; also Foucault 2004; Basham 2018; McSorley 2013; Higate and Henry 2010). This approach is fruitfully articulated in an article by McSorley, 'Towards an embodied sociology of war', in which McSorley urges readers to perceive war as 'fundamentally embodied' (2014: 109).[19] Essentially, *war is carried out by bodies*: war-making, for McSorley, is 'politics incarnate' (2014: 109; also Tidy 2019; MacLeish 2016). The soldierly body is reconceptualized not just as an *instrument* of war, but as the conduit through which conflict between states is expressed. McSorley's concept steers us to use the body and bodily experience as our methodological starting point for studying war.[20] Every chapter in this book demonstrates how the idea of war as 'fundamentally embodied' can reframe source materials from antiquity.

As historians, we are naturally working with data and texts that differ in many respects from those in modern war studies. We do not have the luxury of interviewing or examining living bodies, so instead we read our extant textual and archaeological evidence carefully, and attempt to fill in the gaps where human lives are overcoded as numerical victories or losses, as exemplified by the extract from Livy above. Despite the scarcity of source materials, however, the scholar who seeks to uncover the function and feeling of the fighting body, to read combat as 'politics incarnate', need only look closer at what we do have extant – for example, Roman historiography. Livy's account of the 216 BCE skirmish described a Roman victory, but when we reach the battle of Cannae, a famously disastrous Roman defeat, the body's role in the conflict becomes far more palpable. Although much of the narrative is taken up with quantitative data concerning troop numbers and array (22.45-6), once battle proper begins its embodied nature is painfully evident.[21] First, the Romans struggle to fight owing to physical exhaustion (22.47.10); then, Numidian soldiers are seen hunting for new weapons amongst the bodies of those already dead, in order to 'hack at the soldiers' backs and slit their hamstrings' (22.48.4). The gory aftermath of what Livy terms 'outrageous butchery' (*stragem ingentem*, 22.48.4.4) depicts a battlefield through damaged body parts – *cruenti* (22.51.6.4, bloody men), *volnera* (22.51.7.1, wounds), *femur* and *poples* (22.51.7.2, thigh and knee), *cervix* and *iugulus* (22.51.7.2, neck). The most striking reminder of the human cost comes when those picking for spoils are confronted by horribly wounded men pleading for death and, in a harrowing *exemplum* of combat disintegration, the corpses of men who had sought suffocation by burying their heads in the earth (22.51.8).[22] A political clash is made starkly corporeal.

Narratives like this provide insight not only into the experience of combat bodies, real or diegetic, but importantly their symbolic function. Livy's amplification of the incarnate in war continues to be shocking because it reminds us of the only battle reality to which we as civilian readers can relate: bodily presence.[23] The devastation described here is what Livy needs to convey the horror and loss of the defeat at Cannae; to downplay a loss, the dead are listed as statistics. Combatants' bodies took on new meanings in Roman war narratives of the imperial period, when pitched battles became less common and soldiers functioned largely as 'keepers of the peace' in the provinces (see Redfern and Pearce's chapter). The highest tally of Roman losses in a so-called pacified province occurred in Jerusalem in 66 CE, as recounted in the historian Josephus' *Jewish War* (66–70 CE). The start of the Jewish resistance saw Roman commander Cestius Gallus forced to retreat from the city (*BJ* 2.540-58) and ambushed by the Jews, in the narrow passage between Upper and Lower Beth-horon. Unable to escape cleanly, Cestius stationed 400 soldiers in the camp overnight, with orders to continue calling the watchwords while the rest of the legion fled. 'At dawn the Jews discovered the desertion of the camp, stormed in to make quick work with their javelins of the four-hundred decoys, and set off in pursuit of Cestius' (*BJ* 2.552, trans. Hammond).

Josephus reports that 400 men were killed, and later 'despoiled', by the Jews who pillaged the camp. In this narrative, the bodies (corpses) function as a plot twist, a clever, strategic trick by a desperate man.[24] Their corporeality is not emphasized, nor is the loss of those non-human lives 'jettisoned' for slowing down the retreat (*BJ.* 456, trans. Hammond: 'mules, donkeys and even the wagon teams were slaughtered'). In both Livy's and Josephus' texts, bodies function both to fulfil political imperatives and as dramatic devices – sometimes mourned, sometimes reduced to digits, but either way providing us with examples of how the body is codified, its meaning modified, in the ways we receive and understand war.[25]

To return to Cannae: in the deliberations following the Roman defeat, with 10,000 refugees adrift, we find further evidence for the politicization of the combatant's body. Livy tells us that one Furius Philus believes the debate hopeless because the death of so many equates to the death of the Republic itself (22.53.4.2-3). The young tribune Scipio disagrees, urging *action*, not discussion (22.53.7.1): 'those who wished to save *res publica* should immediately arm themselves and go with him' (22.53.7.1-3).[26] These opposing reactions to the defeat nevertheless both demonstrate how the state of the soldierly body is directly connected to the state of *res publica* – 'the matter of people' – and Rome's fortunes in the war. In this way, we read the combatant's body as both symbol and perpetrator of governing power: this is politics incarnate. Scipio's response indicates that if the state is threatened, the anthropomorphization of its power must be able to withstand this threat *and* to embody threat in response.

As this brief reading of Roman war narratives demonstrates, by using concepts like that of war as 'politics incarnate', we can construct a comparative framework for reading ancient materials that draws attention to and critiques the transhistorical nature of war as an embodied phenomenon. 'Brutality does not explain war in the least,' writes the philosopher Gilles Deleuze, 'and yet brutality discovers in war its best means' (2004: 20, trans. Taormina). In Chapter 2 of this volume, Fiona McHardy explores the brutality of warfare by examining literary descriptions of defacement and decapitation in ancient Greek war narratives. McHardy aims to unearth the significance of this kind of injury for Greek populations and therefore the emotional impact of wounding, being wounded, or witnessing wounding on the face or head. McHardy's subject matter provides a stark example of how the soldierly body functioned as a medium of interstate violence, and how damage to the head, specifically, carried a particular cultural meaning: martial bounty incarnate.

Reading for the body involves understanding the way that bodily activity – in both the preparation and the waging of war – is as immediately political as any strategic, tactical or diplomatic endeavour. By foregrounding the body in the study of war, we can confront the way in which human injury and loss for political purpose has been normalized and overlooked for millennia. Responding to McSorley's work, Joanna Tidy writes: 'Correcting the erasure of people in the theorization and study of war (previously understood and analysed purely as a domain of state institutions) involves putting bodies "back in"' (2019: 221).

'Putting bodies back in' to the study of war means, firstly, looking for the body in existing materials; secondly, using this evidence to expose bodily experiences of warfare, seeking out *all* bodies involved in war-making, including 'collateral damage'; also the gender and ethnicity of bodies affected by conflict, and by extension the logics and spaces of war that shape state power into flesh, blood and boots. By focusing our attention less on the tactical deployments of bodies and instead on the fleshly repercussions of these deployments, the mechanisms that prepare them for war and normalize their mobilization and their signification in sociocultural and political spheres, we are 'putting bodies back in': we are working towards an embodied history of ancient warfare.

Formations, Fortifications and Fallout

As the chapters in this book demonstrate, war is a many-headed monster, its force permeating far beyond the battlefield, the camp walls, or the besieged city, with causes and consequences rooted in multifarious sociocultural and political phenomena. Therefore, the study of war should similarly stretch across these varied manifestations, as advocated in a special edition of *Security Dialogue*, 'Becoming War' (2020), in which Antoine Bousquet, Jairus Grove and Nisha Shah encourage multimedia, multidisciplinary approaches to war studies.[27]

Researching, teaching and thinking about *ancient* war should similarly embrace multiple perspectives, methodologies and source materials. The case studies in this book are embracingly and necessarily broad, providing a rounded introduction to research questions concerned with the combatant's body. While this book concentrates on the Graeco-Roman world as a starting point, due to the expansive nature of Greek and Roman imperialistic activity, the chapters necessarily incorporate neighbouring states and auxiliary troops, to consider how soldierly bodies were perceived by the multifaceted populations of the ancient Mediterranean and bordering territories. The dialogue between these chapters will hopefully inspire readers to revisit histories of professionalized violence, keeping vigil for the somatic, while also revealing the fruitfulness of looking further afield to lesser-studied territories like Parthia (Chapter 4) or colonized Dalmatia (Chapter 5).

One viewpoint – like one human sense – is never enough. As the contributors to this volume stem from a range of disciplinary backgrounds, within and beyond academia, their variant perspectives go some way towards reflecting the diverse interactions between the body and war-making, and the changing perceptions of these bodies in antiquity. All are *writing the body back in* to ancient warfare by considering the warring human, rather than 'war machine'.

When it comes to thinking about the effects of war-as-fighting on the body, archaeological evidence can prove especially useful for accessing the physicality of embodied warfare. As Redfern urges us to do, the interpretation of paleotrauma, for instance, should lead us to consider 'the physical and psychological consequences of these traumas' and not solely what trauma can tell us about fighting tactics or medical care (Redfern 2010: 443; see also Liston 2020). Redfern and Pearce contribute to this volume (Chapter 3) a bioarchaeological study of life on the frontiers, asking what it meant to have *a military body* in the unruly province of Roman Britain. Redfern and Pearce examine skeletal condition and the types of injuries sustained to question whether a combatant's body can be determined against other humans living with the habitual violence of borderland occupation. Bioarchaeological datasets are paired with epigraphic evidence in this chapter to demonstrate how cross-disciplinary collaboration can create a fuller picture of what it was like to inhabit a combatant's body: the application and feel of armour, the swing and strike of a weapon, the wounds it inflicts.

By concentrating on one small but crucial facet (the body) of a global phenomenon (war), such a reorientation allows for a reconsideration of how we conceptualize combat in the ancient Mediterranean. Lucia Nováková's rereading of the trauma of battle depicted on Greek war memorials (Chapter 7) surveys the way that bodies crowding, twisted together, are a primary symbol of conflict in iconography. As Nováková shows, however, in these sculptures the combatant's apparent pain can be construed as misleading: what is being commemorated is the highly anthropocentric depiction of conflict, with a focus on dynamic, idealized imagery rather than realistic renderings of the battle experience.

The visce-realities of war-making are brought to the fore in this volume to prompt discussion about how bodily attributes and conduct revealed the status of combatant. What made a body soldier, rather than athlete, or politician? Archetypal warrior bodies celebrated in ancient Mediterranean artworks are often instantly recognizable through poise and perceivable menace, equipped with distinctive shields, javelins and helmets. These associated pieces of paraphernalia were not merely markers of status but essential features facilitating the body's functioning in battle. Scholarship concerning military kit throughout the ancient Mediterranean has predominantly focused on the manufacture and supply mechanisms of uniform and equipment (Rossi et al. 2015; Goldsworthy 2003; Coulston 1998), while extant literature describing battlefield technology tends to focus on the economic or the technical, exemplified in Vitruvius' *On Architecture* 10. However, the somatic experience of using this equipment and its bloody, bodily purpose merits more critical attention. Lloyd Llewellyn-Jones (Chapter 4) uses the Parthian 'Shami Prince' to consider how the soldier's Personal Clothing System has become inseparable from conceptions of the body. Drawing on theoretical models for understanding the relationship of personal space to clothing, Llewellyn-Jones explores how Parthian armour was designed and functioned symbiotically with the body, rendering human, weapon and steed one supple fighter.

Such an approach echoes modern discussions concerning the combatant as 'cyborg', where flesh, clothing and kit work together as one *technology of war*. 'To the outside observer,' writes MacLeish, 'weapons and armour signify soldiers' invulnerability and lethal capacity. But for soldiers these objects are material, bodily environments through which they understand their vulnerability to violence just as much as their ability to produce and withstand it' (2012: 52; see also Basham 2013: 11; Woodward and Winter 2007). At the level of discourse alone, the Greek, ὁπλίτης (hoplite, heavy-armed soldier), and the English 'cavalry' (from the Latin *caballus*, horse), convey that same symbiosis of soldier and war-tech latent in MacLeish's statement.[28]

It is not only the body's interactions with objects that established the status of combatant, but also the co-construction that takes place in whichever space it inhabits.[29] Although all sorts of bodies entered the military collectives of the ancient Mediterranean, military activity of any kind forces the body into a new realm of codes and logics that shape the soldier's sense of self and perception of others. There is by now a widely recognized imperative to decentre and dismantle monoethnic representations of living and dying in the ancient Mediterranean (Hingley 2014; Tolia-Kelly 2011; Woolf 1998; Webster and Cooper 1996), and the same goes for war-making and its embodiment (Sidebottom 2004: 1–15). Chapters in this volume explore the question of how we understand not only those bodies populating Greek or Roman armies, but also those inhabiting territories colonized or provincialized as a result of war.

Military activity and violence can be shown to work together to manifest logics of nationship and status (see Lee 2013), especially in opposition to enemy soldiers and Othered communities (Tac. *Ann.* 15.4; Virg. *Aen.* 9.617; Webster and Cooper 1996). It has been argued that enlistment in the 'brotherhood' of ancient imperial war machines almost programmatically forced those whose lives had been overturned by foreign invaders to inculcate a new persona – as soldier (Whittaker 2009: 193–5; Phang 2008: 79; Mattingly 1999: 48). For instance, in Roman military institutions the soldierly *habitus* is inculcated through daily routine and drill (the 'military step' in Veg. *Mil.* 1.9; Polyb. 10.20; see Chidwick 2018). Physical exercise, along with uniform and the prospect of pay, worked to construct what Dick Whittaker describes, in the context of the imperial Roman army, as a 'corporate identity' (2009: 195–6) which overcoded pre-established social and ethnic identities. As critiques of cultural 'hybridity' (Hardwick and Gillespie 2007) remain central in postcolonial approaches to ancient history, they seem especially pertinent when we consider that those joining military communities had to renegotiate their innate sense of self while embodying the very mobilization and military praxes that dismantled native identities in the first place. Ewan Shanks Coopey and Danijel Džino (Chapter 5) contribute to this discussion by examining local military dress and war monuments to ask how colonized communities in Dalmatia responded to the Roman occupation. Material evidence indicates that soldiers in veteran communities and Roman provinces across Dalmatia used their bodies to express their ethnicity through visible, corporeal means. Bodily decoration and the depiction of bodies in funerary artwork reveal how soldiers distinguished themselves not just from their enemies, but also from the civic community around them. Adopting the mantle of 'soldier' meant embodying numerous heterogeneous identities simultaneously, including perceiving one's body as 'combatant' rather than 'civilian'.

This book further engages with debates concerning how bodily actions perpetuate praxes of 'othering' (Laforse 2013), by considering how gender identities were embodied in combat. In both modern and ancient thought, the ancient warrior's body is almost always configured as masculine. Jacqueline Fabre-Serris and Alison Keith brought to light the gender politics of war in the Graeco-Roman world in *Women and War in Antiquity*, describing battle as a 'quintessentially masculine pursuit' (2015: 1; see also Walde and Wöhrle 2018). The female form and notions of femininity were entirely in contrast with the exemplary soldierly body, one hardened and unbreakable like the outer walls of a frontier fort (Alston 2010: 205). As this author's Chapter 6 explores, rejection of feminine or nonbinary bodies – descriptions characteristically reserved for social deviants or the enemy – reflect attitudes (if not official policies) in today's Western armed forces (Sharrock 2015: 158; Woodward and Winter 2007). The idea of combat itself is often, as Cynthia Enloe has pointed out, considered to be solely the activity of two masculine bodies in opposition.[30] A focus on the features and behaviours of the combatant's body itself can lead to critique of this narrow conception. Narratives of bodily action in battle offer opportunities to consider how bodies were gendered in the combat zone: in poetry or myth, women *could* participate in conflict, and the embodiment of gender norms could be as changeable as battle itself. Chapter 6 considers the kind of 'military masculinity' discernible in war narratives by discussing one of few female fighters in extant Graeco-Roman literature, the warrior princess Camilla in Virgil's *Aeneid*. As *dura femina* (*Aen.* 7.807, hardened woman), Camilla's prowess in combat not only flouts sexed-body stereotypes, but prompts us to consider the interrelation between masculinity, femininity and the construction of a specifically military body. Camilla's performative masculinity is a product of preparing for and participating in combat, rather than biological make-up.

Furthermore, following archaeological evidence of women across ancient military zones,[31] scholars have called for 'a serious re-imagining' (Enloe 2000: 35–6) of ancient warfare, and contributors to this volume also encourage recognition of the presence and function of female or nonbinary bodies in war-making. Hannah Sorscher (Chapter 8) urges us to reconsider who counts as a 'combatant' by discussing the phenomenon of second-phase warfare as it features in the *Iliad*, arguably the most famous and influential 'poem of death' extant from the ancient Mediterranean (Edwards 2007: 21; see also Scarry 1985: 123). Second-phase warfare occurs *after* the fighting between male collectives, where bodily violence shifts from the soldier to those raped, murdered, or enslaved by conquering forces. Drawing on the work of Kathy Gaca, Sorscher argues that these bodies become unwilling combatants themselves, in a phase of conflict that might have been more relatable to Homer's audience than the sweeping battle scenes between Trojans and Achaeans.

Sorscher's contribution works towards a need 'to get to the bottom of war as an experiential phenomenon', as described by Christine Sylvester (2012: 490). To study war requires traversing all aspects of its manifestation, including the way it can transform bodies that do not enter the warzone freely into warriors. Even those soldiers who manage to survive the physical hazards of war-making cannot escape the warzone forever, but embody that experience for their remaining years. Jane Draycott's Chapter 9 closes this volume by exploring what happened to fighting bodies that were no longer fighting fit. Draycott surveys the techniques for rehabilitating those bodies irreversibly transformed by the injury fundamental to their profession. Through assistive technology, including prostheses and mobility aids, the soldierly corpus could be remade. In the ancient Mediterranean, as today, the effects of post-combat bodily reconstruction served as reminders of the corporal realities of combat, beyond the battlefield.

Finally, a somatic study of warfare ought also to touch on the afterlife of the military body. 'To read war literature is constantly to stumble over corpses,' writes Kate McLoughlin in *Authoring War: The Literary Representation of War from the Iliad to Iraq* (2011: 73). If injuring is 'war's main purpose and outcome' (Scarry 1985: 63), then disfigurement and death are readily accepted by-products. The 'noble death' that dominated poetic narratives of war up to the Augustan period was perhaps so prized because the alternative did not bear consideration. Yet, as Annemarie Ambühl notes in her Conclusion, the texts discussed throughout this volume address questions concerning the soldierly body's capacity to endure the physical trials involved in war-making. What they do not address, however, is *why* the body must endure these ordeals; why the body has been consistently central to war, whether it is the harsh preparatory regimes of the Roman camp, or in the simple *digital* act of pushing a button that will destroy an Afghan settlement. If war is a political contest, then why does it continually manifest itself in physical brutality?

Notes

1. The material for this Introduction was first presented at the Warfare in Antiquity Conference 2019 at King's College London, and subsequently at the Institute for Classical Studies Summer 2021 seminar series. Many thanks to the conference organizers and delegates for informing the writing of this chapter.
2. A reader need only consult Campbell and Tritle (2013) to find plenteous discussion of the 'embeddedness of warfare' in the Greek *polis* (Millett 2013: 55), and the ubiquitous labelling of Rome as a 'militaristic society' (Lovano 2013: 74).
3. *nostri exercitus primum unde nomen habeant, vides; deinde qui labor, quantus agminis: ferre plus dimidiati mensis cibaria, ferre, si quid ad usum velint; ferre vallum.* All translations from Latin are my own.
4. See Ambühl (2019: 130–1). On the power and effect of wounding in Roman battle, see especially Leigh (1995), and more from Cic. *Tusc.* 2.16. Cicero goes on to say that veterans, despite their physical inferiority, perform better in conflict scenarios because their bodies are more accustomed to the corporeal realities of war-making.
5. On bodies and body parts in the ancient Mediterranean, see, for example: Draycott (2022 and 2023); Bradley, Leonard and Totelin (2021); Surtees and Dyer (2020); Walters (2020); Bodiou and Mehl (2019); Garland (2010); Porter (2005); Wyke (1998).
6. On the metaphor of the 'body politic', see especially Walters (2020): in the fable of Menenius Agrippa (Livy 2.32.8-12), 'The image of the body invoked is that of a unified organism composed of discrete organs with individualized functions, and only by working together do the parts ensure the health of the whole' (11).
7. Vegetius specifies: 'So let the adolescent who is to be selected for martial activity have alert eyes, straight neck, broad chest, muscular shoulders, strong arms, long fingers, let him be small in the stomach, slender in the buttocks, and have calves and feet that are not swollen by surplus fat but firm with hard muscle' (*Mil.* 1.6, trans. Miner).
8. *The Oxford Essential Dictionary of the U.S. Military* (2002).
9. See also Quint. *Decl. Min.* 339.15; Val. Max. 2.7; Livy 8.6.14, 8.35.4, 9.17.10. On the discrepancies in accounts of Roman *disciplina militaris* and its exaggeration, see Machado

(2021). While 'unquestionably brutal' (394) military punishments were not systematized and did not result in uniform obedience.

10. Also, MacLeish (2016); McSorley (2014); Scarry (1985). On biopolitical regulation and 'resiliency': soldiers are 'exposed to harm and subject to a huge range of measures to keep them healthy and alive [. . .] Soldiers' health is constantly monitored and intervened upon; soldiers themselves are both deeply invested in their own fitness and keenly sensitive to the ways that the Army attends to it' (MacLeish 2012: 6, citing Agamben 1998; Foucault 2004; Foucault 1991).

11. See Chrissanthos (2007: 225); Barkawi and Brighton (2011: 128); Kümmel (2003). See especially Ambühl's special edition of *Thersites* (2016a) 'War of the Senses – The Senses in War: Interactions and Tensions between Representations of War in Classical and Modern Culture', in which Thorne (2016: 80) describes reading trauma narratives comparatively as 'an eye-opening corrective to the academic temptation toward abstracted scholarly musing on the past, distanced from the real terrors and hardships of life (ancient or modern).'

12. See especially Caston and Weineck (2016); Chrissanthos (2007); Krebs (2021); all Gaca listed in the References. A particularly popular approach in literary studies, see: Pope (2020); McClellan (2019); Gaca (2015); Bartsch (1997).

13. See Heidenreich and Roth (2019); van Lommel (2013); Melchior (2011); Shay (2003). Caesar understood 'the psychology of war' (Lovano 2013: 75, Caes. *BCiv.* 1.21).

14. On the 'sensory turn' see Ambühl (2019: 120–1); Ambühl (2016a); Butler and Purves (2013).

15. This pattern is replicated for the most part in modern literature on the subject (for critique of this practice, see Basham 2013; Barkawi and Brighton 2011).

16. 'Wars, soldiers, battles: literature is full of these. Armed conflict is among the earliest and most powerful subjects in literature and continues to preoccupy writers across the spectrum of genres and forms' (Beck 2016: 84).

17. *ad mille et septingenti caesi, non plus centum Romanorum sociorumque occisis*. Livy follows, *Hannibal id damnum haud aegerrime pati* (22.41, Hannibal was not troubled by the loss), instead judging it strategically valuable for making the Romans overly confident.

18. Scarry continues: 'Though this fact is too self-evident and massive ever to be directly contested, it can be indirectly contested by many means and disappear from view along many separate paths' (1985: 63–4).

19. Also, MacLeish (2016). This idea dominates any discourse which deals with the 'biopolitical' (Foucault 2004; Agamben 1998).

20. In this way we can 'develop the understanding of different conflicts via the comparative analysis of their distinctive methods of injury [. . .] modes of somatic engagement and flow, affective resonances, body pedagogics and the inculcation of militarist structures of feeling' (McSorley 2014: 119).

21. Just prior, Hannibal is coerced into confrontation owing to dwindling resources leading to hunger (22.43).

22. Nevertheless, the Cannae narrative concludes with another list of numerical losses (22.49).

23. This volume focuses on the soldierly body but, of course, numerous other bodies fell victim in ancient war-making – animal, vegetable, mineral. During the battle of the River Metaurus, 207 BCE, Livy describes the slaughter of elephants using chisel and hammer, should the beasts become uncontrollable (27.49). Before the establishment of official military hospitals, there was little difference between treatments for sick soldiers and sick animals. Roman military doctors were not officially installed until the principate of Augustus (Wesselingh 2008). See

Chrissanthos on military living conditions, including all the messy realities of being human (2007: 228).

24. The legionary eagle was then captured (Suet. Vesp. 4.5); Cestius' death soon afterwards may have been the result of 'vexation' at the defeat (Tac. *Hist.* 5.10).
25. König (2022) explores the interactions between war narratives in diverse forms.
26. *audendum atque agendum, non consultandum ait in tanto malo esse. irent secum extemplo armati qui rem publicam salvam vellent.*
27. Kümmel (2003: 417) introduces a chapter arguing for more holistic approaches in military sociology with Ovid's story of soldiers emerging from dragon's teeth (*Met.* 3.102-30): 'the simple truth that the military is a highly complex social phenomenon in itself and one that cuts through various levels and touches several different contexts and is thus subject to multiple processes of interpenetration'.
28. Identified by Spinoza, who uses the figure of the soldier to demonstrate how memory-making involves remembering both one thing and its associated interacting bodies: 'And in this way each of us will pass from one thought to another, as each one's association has ordered the images of things in the body. For example, a soldier, having seen traces of a horse in the sand, will immediately pass from the thought of a horse to the thought of a horseman, and from that to the thought of war, and so on' (*E*2P18, trans. Curley).
29. 'Our ability to make sense of our surrounding environment, and of ourselves in the world, presupposes a set of bodily actions – turning the head, moving, reaching, picking up, tasting and so on – through which the perception of objects in the external world and the perception of one's own body are necessarily intertwined' (McSorley 2014: 108). On this phenomenon in classical Athens, see Pritchard (2013). The soldier is a paradigmatic figure with regard to the inculcation of a particular institutional *habitus* (McSorley 2014: 115); see Petr. *Sat.* 82, for a parodical example of the dangers of the wrong uniform.
30. 'Combat, in other words, is an activity narrowly reserved for only those occasions when there is a two-sided militarized violence, in which both sides are ready for the violence and are equipped with weapons' (Enloe 2013: 260).
31. Enloe (2000: 35–6) cites Carol van Driel-Murray's discovery of women's shoes in military garrisons: 'the Roman conquest of Britain was about to undergo a serious re-imagining. So was soldiering. So was military strategy.'

PART I
BRUTALITY IN THE FIELD

CHAPTER 1
FIGHTING FIT: VISUALIZING THE BODY OF THE COMBATANT

Helen Snell (artworks)
Hannah Marie-Chidwick (translations and introductory text)

While military history is rife with sinister variety – surprise attacks, innovative insurgencies, never-before-seen modes of destruction – one material persists as the primary technology and currency of war-making, whether in ancient Parthia or the Portsmouth docks: the human body. By centring her representation of war on *bodies*, Helen Snell's work reveals, in a manner more immediate and shocking than a written study, the intimate entanglement of human and post-human in warfare.

Snell's images set the tone carried through every chapter of this book, encouraging a holistic reflection on the transformations, multiplicity, myth-making and horror of combat throughout history, fit for a modern readership. Like the other chapters of this volume, Snell examines wartime evidence to consider the corporeal, drawing from the extraordinary archives at the National Museum of the Royal Navy where she was the inaugural artist in residence in 2014–17. Her exhibition at the NMRN in 2016–17, entitled *A Long Shadow*, alluded to the lasting effects on body and mind of any military experience. In the images to follow, she pairs artwork and photographs with extracts from the fourth-century CE Roman military historian Vegetius' didactic *Epitome of*

Military Science, throwing stark light on the expectations that have long shaped the call of duty. 'Military matters,' writes Vegetius, 'consist of arms and men' (*Mil.* 2.1, evoking Virg. *Aen.* 1.1). Snell's work epitomizes the aims of this volume by showcasing the various perceptions, receptions and constructions of the soldierly body, working with digital multiples as a way of exploring, through artistic media, themes of mass production and the relationship between the group and the individual. For her, this approach has a particular resonance with military themes.

In this collection, Snell portrays living bodies and technologies embroiled in a war still prevalent in the Western imagination today, for instance in the piece on the previous page, *Red Dots*, a digitally manipulated photograph in which every naval recruit is anonymized by their personal equipment. They are faceless, dehumanized, like the statistics listed in Livy's histories, or the records of veterans laid up in Vindolanda's military hospital. Similarly, *Meet the Marines* and *Liferaft Goggles* show how entering combat can mean adopting body-modifying technologies, reminiscent of those described in Jane Draycott's and Lloyd Llewellyn-Jones' chapters. Meanwhile, *ILLUSTRIOUS* and *HMS Broke* transplant somatic features onto an inhuman vessel, evoking the wide reach of war beyond the anthropological. The damage to the ship wall appears so much like an injured combatant's torn skin, that when *ILLUSTRIOUS* appeared on screen at the colloquium which inspired this volume, the council chamber fell eerily silent. Vegetius' text serves to accentuate the near-universality of Snell's themes: the combatants themselves must stand *tamquam murus ferreus* (*Mil.* 2.17, like an iron wall; also Luc. 6.201; Hom. *Il.* 4.299), in terms of both function and physical repercussions. Ancient military strategists devoted copious attention to engineering and tactics for obliterating defence structures, particularly in siege warfare, but never forgot that men were as key to these structures as timber and stone.[1]

The preparations, violence and loss of individuality permeating battle narratives throughout this volume are here presented in glaring, glossy actuality. The *conditioning* of the body for war is Snell's presiding subject: the laser cut *Fighting Fit Triptych*, part of a 2013–14 touring exhibition, splices soldierly bodies with war technology, alongside Christian iconography evoking ideas of devotion and sacrifice. Ancient Mediterranean military cultures incorporated a strong sense of duty and dedication in their attitudes to the body, where drill and close-combat experience made a fighter 'fit' for service. 'Neither the arena nor the field proved a man invincible in war, unless he was shown to be diligently trained at the post' (Veg. *Mil.* 1.11).[2] Like other military writers before him, Vegetius focuses on recruiting, training, disciplining and deploying soldierly bodies (*Mil.* 1.1; Polyb. 6; Sall. *Jug.* 45; Val. Max. 2.7; Front. *Strat.* 2.3, 8, 4.1-2). Snell's art research focused on comparable military propaganda and evocative post-combat imagery in the NMRN archives, which offer visual reflections on the process from recruitment to discharge.

There were, of course, artistic representations of warring bodies in the ancient Mediterranean: Lucia Nováková surveys the varied depictions in Greek and Hellenistic art, while Ewan Shanks and Danijel Džino examine how conquered combatants function in military monuments. But these are crafted portrayals; we must imagine for ourselves the wounded soldiers stumbling across a smoking battlefield (in whatever form), amid ruined war machines. Archive photos provide real snapshots of these messy experiences, here offset by Vegetius' orderly manual – written by a strategist, for strategists.

Pieces like *Meet the Marines* and *Fighting Fit* remind us of how attitudes to medical care in the forces differ from those of civilian medicine. Chapters in this volume by Draycott, Rebecca Redfern and John Pearce, and this author, explore attitudes to battle-ready and post-combat bodies where 'fitness' means perceiving the body as war-tech, closely controlled and reparable (Veg. *Mil.* 3.2). Still today, the somatic criteria for enlistment to the UK armed forces are as important as at the Roman Empire's height: medical and fitness tests determine a recruit's body 'ready to fight', fit for exposure to injury. The UK Royal Navy's application pages list advice and training plans to prepare for the assessments, let alone for service itself (Royal Navy 2022), required even of those applying for the Royal Marines Band Service.[3] While a military health regime can engender positive lasting effects, the findings of a recent study of UK national electronic healthcare records suggest combatants' bodies and minds can still bear distinctly veteran hallmarks long after leaving the forces.[4]

Readers of this volume may find themselves flicking back to Snell's chapter as a visual companion to the rest, much as the colloquium delegates frequently recalled her striking viscerality in roundtable discussions. As the temporal distance between the twentieth century CE and the fifth century BCE ruptures, like the hull of a bombarded ship, Snell's powerful images leave us reflecting on the corporeal in conflict beyond what has been recorded on paper and screen.

Notes

1. Veg. *Mil.* 2.25 and 4; Vitr. *De arch.* 10–16. See Levithan (2019: 3): 'writers began to lavish attention on artillery and wall-damaging techniques, fetishizing technology and obscuring the centrality of the assault to the siege and of human behavior to the assault'.
2. *nec umquam aut harena aut campus invictum armis virum probavit, nisi qui diligenter exercitatus docebatur ad palum.*
3. Roman war-trumpeters were also trained for combat, though their role was more acutely militaristic (Veg. *Mil.* 3.5).
4. Leightley et al. (2020): military or ex-military personal histories were identified through patterns of key words in the records' free text.

Neither the arena nor the field proved a man invincible in war, unless he was shown to be diligently trained at the post.

Veg. *Mil.* 1.11

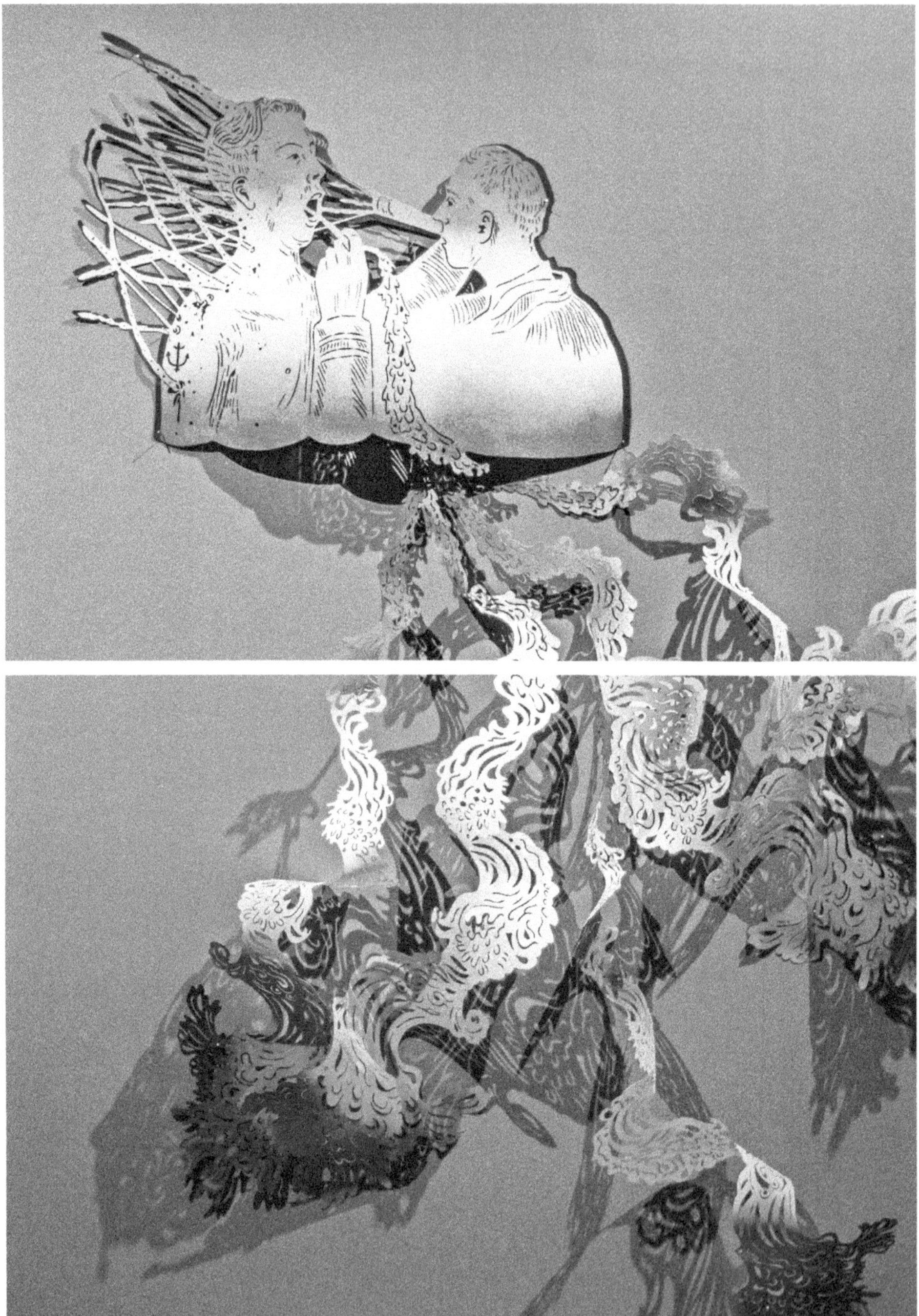

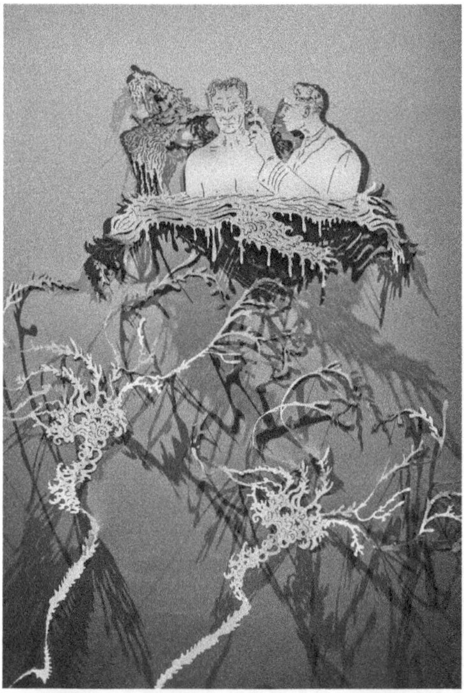

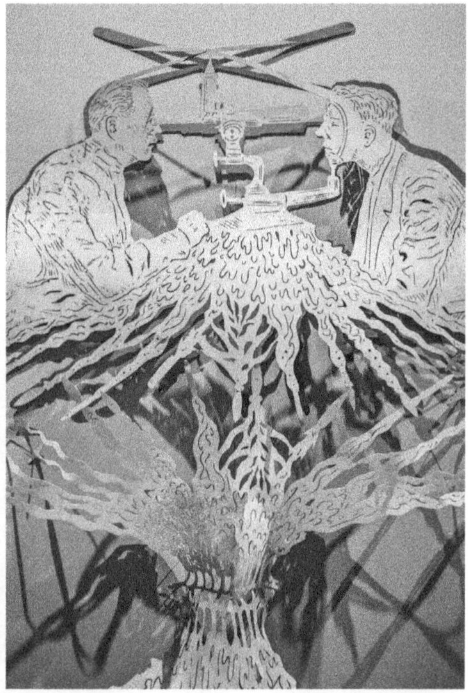

FLIGHT DECK
PORT AFTER ELEVATOR BHD.

Then the heavy infantry took up the fight and stood just like a wall, so to speak, of iron, and fought not only with javelins but also swords at close-quarters.

Veg. *Mil.* 2.17

Damage sustained in
collision, also shell damage
in side plating near the bow.

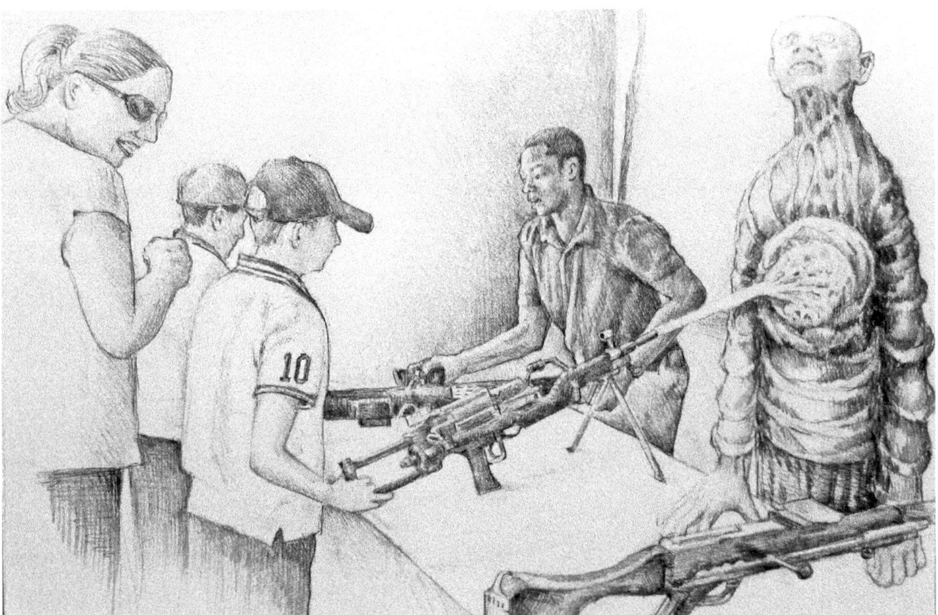

1.1, 1.2 RED DOTS

Digitally manipulated archive photograph.

Crew in Rees-Hall escape apparatus *c.*1910. Early submarines often sank. Several accidents in the run up to the First World War prompted the Admiralty to develop an escape apparatus for trapped crew members. The Rees-Hall escape apparatus briefly went into production. It incorporated a canister containing 'Oxylite', a chemical compound which gave off oxygen and absorbed carbon dioxide. Original archive image courtesy of The National Museum of the Royal Navy.

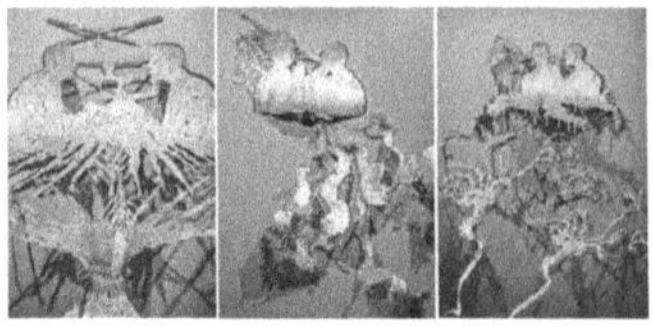

1.3, 1.4, 1.5, 1.6 – FIGHTING FIT TRIPTYCH

Laser cut paper, heat formed laser cut acrylic, steel pins, spray paint.

Weeping Wound
Hearing test
Sight test
Image credit © Nisha Haq Photography.

1.7 – FIGHTING FIT TRIPTYCH

Laser cut paper, heat formed laser cut acrylic, steel pins, spray paint.

Installation view from *Burnt Out*, 20–21 Visual Arts Centre, Scunthorpe. 2013–14.

A national touring exhibition funded by the National Lottery through Arts Council England. 'Fighting Fit Triptych' relates to the tragic irony of the Army or Navy medical examination, referencing Christian iconography and the parallels between contemporary and classical notions of sacrifice, devotion and the substitution of the individual for the collective.
Image credit © Dominic Mason.
Original archive image courtesy of The National Museum of the Royal Navy.

1.8 – ILLUSTRIOUS

Digitally manipulated archive photograph.

HMS Illustrious Damage to Aft Lift 1941. One of a series of six large scale photographs, of blast holes and torpedo damage to ships. Exhibited as part of *A Short Fuze*, a series of site-specific installations at Explos in Museum of Naval Firepower, Gosport in 2014 during Snell's residency at The National Museum of the Royal Navy. Acquired by NMRN for the collection. Archive image courtesy of The National Museum of the Royal Navy.

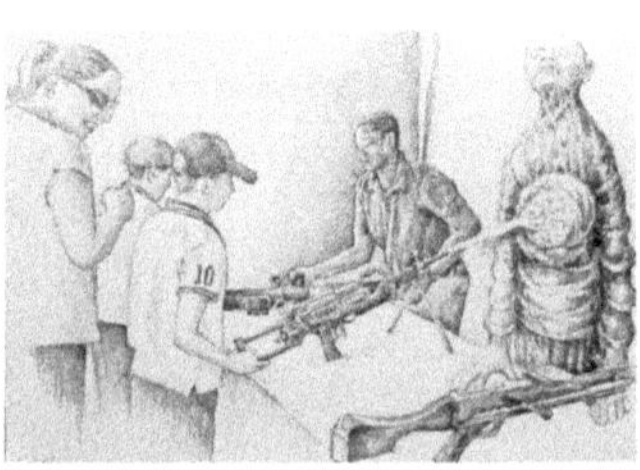

1.10 – Meet the Marines (First Blood)

Pencil sketch.

From a series of drawings prompted by a 'Meet the Marines' event at Bristol Harbour Festival, July 2014. Image credit © Helen Snell.

1.9 – *HMS Broke* Damage sustained in the Battle of Jutland

Pencil drawing.

Illustration of battle damage from the Battle of Jutland, 31 May 1916.

From a small sketchbook containing *c.* 80 drawings showing the damage to the ships of the British Fleet at the Battle of Jutland (RNM Manuscript 2002/14). The volume is a rich source for research into the battle, and shows in unflinching detail the dangers faced by crews. The drawings were made in secret as the ships were in dockyards for repair. To start, there is an aerial view of each ship showing the configuration of hits to both port (marked in green watercolour) and starboard (marked in red watercolour). These are accompanied by vivid illustrations of close-up sections of damage caused by shell fire, torpedoes, collision, fire and internal explosion with a precise description of the cause of the damage. Archive image courtesy of The National Museum of the Royal Navy.

1.11 – Liferaft Goggles

3D nylon polymer print and laser cut paper.

This work was inspired by the need for invisibility and silence in modern warfare and the cowardly advantage of subterfuge. Snell's design represents a counter-intuitive response to a set of Second World War night vision binoculars in the collection at the Royal Navy Submarine Museum; it is a conspicuous, life-affirming prosthesis. Each eyepiece becomes a searchlight. Each layer in the build represents a placenta–liferaft hybrid. The structure was informed by the application of 3D printing in orthopaedic surgery to create bespoke prostheses, as facial implants.

CHAPTER 2
BATTLEFIELD DECAPITATION AND MUTILATION IN THE ANCIENT GREEK IMAGINATION

Fiona McHardy

This chapter explores the representation of mutilation in Greek texts with particular focus on the violent and abusive treatment of the head of those killed in battle.[1] The research forms part of the 'corporeal turn' in warfare studies (McSorley 2014: 121), reflected in the close attention paid to bodily mutilation by classicists in recent years.[2] Much of this recent discussion has focused on ethical considerations, contemplating the extent to which bodily mutilation could be seen as acceptable and appropriate in the context of the ancient battlefield, or whether it was deemed excessive and barbaric, not proper behaviour for Greek warriors. This debate has focused in large part on the treatment of the body of Hector by Achilles in the *Iliad*, as well as Pausanias' pronouncement about the appropriate way to treat the corpse of Mardonius following the battle of Plataea (Hdt. 9.78-9). In this chapter, I delineate some of the complexity involved in the ancient Greek understanding of battlefield decapitation, looking especially at symbolic, psychological and moral aspects, before turning to discussion of the example of Tomyris, Queen of the Massagetae, who, in Herodotus' account, pushes the decapitated head of her opponent Cyrus into a wineskin filled with blood after conquering him on the battlefield (Hdt. 1.214). The example provides a useful comparator to the Iliadic Achilles, in that the target in each case is a military leader whose bodily mutilation symbolizes defeat and the loss of status that accompanies it. In both examples, the mutilator is driven by heightened emotions and a strong desire for revenge, but their mutilation of the enemy is also coolly calculated and well-matched in response to the intended aggression of their opponent.[3] Rather than recognizing this aspect of their vengeance, critics have condemned their actions as indicative of savagery.[4] Although different in terms of gender, ethnicity, time and place, as well as the genre in which they are featured, in the examples of both Achilles and Tomyris, the infliction of bodily abuse on a corpse is a manifestation of the intense emotional pain experienced by the inflictor of violence. Rather than appearing as a significant systematic approach to warfare, battlefield decapitation and mutilation functions as a bodily code.[5] This code is understandable in an ancient Greek context because of the way in which the body of a soldier can be read through knowledge of attitudes towards different parts of the combatant's body as well as the meaning underpinning injury to these parts. Greek authors use the image of the mutilated corpse of a leader as an outward expression of the humiliation of loss, as well as of the emotions that underpin Greek notions of warfare, especially pain, anger and the desire for revenge.[6]

Bodily Mutilation

Commentators working in the fields of international relations and sociology have noted that, while warfare is performed by human participants whose bodies experience war physically and emotionally through attacks, mutilations, rapes and killings, nevertheless, warfare is often discussed at a strategic level with attention given to the outcome for cities and nations rather than individual bodily experiences.[7] While ancient narratives can exhibit similar tendencies, there is also within them some degree of focus on the experiences of the individual. Nevertheless, historical accounts often incline towards the glamorization of death in battle and eschew vivid depictions of the realities of warfare, including by referring to the systematic violence inflicted on women and children captured in war with stock phrases, as Kathy Gaca has detailed.[8] The concept of warfare as a rule-orientated exchange between men fighting on a battlefield for supremacy is undermined by the realities of what occurs. In discussion of ancient hoplite warfare, informal rules supposedly governed this fighting style, including the immunity of certain people and places (such as heralds and temples), the right to reclaim your dead and the ransoming of prisoners, although compliance with these unwritten rules could not be assured.[9] Combatants could elect tactically to ignore these unwritten rules of war to give themselves military advantages. By perpetrating excessive acts of violence on the bodies of enemies, combatants make use of what Kevin McSorley terms 'corporeal tactics', where the aggressive treatment of opponents' bodies is advertised in order to create terror (2014: 107).[10] This interpretation of combatants as making rational, tactical decisions is supplemented by considering their irrational, emotional responses during warfare,[11] leading to a combination of rational performance of violence by soldiers who are trained to inflict damage on the enemy and socially expected to do so, and their instinctive, aggressive responses to their painful experiences of battle, including suffering injury to themselves or loss of comrades.[12]

Simultaneously, as McSorley outlines, targeting specific body parts for abuse is 'expressive and productive of a dehumanizing discourse' (2014: 120). Combatants can engage in literal and metaphorical conversation by verbally threatening and physically inflicting bodily wounds, which are translatable in the context of the battlefield through combatants' knowledge of this bodily language within the larger social context. The network of associations between bodily mutilation, punishment, defeat and enslavement indicates issues underpinning infliction of bodily mutilation on living foes, as well as violence done to enemy corpses in the Greek imagination. According to Aristotle's definition, mutilation involves loss of bodily integrity to the extent that a significant external body part that cannot regrow is removed (*Metaph.* 1024a12-29). This conception of mutilation is not specific to warfare, so can cover mutilation that occurs in sporting contexts, accidents, brawling, or as a form of corporal punishment. Disfigurement and mutilation at the hands of others can be viewed as a sign of denaturalization of the body synonymous with the experiences of enslaved people.[13] The susceptibility of enslaved people to corporal punishment is alluded to in Attic oratory (Dem. 21.180, 22.55, 24.167). Marks, scars and mutilations of the body are used by the viewer as an indicator of identity and status (Lee 2015: 82). According to Socrates in Plato's *Gorgias*, such bodily markings remain on the dead as identifiers of their status and character in life (524c). These ideals of bodily perfection present difficulties for soldiers for whom it is realistic to experience wounding, mutilation and scarring as part of active service.[14] While battle scars, especially those received on the front of the body rather than the back,[15] demonstrate that the soldier has participated in warfare and is not a coward who has shirked the front line, at the same time there are opportunities for combatants to inflict injury on their enemies in ways that will humiliate or dishonour them. So, while the coward who avoids battle or runs away is *atimos,* without honour legally and socially,[16] the wounded veteran, who is by rights worthy of praise earned via his military contributions, is also pitiable for his loss of bodily perfection or unity,[17] and, depending on the wounds he has received, can be subject to a perception that he is less honourable because his body has been mutilated.

The movement of status from warrior to slave can also be actualized in cases where the victorious elect to enslave the vanquished rather than killing or ransoming them (Muller 2014: 56). In these cases, the infliction of corporal punishment or bodily maiming, such as branding, signifies the mastery of the conqueror over the body of the conquered and the humiliation of the defeated. Where mutilation is performed on the living, their bodies continue to bear the evidence of their defeat. In the case of Eurytion, who has his ears and nostrils cut off in punishment for his drunken behaviour in the house of Peirithous, he is left alive with the visible markers of his wrongdoing and subsequent chastisement (*Od.* 21.300-2). Similarly, Malcolm Davies argues that the disloyal goatherd Melanthius is left alive when his extremities are cut off following the defeat of the suitors (*Od.* 22.475-7) and he is forced to exist (perhaps not for a very long duration given the extent of his injuries) in the sight of onlookers as a 'living corpse'.[18] Antinous threatens the beggar Irus with severing of ears, nose and genitals by the renowned maimer of men, King Echetus (a similar punishment as that inflicted on Melanthius), if he cannot defeat the disguised Odysseus in a fight (*Od.* 18.86-7). Here again the associations with low status and dishonour are intended to be carved into the body as signifiers of humiliation following defeat.

The role of honour and humiliation attached to bodily mutilation, as well as the tactical employment of mutilation albeit, in an extraordinary way, can be seen clearly in Herodotus' story of the capture of Babylon. The Persian noble Zopyrus is said to have mutilated himself by cutting off his own nose and ears, as well as shearing his hair and scourging himself, so that it will look as if he has been punished by the Persian ruler Darius, and it will be credible that he is now an enemy of the Persians (Hdt. 3.154). After he cuts off his own extremities,[19] he explains his devious plan to Darius (3.154-5). He then supposedly defects to the Babylonian side and demonstrates his trustworthiness to them through actions, by attacking groups of Persian soldiers who have been sent out for this purpose (3.157), but most significantly through his appearance, which evokes the pity of the Babylonians (3.155.2).[20] As the story shows, no-one is able to anticipate that a man would perform such a severe act of bodily mutilation on his own facial features. His ostensible desire to join the enemy in order to achieve vengeance against the ruler who supposedly punished him (3.156.3) is conversely an expected response and psychologically persuasive. In this tale, Herodotus makes clear that Zopyrus mutilates himself in a way that will not heal (3.154),[21] demonstrating his commitment to achieving military victory, even though it entails the life-long cost of having to bear the visual signs of punishment and dishonour outwardly. That this is an extreme decision is highlighted in Darius' exclamation that Zopyrus must have been out of his mind to have mutilated himself in this way (3.155). The magnitude of his commitment is further highlighted by the repeated reference to his status in Persian society. He is referred to as being 'most well respected' (δοκιμώτατος) both in the perspective of Darius (3.155.1) and the Babylonian enemy (3.157.1). This contrasting of the status and achievements of Zopyrus with the identity as low status and criminal that his facial mutilation suggests is emphasized in Darius' shocked pronouncements, as well as in his reported repeated (but ultimately hopeless) desire to see Zopyrus free of mutilation (3.160) so that his appearance might fit his remarkable military success, rather than being a bodily manifestation of defeat and humiliation.

Mutilation of Corpses

Similar associations are applicable to the mutilation of dead soldiers on the battlefield.[22] As Jean-Pierre Vernant has argued, mutilation and disfigurement in death was dishonourable, the 'sinister obverse' of the desired 'beautiful death' in battle (1991: 67; cf. Kucewicz 2020: 108). This 'sinister obverse' is captured by Tyrtaeus in his depiction of the shameful sight of the old Spartan warrior's corpse with his genitals mutilated (fr. 10.21-5).[23] Some of the terminology of mutilation has embedded within it a concept of shame that impacts upon the victim of the attack, his family and comrades, as Cezary Kucewicz has demonstrated (2016: 433–4).[24] So, enemy combatants could aim tactically to destroy the prospect of honourable and beautiful death for their opponents by mistreating their corpses and inflicting shame upon them.[25] The most common way in which this is threatened in Homeric epic is the taunt that a warrior will be killed and his body left to be devoured by wild dogs and birds.[26] For Vernant, this fate signifies the descent of the human form into chaos where the corpse is dehumanized to such an extent it is no longer distinguishable, but instead becomes part of the beasts that have consumed it (1991: 71–2; cf. Holmes 2010: 34, 2015: 31–2). The onus is upon the fallen man's comrades to carry away his body from the battlefield to avoid this fate. While these threats of being left to be eaten by wild animals do not involve active mutilation of the corpse by a fighter, the impact upon the dead man's body of this treatment is related to other forms of bodily mutilation in death.[27] Further, combatants in the *Iliad* become assimilated to wild animals in their attacks on the enemy through similes and take on some of their aggressive and bestial features (O'Brien 1990: 106). Both Agamemnon and Menelaus are connected to wild lions in the passages where they perform bodily mutilation in battle (*Il.* 11.122-42; 17.46-69), manifesting impressive displays of bodily strength over their foes whom they defeat by inflicting devastating injuries. The connection of warriors to wild beasts that devour raw flesh expresses their more instinctive level of aggression towards the enemy, alongside their training as soldiers that allows them to overpower their enemies.[28]

While there is some evidence to suggest that bodily mutilation could be used to terrify opponents (*Il.* 14.506-7; Xen. *Anab.* 3.4.5),[29] or as a retaliatory strategy where combatants employed tactics used by their enemies against them (Muller 2014: 60, 69), Greek texts typically closely associate corpse mutilation with the heightened emotions of combatants rather than with military tactics.[30] The imagery of bodily injury and mutilation is used by Greek authors to indicate the feelings of grief and anger generated by the pain of injury or the loss of a comrade.[31] The combination of a more tactical attack in which the attacker desires to inflict dishonour and pain upon his opponent is enmeshed with his emotional response, generated out of feelings of anguish and rage that lead to the 'overkill' reaction where the attacker continues to abuse the enemy's body after it is dead.[32] This combination of elements can be perceived most clearly in the *Iliad* in the treatment of Hector by Achilles. His use of speech to outline his motivations sits alongside his unrelenting wish to maltreat the body of Hector, driven by his passions.

Achilles' initial emotion on hearing of the death of Patroclus is an overwhelming grief (*Il.* 18.22). This morphs into anger that fuels his aggressive assault on the Trojans,[33] depicted as a relentless bloody onslaught where the river is clogged with corpses (21.218-20).[34] Apollo, who lures Achilles away so that more Trojans can escape into the city, observes that Achilles' anger is so intense that he cannot recognize that he is chasing a god (22.10). In his exchange with Hector prior to their duel, Achilles highlights his desire for revenge and associates this with his refusal to pledge that he will not abuse Hector's corpse (22.260-72).[35] He likens himself to a lion or wolf when he suggests that Hector's body will be subjected to mistreatment upon death. Upon fatally wounding Hector, Achilles once again reiterates that this killing is in revenge for Hector's slaying of Patroclus, and he threatens to expose Hector's body to be eaten by wild animals and birds rather than returning it for a proper funeral such as the one Patroclus will receive (22.331-6). He manifests his bestial desire to eat raw the flesh of Hector, taking on the role of the animals that devour the bodies of the dead on the battlefield in order to enact his revenge (22.345-54).[36] This murderous desire is expressive of the intensity of Achilles' rage and desire for revenge, symbolizing his psychological state of pain and anger, but at the same time the image of consumption of raw flesh is representative of the natural process that occurs when dead bodies are left exposed on the battlefield.[37]

These emotions are further intensified in the reactions of Hector's family. Priam's ardent wish that Hector had died in the arms of his parents and hope that he can ask Achilles to return the body (22.427-8) indicates the significance to family of a proper funeral and death without bodily mutilation.[38] He imagines that his own body, like his son's, will be feasted upon by dogs, but in his vision, the dogs he has reared in the palace will drink his blood when he is killed. For Priam, his aged, mutilated body will be the height of dishonour (22.66-76). The motif of dogs consuming a corpse is both symbolic of Priam's fall from high status to utter humiliation upon the destruction of his city, and prophetic of the natural end for those whose bodies are abandoned with no kin to bury them.[39] Hecuba's focus is different. Her grief for her dead son is accompanied by intense anger targeted at her son's killer, whom she calls an 'eater of raw flesh' (24.207, ὠμηστὴς). Her emotions lead her to desire revenge in kind by eating Achilles' liver raw (24.212-14). This wish is not a realistic one that Hecuba plans to execute; rather, the imagery of cannibalism conveys the intensity of her emotions.[40] Hecuba manifests the same powerful rage and grief as Achilles and reflects her similar psychological state in her speech. Both take on the animalistic desire to consume their enemy in revenge that is suggestive of the doglike, devouring characteristics of the Furies, personifications of revenge, who within Aeschylus' *Eumenides* threaten to suck the blood from Orestes while he is still alive (*Eum.* 264-8).[41]

In killing Hector, Achilles is depicted using cool calculation in aiming his shot as Hector charges towards him. He locates the gap in the armour where Hector's neck is exposed and drives his spear right through (22.326-8).[42] This part of the description highlights his experience and training as an elite warrior, calmly finding the best location to kill his enemy swiftly. At the same time, his speeches to Hector make clear his intention to dishonour him by maltreating his corpse. Following the killing of Hector, the other Achaeans gather round to attack Hector's corpse (22.369-74). As Lawrence Tritle has noted, this irrational, brutal attack on a man who is already dead is indicative of an outpouring of emotions caused by the painful loss of comrades that Hector had killed (1997: 129–30). Achilles' treatment of the corpse is also driven strongly by his emotions of grief and anger (24.22).[43] He first attaches Hector's body to his chariot and drags it through the dust in front of the city of Troy (22.396-404; cf. 22.464-5), then persists in dragging Hector's body around the tomb of Patroclus over the following days (24.14-16). Achilles is once again likened to a lion attacking flocks in his ferocity towards Hector's body, this time by Apollo, who decries Achilles' lack of pity and his relentless fury towards Hector even though he is dead (24.42-5).[44] Not all the gods agree with Apollo, though. Rather, each god's assessment of the appropriateness of Achilles' treatment of Hector's body is determined by their loyalties in the war (24.23-6), indicating that Apollo's view is not a general objection to corpse mutilation, but is a specific objection because of his love for Hector and allegiance to the Trojan people.[45]

Battlefield Decapitation

Achilles' mutilation of Hector does not include decapitation, though he promises Patroclus he will bring him Hector's head and armour (18.334). This fate is the anticipated outcome, and justifiable within the socially expected performance of reciprocity on the battlefield,[46] as Hector's plan had been to drag away the body of Patroclus, sever his head and throw it to the dogs (17.125-7), though he does not succeed because of the intervention of Ajax.[47] Decapitating the corpses of leading figures is threatened several times in the *Iliad*, though it is not enacted. The threat that is expressed therefore functions not as a remnant of uncivilized warfare practices that the poet has failed to erase completely, as Gilbert Murray proposed (1967: 10), but rather as an expression of the combatants' emotions, a verbalization of their desire for revenge in threat form, that is ultimately not performed in these cases. The act is intended to restore the balance of honour in favour of a killed warrior, and it is also suggested that it would provide comfort to the dead man's family, as can be seen from Euphorbus' threat to bring Menelaus' severed head home to his parents to console them for the loss of his brother (17.39-40).[48] At the same time, the proposed removal of the head is symbolic of the loss of leadership and status. The threatened display of Patroclus' head on the palisade that Iris uses to goad Achilles into action (18.175-7) further indicates that the purpose of the removal of the head is to demonstrate supremacy over an enemy leader, and with it your threat to other warriors, while humiliating the defeated man.[49] Cutting off the head for display is significant in that it is identifiable in a way other limbs are not, as well as indicating that the enemy is dead (Vermeule 1979: 107; cf. Fields 2005: 69). At the same time, removing the head from the body means the corpse is deprived of its identity and becomes unrecognizable, dishonouring the corpse (Muller 2014: 47). The instances of threatened decapitation that occur in the *Iliad* involve significant characters (Fantuzzi 2020: 259) and are separated from examples in which a man is mutilated as part of fighting or directly following his death without the head being retained as a trophy, such as when Agamemnon shames the sons of Antimachus because their father dishonoured Menelaus (11.122–7) or avenges himself for being wounded (11.238-40).[50]

Similar motivations for targeting the head as the subject of mutilation are expressed in later Greek sources also. While decapitation for the purposes of trophy-hunting is mentioned in some cultural settings by Herodotus and Xenophon,[51] their examples of decapitation in warfare involving Greeks and Persians are instances concerning the removal of the head of a leader who has been killed on the battlefield for display (Muller 2014: 70). In these instances of decapitation, defeated leaders are targeted specifically, rather than the practice being generally undertaken as part of warfare. Within Greek sources, the head can be read as part of a 'cultural coding' of the body wherein heads, especially the heads of rulers, signify control and leadership.[52] Removal of the head in warfare is a graphic illustration of the removal of the leader from his position of power and status at the top of the chain of command, as well as being indicative of the humiliation of defeat for the commander. Xenophon recounts how Cyrus the Younger's head and right hand are severed after his failed attempt to seize power from his brother Artaxerxes (*Anab.* 1.10.1).[53] The generals of the Greek troops are also killed by beheading (2.6.1). Histiaeus of Miletus, instigator of the Ionian revolt, is also decapitated after his capture and his pickled head sent to the Persian king Darius (Hdt. 6.30.1). Another leader who is involved in rebellion, Onesilus of Salamis in Cyprus, has his head cut off and displayed above the gates of the Cypriot city of Amathus, according to Herodotus (5.114.1) after he is killed in battle (5.113).[54] Herodotus similarly maintains that Xerxes' rationale for decapitating Leonidas after the defeat of the Spartans at Thermopylae and sticking his head on a stake was because Leonidas was a king and general (7.238.1). Herodotus goes on to explain that Xerxes' behaviour was not motivated by tradition but by emotions, when he states that Leonidas made Xerxes furious (7.238.2). In each of these instances, the leaders who were decapitated were felt to have generated particularly strong emotions in their opponents through their military rebellions. The display of the head confirms the defeat of a rebellious foe, as well as being demonstrative of the strong anger generated by his actions.

Similar feelings underpin the Herodotean tale of Lampon of Aegina, who suggests to the Spartan king and general Pausanias that the head of the Persian general Mardonius should be cut off and displayed on a pole following his defeat at the battle of Plataea (9.78). According to Herodotus, Lampon explained that treating Mardonius' head in this way would be appropriate vengeance because Mardonius had been involved with Xerxes in maltreating the body of Leonidas. However, Pausanias maintains that the actions proposed by Lampon are un-Greek (9.79).[55] He rejects the idea, saying that he does not crave personal honour and that Leonidas has been avenged sufficiently (9.79; cf. 9.64). This rejection of corpse mutilation coheres with classical Greek ideals about treatment of the fallen in warfare. Notably, Pausanias also tells the men not to touch the spoils, but asks the helots to gather it for a special tribute to Delphi (9.80), further indicating the way in which he distinguishes his approach from the Homeric style where booty confers honour upon an individual warrior. However, Herodotus states that the men actually stole as much as they could of the booty and sold it for profit, revealing his expectation of a gap between ideology and practice. Although Herodotus represents mutilation of corpses as un-Greek, he simultaneously shows at least one Greek encouraging the practice because of a belief in appropriate vengeance, making clear that this issue is not straightforward. Elsewhere Greek troops are said to make use of battlefield decapitation when it is strategically beneficial, even though the practice is not usually associated with hoplite warfare. The Greek mercenaries returning from their campaign serving Cyrus the Younger are said by Xenophon to have severed enemy heads and displayed them as a tactic to terrify their opponents (Xen. *Anab.* 3.4.5).[56] While examples of this kind are infrequent in Greek texts, it is possible that bodily mutilation in warfare featured more significantly than Greek authors typically suggest, driven by emotions such as anger and fear and the desire for revenge, as well as being employed tactically when troops perceived they could gain military advantage by its use.[57]

Lampon too mentions military advantage in his attempt to persuade Pausanias to decapitate Mardonius. He maintains that the mutilation would also serve as a deterrent to any others who might think of attacking the Greeks. The utilitarian notion of display of body parts generated in warfare to make aggressors think twice about engaging further in warfare again sits alongside the concept of revenge driven by strong emotions, especially anger. These explanations of Lampon indicate that a Homeric sensibility towards battlefield decapitation, where bodily mutilation on the battlefield is said to be praiseworthy when it delivers just revenge, was thought to persist for some Greeks in particular situations,[58] although Herodotus deems the suggestion 'most unholy' (9.78, ἀνοσιώτατος).[59] This type of thinking about the burial of dead bodies that have fallen in warfare through proper ritual, rather than leaving them exposed or mutilated, was debated on stage in Sophocles' *Antigone,* which features a difference of opinion regarding how the body of a traitor fighting against his own land should be treated: whether the treachery merits exposure, or whether proper burial should be undertaken because of the need to follow appropriate rituals. In these authors, as in Homer, the decapitated or mutilated body is used symbolically to illustrate the loss of status and leadership of the rebel who is punished for his insurrection. The head or body is displayed to demonstrate the strength of the conqueror, as well as to deter further acts of rebellion. At the same time, the heightening of emotions in these particular examples, which in several of the cases involve fighting between brothers and acts of treachery, is expressed in the texts though the winner's urge to mutilate the corpse.

Tomyris and Cyrus

The motif of battlefield decapitation and corpse abuse that features predominantly in Herodotus' account of the death of Cyrus the Great encompasses a similar network of associations as discussed in previous examples, and seemingly draws inspiration from other literary sources in its use of imagery.[60] In Herodotus' telling, it appears to be heavy in symbolism and to depict a mythologized version of the death of Cyrus,[61] just as Cyrus' birth and early life are highly resonant of a folktale (1.107-22). Herodotus notes that other versions of Cyrus' death were in circulation,[62] but indicates that he believes this version to be the most likely (1.214.5), despite its folktale qualities.[63] The tale of Cyrus' death slots into the story-arc of his life, where he starts off life in poverty, grows to power and success through his innate qualities, but is ultimately destroyed through overconfidence and the desire to extend his power too far.[64]

Cyrus is said to have died as part of a campaign against the Massagetae because he angered their queen Tomyris. That his intent was to add to his power and lands is made clear by Herodotus, who maintains that Cyrus initially aimed to gain control of the kingdom by marrying Tomyris. She, however, understands Cyrus' sly intentions when he proposes marriage, and rejects his offer (1.205.1). So, Cyrus advances with his army to achieve his goal. In the early part of the story, Tomyris is cast in the role of wise counsellor, intelligently recognizing what Cyrus intends and attempting to persuade him not to attack (1.206).[65] Croesus also offers Cyrus tactical advice, including one point based on the gender of Tomyris, suggesting that it would be shameful for Cyrus to retreat before a woman (1.207.5).[66] Croesus advises that the Persians make use of deception to achieve victory (1.207.6-7). That the men were thought to make tactical military decisions based on thinking about the concepts of honour and shame is suggestive of the impending decisions about the treatment of the dead body of Cyrus, based on similar concepts.

At the same time, the men are said to use tactics involving trickery, employing the type of deception associated with female leaders elsewhere in Herodotus,[67] as well as being reminiscent of Odysseus and Diomedes attacking Rhesus as he slept (*Il.* 10.474-97; cf. *Rh.* 611). They leave a weak portion of the Persian army in the Massagetae territory, along with a large banquet and a third of the Massagetae troops, led by Tomyris' son Spargapises, and believe they have achieved an easy victory over their enemy when they have defeated just this small group. Their decision to feast on the banquet rather than maintaining their guard indicates their naivety.[68] Off-guard and drunk from the wine, they fall asleep, leaving them open to attack by the remainder of the Persian army.[69] Many are killed in their sleep and Spargapises is taken captive (1.211).

Tomyris' response on learning what has happened indicates the associations of honour with victory through force of arms in battle, but that use of trickery and especially overcoming the men with wine rather than through military superiority is less honourable (1.212). She emphasizes that Cyrus is unreasonable and has an insatiable desire for blood, indicating with this image the extent of his quest for power over others.[70] The association of insatiable desire for blood with empire-building turns at the end of Tomyris' message into an association of desire for blood with vengeance. This set of associations is evocative of Aeschylus' Clytemnestra, who, similarly, is characterized as vying for supremacy and caught up in a cycle of bloody revenge because of the ongoing feud over power in the house of Atreus.[71] Her famous speech that spells out her desire for control over Agamemnon while speaking metaphorically of his death (*Ag.* 958-74) is redolent of the potent destructive powers of female speech.[72] Her oath (1431-7) alludes to concepts of justice and revenge connected to her child (cf. Fletcher 2007). Similarly, Tomyris' oath, which Cyrus ignores, is also related to concepts of justice and revenge in connection to her child and, like Clytemnestra's speech before the death of Agamemnon, portends Cyrus' doom.

Since Cyrus elects not to follow Tomyris' wise counsel,[73] he does not release Spargapises, who chooses to kill himself because of the dishonour of his capture while drunk and his failure to recognize the ruse (1.213). Tomyris determines to enact the oath she swore by advancing against Cyrus in full military strength (1.214.1). Cyrus is said to have been killed in battle, along with much of the Persian army. Herodotus maintains that Tomyris sought out the body of Cyrus among the Persian dead, indicating her desire to target this particular Persian for revenge.[74] She places blame on Cyrus as the leader and determines to fulfil her oath by making Cyrus drink his fill of blood as she had promised. This image of blood-drinking as a metaphor for insatiable desire for power is transformed into a visual image when Tomyris is said to place the severed head of Cyrus into a wineskin filled with human blood (1.214.4).[75] Though Tomyris' oath is metaphorical in quality, she fulfils it literally. She explains her behaviour through a revenge speech that outlines how her act is not a random one, but one that she had promised to undertake as part of her oath of vengeance. This speech indicates that Cyrus' behaviour was dishonourable, in that he defeated her son using guile, and hence his treatment at her hands is now appropriately dishonourable through her maltreatment of his corpse.[76] As seen in previous examples, Cyrus' severed head signifies the end of his reign, as Herodotus makes clear when spelling out the length of Cyrus' rule over the Persians (1.214-3). The loss of his head is symbolic of the loss of his rule and status. The maltreatment of the head indicates the strength of Tomyris' emotions, enacting this revenge in a way that simultaneously replicates how her son had been defeated by drinking wine and captures the imagery of eating raw flesh discussed earlier in this chapter.[77] Tomyris herself does not express the wish to eat Cyrus raw. Instead, she makes Cyrus into the drinker of blood, characterizing him as the bestial aggressor who has killed her son through his power-hungry actions. Here, Cyrus is not the powerful wild beast of Homeric simile defeating his enemies. Instead, with this image of Cyrus' blood-drenched head, Tomyris reveals

Cyrus' flaws as a leader, in that his severed head indicates the failure to meet his military goals and his loss of leadership, while its coating of blood indicates the negative consequences of his expansionist policies.

Like the Homeric Achilles, Tomyris is shown abusing the corpse of her defeated foe out of a desire to avenge a loved one. Just as Achilles promises Patroclus that he will undertake bodily mutilation of the enemy before joining battle against the Trojans, Tomyris promises bloody vengeance metaphorically, though the reader only understands her full meaning after she has completed her revenge. Tomyris is shown to be rational and able to offer wise advice, while also exhibiting heightened emotions that lead her to punish Cyrus through bodily mutilation and humiliation. Achilles too shows cool calculation in his attack on Hector, as well as extreme pain and anger exhibited through his mutilation of Hector and especially his declaration that he wishes he could eat him raw. Further comparison is possible between Tomyris and the Iliadic Hecuba, who wishes to avenge her son by eating the body of his killer raw.[78] The motif of drinking blood is used in this example, though the dead man is forced to 'drink blood' to fulfil his alleged desire to drink the blood of his enemies. As such, Tomyris' actions, like Achilles', can be seen to fit into a pattern of delivery of justice where an aggressor is killed in retaliation for his treatment of those close to the avenger.[79]

Conclusions

This chapter has investigated the complex ways in which the mutilated body of the combatant can be read and interpreted by internal and external audiences of Greek texts about warfare. Mutilation of the body and decapitation on the battlefield involve injuries to the body, which convey a number of intertwined meanings involving concepts of status and assumptions of psychological significance. The threat of corpse exposure is feared because of the way that the body will be mutilated by wild animals after death, dehumanizing it. Warriors who take on the attributes of wild animals to inflict mutilation on their enemies seek to destroy their opponents' honourable death in battle. For those who are mutilated, especially where the head of a leader is removed, the action conveys his demise and is imbued with shame. The head is coded to signify power and leadership, as well as military decision-making. The removal of a head is symbolic in nature, indicating the loss of power of a leader and his defeat in battle, as well as his failure in military strategizing. The maltreatment or public display of the head serves to humiliate the former leader as well as to deter others from attacking. For aggressors electing to mutilate or decapitate the corpse of an enemy, calm strategizing and an out-pouring of emotions are often situated together. The evidence suggests that ancient authors are especially interested in situations where combatants seek out a corpse to decapitate it. Many of the examples here indicate that decapitating a corpse was undertaken in special circumstances where the leader in question had generated intense emotions of anger or pain in his opponent, including through betrayal (such as the examples of rebellions by brothers) or through having harmed someone close to the mutilator, generating the desire for revenge. The image of the mutilated body or decapitated head in

Greek accounts of warfare conveys the intense emotions of pain, anger and loss that have been written into the enemy's body through bloody violence.[80]

Notes

1. See Kucewicz (2016: 428 n. 14); Muller (2017: 211 n.13); van der Plas (2020: 460 n. 6, n. 9) for discussion of the distinction between attacks on the head and neck during fighting and removal of the head after death. My discussion focuses on textual rather than artistic representations, as although there are a number of archaic artistic depictions of mythical mutilations such as the treatment of Hector by Achilles and the decapitation of Troilus (see Gantz, 1993: 597–603), decapitation is not a common theme in classical vase painting. On the black-figure painting of three soldiers seemingly carrying decapitated heads on a lekythos by the Beldam Painter (*c.* 490 BCE; BA331227), see Kucewicz (2020: 203–4 n.68); Vermeule (1979: 107, with 236 n. 30).
2. See esp. Muller (2014); cf. Kucewicz (2016), Kéi (2019); McClellan (2019); Kucewicz (2020); van der Plas (2020). See Ambühl (2016a) for discussion of sensory perceptions and emotions in connection to warfare.
3. Cf. Ambühl (2018) who emphasizes the strategic approach of Tomyris towards revenge from a perspective of gender and ethnicity.
4. Segal (1971b) argues that instances of mutilation increase in intensity in the *Iliad* up to the mutilation of Hector by Achilles. He says that Achilles' thoughts about the treatment of the body of Hector are savage and surpass all others in the epic (ibid., 39). Fields (2005: 58) says that 'the warrior driven by vengeance symbolises the epitome of the savage warrior'. Amaral (1994: 28) calls Tomyris' act savage, macabre and uncivilized. Desmond (2004: 25) compares Tomyris' treatment of Cyrus to Xerxes' treatment of Leonidas at Thermopylae and says that their violent behaviour typifies 'the unnatural excess characteristic of such rulers'.
5. Felton (2014: 52) discusses the ways in which wounding is coded in Herodotus. Cf. Salazar (2000).
6. Felton (2014). For analysis of how wounding can lead to a range of emotions from fear and pain to desire to revenge, see McCoy (2013), especially 6–7 for discussion of Agamemnon's emotional response to the wounding of Menelaus (*Iliad* 4.148-82).
7. McSorley (2014: 113); Sylvester (2012: 493; 2013: 5, 13); cf. Butler (2009); Scarry (1985: 12).
8. Gaca (2010) on rape and enslavement as integral to ancient warfare. Cf. Raaflaub (2014), especially on Euripides' *Trojan Women*. See Hannah Sorscher's chapter in this volume.
9. Ober (1994). He notes that many of the rules broke down during the Peloponnesian War (ibid., 18). See Krentz (2000) on the significance of duplicity in Greek warfare.
10. Cf. McSorley (2013: 9) on the use of televised beheadings in recent campaigns. On feelings of terror in ancient warfare, see Tritle (1997); (2013).
11. McSorley (2013: 2–4); Shaw (2003: 271). See Frembgen (2006: 256–7). Cf. Shay (2003) on the stress response known as 'fight or flight'.
12. Runciman (1998: 736–7) on hoplites' evoked, acquired and imposed behaviour and spontaneous reactions to loss of a comrade.
13. Vlahogiannis (1998: 24). See Lee (2015: 86) on the social impact of body markings.
14. See Jane Draycott's chapter in this volume on disability and prostheses.
15. See *Iliad* 22.283-4, where Hector states that he does not intend to allow Achilles to hit him in the back, but Achilles will have to hit his front as he is charging towards him.

16. On laws of cowardice and the state of *atimia*, see Aeschin. 3.175-6; Lys. 14.6-7; Andoc. 1.74.

17. On the significance of bodily unity, see, e.g., Lee (2015: 82–6); Muller (2014: 41; 2017: 213); Mouton (2020). Kéi (2019) argues for the appreciation of the aesthetics of mutilation alongside the appreciation of the ideal youthful body. On the vulnerability of the wounded hero, see McCoy (2013).

18. See Davies (1994) for discussion of the debate concerning whether Melanthius was mutilated when alive or dead. See Dronke (1969: 173) on the 'living corpse' concept.

19. This punishment occurs elsewhere in Herodotus performed by an Egyptian king (2.162), a Persian noble (3.118) and the wife of the Persian king (9.112). Cf. Muller (2014) 46; Felton (2014: 54 n. 39). Cf. Desmond (2004); Strid (2006: 393) on violent punishments exacted in Herodotus.

20. Muller (2017: 219–20) notes the vividness of the image of blood in the description of Zopyrus' injuries.

21. λωβᾶται λώβην ἀνήκεστον (3.154.2, he mutilated himself irreparably).

22. See Muller (2014) for instances of battlefield mutilation in Greek sources. He notes that mutilation occurred across the ancient world inhabited by the Greeks and was perpetrated by Greeks as well as non-Greeks (ibid., 47).

23. Adkins (1977: 91–2) discusses the allusion to Priam (22.71-6) in these lines. See further below on this example.

24. Cf. Adkins (1960: 43) on shame. On terminology, see Segal (1971b: 15); Vernant (1991: 67–8).

25. Vernant (1991: 70, 74); Lendon (2000: 8–9). See van Wees (1992: 67–8, 115) on the strategic competition for honour among Homeric warriors who can seek to gain advantage over others by inflicting shame on them.

26. Bassett (1933: 47–9); Kucewicz (2016: 430; 2020: 37–8); Muller (2014: 53; 2017: 214); Martin (2020: 122). Vernant (1991: 73) notes that the bodies of Sarpedon, Patroclus and Hector are preserved by the gods and avoid this fate. Following the death of Hector, Priam foresees his own shameful death devoured by his own dogs because his son can no longer protect him (22.74-6). Cf. Vernant (1991: 64). Fear of this fate persists in the classical period and is explored by tragedians in plays about the exposed corpse of Polynices that his sister Antigone feels compelled to bury properly.

27. Van der Plas (2020: 460) notes the distinction from the perspective of the perpetrator. From the perspective of the dead man, the affect is less distinct.

28. See, e.g., *Il.* 16.155-67, where the Myrmidons are likened to wolves that eat raw flesh. Cf. 5.782-94, 7.256, 11.473-84, 15.592.

29. Lendon (2000: 4); Tritle (1997: 127; 2013: 289); Kucewicz (2020: 5); van der Plas (2020: 465).

30. Bassett (1933: 41); van Wees (1996: 51–3); Lendon (2000: 3–11); Kucewicz (2016: 427, 429, 436; 2020: 35); van der Plas (2020: 460).

31. Van der Plas (2020: 460) notes that of the four types of active mutilation she delineates (defensive, aggressive, offensive and necromantic), the one most commonly found in the *Iliad* is 'aggressive mutilation' that concerns the same strong emotions that led to the killing being exhibited in the mutilation.

32. Cf. Frembgen (2006: 256–7) for the complex blending of primal emotional motivations and rational, culturally motivated retaliation.

33. Shay (2003) analyses Achilles' actions using a comparative psychological approach. Achilles' grief at the loss of a comrade (ibid., 41–54) leads to his furious 'berserker' state (ibid., 77–99),

in which he enacts high levels of brutal violence before finally perpetrating corpse mutilation (ibid., 117–19). Cf. Tritle (1997).

34. Achilles' disposal of the corpse of Asteropaeus in the river where his fat will be devoured by fishes and eels (21.200-4) distresses the river god Scamander. See Segal (1971b), ch. 4, on the violence in this episode. See Homes (2015) on the personification of the river and his identity as a 'fluid body' (2015: 34 n. 12).

35. Hector's request for a pledge that they will not mutilate one another's corpses (22.256) is suggestive of the fact he anticipates that this is Achilles' intention. His further request to avoid this fate (22.338-43) shows how negatively it is perceived and how important proper burial is to dying soldiers.

36. Van der Plas (2020: 468, n. 31) categorizes this as 'offensive mutilation' where the primary goal is not to kill, but to mutilate the enemy, in this case through cannibalism.

37. See Muller (2017: 215) for discussion of Tydeus' eating of his enemy's severed head and Athena's disgust. Eur. *Meleager* fr. 537, ὁρᾷς σὺ νῦν δή μ᾽ ὡς ἐπράϋνας, Τύχη (he will come to cannibal pleasures and tear the head of Melanippus with gore-red jaws, trans. Collard and Cropp). Cf. Apollodorus, 3.76-77. Schol ad *Il.* 5.126 notes that Tydeus behaved like a beast.

38. See Kucewicz (2020: 41) on the significance of proper burial. The impact on the family members of Promachus is referred to by Peneleos after he severs the head of his opponent and displays it on the end of his spear (*Il.*14.500-505). Euphorbus claims his parents, distraught at the death of his brother, will gain comfort from receiving the severed head and armour of their son's killer, Menelaus (17.33-42).

39. Segal (1971b: 33) interprets this image as 'the destruction of civilized values'. He further notes referring to the death of Hector that 'the outrage of the corpse comes to forebode the outrage of Troy itself' (1971b: 47).

40. O'Brien (1990: 116) sees this as moral degradation in both Achilles and Hecuba. Cf. Harris (2003: 12) for criticism of Hecuba's anger. Segal (1971b: 38) calls Achilles' statement 'savage'. But as van der Plas (2020: 468 n. 31) notes, although Hecuba's desire is shocking, it is a wish that is unfulfilled. Cf. Ambühl (2018: 32).

41. Fowler (1991: 91–3). She discusses Clytemnestra, who is also characterized as a dog-like blood-drinking Erinys (A. *Ag.* 1428–9). Cf. also *Ag.* 1188–90 on the Furies as drinkers of human blood. See Hall (2018) on the Erinyes. O'Brien (1990: 106) notes that Zeus describes Hera as desiring to eat Priam and the Trojans raw to indicate the intensity of her fury towards them and desire for revenge (4.34-6). Zeus also likens her to a dog (8.483) in her wrath (Graver 1995: 48). Euripides' *Hecuba* is transformed into a dog after performing revenge (1265).

42. Felton (2014: 51) on the vulnerability of the body to attack in gaps between pieces of armour. McClellan (2016; 2019: 39–40) discusses the example of Imbrius (*Il.* 13.170-5), who is killed by Teucer and decapitated post-mortem by Ajax, arguing he is a stand-in for Hector. His head lands at the feet of Hector. McClellan goes on to speculate on whether Hector was decapitated by Achilles in other versions.

43. ὣς ὃ μὲν Ἕκτορα δῖον ἀείκιζεν μενεαίνων (24.22, so in his fury he mutilated noble Hector).

44. Apollo protects the skin of Hector so that Achilles cannot tear it (24.18-21), undermining Achilles' attempts to disfigure his enemy's body.

45. Kucewicz (2016: 425–6) argues that there is a lack of evidence for moral objection to mutilation in the *Iliad*; cf. Bassett (1933: 41); van Wees (1992: 129–30); Fantuzzi (2020: 259). Segal (1971b: 13) states that corpse mutilation in Homer generates repugnance and some moral outrage. Lendon (2000: 9) says that mutilation offends the gods. On vengeance and infliction of dishonour as part of Homeric morality, see Yamagata (1994: 144). Plato (*Resp.* 3.391b-c) condemns Achilles' treatment of Hector's body.

46. Kucewicz (2016: 430). Cf. McClellan (2019: 39–40) on the hints that Achilles would decapitate Hector.
47. The Trojans' plans to drag away Patroclus by passing a strap around his ankles (17.289-94) also mirror Achilles' treatment of Hector. Cf. Bassett (1933: 56).
48. Van der Plas (2020: 467) categorizes this as a necromantic mutilation where the head is a trophy.
49. Van der Plas (ibid.) notes that this planned mutilation is a 'strategic and psychological weapon instead of an impulsive act'.
50. Ibid. (462–3). Cf. Lendon (2000: 5, 8); Kucewicz (2016: 428–9).
51. In Herodotus (4.65), Scythians drink from the skulls of their enemies (4.103.3), and Tauri use enemy heads on stakes to guard their houses. In Xenophon (*Anab.* 4.7.16; 5.4.17), local warriors that the Greek army encounter collect heads to parade them while singing and dancing. See Fields (2005: 55–6) for discussion of these alongside examples from Roman sources and archaeology.
52. See Felton (2014) on the significance of the thigh as a bodily code in Herodotus. She notes that thigh wounds are frequently suffered by kings (ibid., 51) and that Herodotus uses the wounding of Cambyses in the thigh as a literary technique to symbolize his demise (ibid., 53).
53. On the different accounts of the death in battle of Cyrus by Xenophon, Ctesias and Deinon, see Bassett (1999).
54. On the possibility that Onesilus' decapitation forms part of heroization ritual, see Petit (2004); Hornblower (2013: 298).
55. See Muller (2014: 41) on the tendency of Greek texts to associate bodily mutilation with barbarians. Cf. Kucewicz (2016: 425), and especially (2020) for discussion of the shifts in thinking about mutilation in classical times.
56. Tritle (1997: 127). See Muller (2014: 60) for discussion of examples in Diodorus Siculus where mutilation is inflicted as a kind of education on armies who use mutilation as a tactic (14.46.3-4; 19.103.5).
57. Fantuzzi (2020: 260) hypothesizes that criticisms of mutilation in warfare might have circulated principally in intellectual circles. Cf. Kucewicz (2020) for discussion of differing ideas about appropriate treatment of the dead connected to elites and ordinary people.
58. Kucewicz (2016: 436) notes that 'the mere fact that Lampon suggested it in the first place indicates that the practice of mutilation was still regarded by some as a viable way of exacting revenge'. Cf. Fantuzzi (2020: 260).
59. Cf. 7.238.2 on Xerxes' transgressive treatment of Leonidas' corpse. Cf. also Hdt. 3.16.2 on Cambyses' 'not holy' (οὐκ ὅσια) treatment of the body of Amasis, which goes against religious and cultural norms (3.16.4).
60. Cf. Ambühl (2018).
61. Payen (1991: 39); Mitchell (2023). Weststeijn (2016) draws comparisons between this tale and those of Judith, Zenobia and Jalila that all feature wine, blood and revenge.
62. In Xenophon's *Cyropaedia*, Cyrus dies peacefully in his bed (8.7.28), while in Ctesias' *Persica*, (fr. 9) it appears that Cyrus is wounded at war, but then returns home to die. In Diodorus Siculus, Cyrus is captured in battle and crucified (2.44.2). See Sancisi-Weerdenburg (1985) on the different versions.
63. The Herodotean version including Tomyris and Cyrus' decapitation following defeat in battle is heavily reflected in later versions (Lucian *Charon* or *The Inspectors* 13; Polyaenus *Strat.* 8.28.1; Amm. Marc. 23; Just. *Epit.* 1.8.13; Val. Max. 9; Paulus Orosius *HAP* 2.7).

64. On the mythologization of the historic figure of Cyrus, see Mitchell (2023). On the rise and fall story pattern for a leader that is typical in Herodotus, see Immerwahr (1966: 75–8); Avery (1972); Amaral (1994: 18); Harrison (2000: 43–5). See Saïd (2002: 128–9) on the resemblance to tragic plotlines.
65. Cf. Romero Mariscal (1998–9: 241). Amaral (1994: 19) compares Tomyris to Artemisia who offers wise counsel to Xerxes that he praises, but ignores (8.68).
66. Cf. Ambühl (2018: 28) for a detailed account of gendering in this tale.
67. See especially Hdt. 2.100 on Nitocris who uses a similar trick involving a feast to lure her brother's enemies to their death in revenge for killing him. See also Artemisia's trick at Salamis (8.87-8). See Lateiner (1990) on Herodotus' appreciation of tricksters. In Polyaenus' *Stratagems*, 8.28.1, it is Tomyris that makes use of deception against the Persians fitting gendered stereotypes about use of trickery in Greek literature. Cf. Ambühl (2018: 24–5, with n. 15, 39), who notes the reversal of gender roles in the purported dispositions of Cyrus and Tomyris.
68. On associations of the banquet with destruction in Herodotus, see Bowie (2003).
69. On legal attitudes to wine and drunkenness, see Dimopoulou (2019).
70. Cf. Hdt. 4.64 for Scythians drinking the blood of the conquered enemy as a symbol of military success. They also take captured heads to the king in return for booty, again a physical demonstration of bodily dominance.
71. Cf. Ambühl (2018: 33–4) on the comparison of Tomyris to Clytemnestra as vengeful mother.
72. See McClure (1996–7) on Clytemnestra's speech as a binding spell.
73. Avery (1972: 537–8) notes Cyrus' failure to follow wise counsel here and Croesus' errors in his advice to Cyrus at this stage in his career.
74. Herodotus does not specify whether Tomyris herself severed Cyrus' head, though she is said to do so in some other sources (e.g. Lucian *Charon* 13, Τόμυρις ἐκείνη ἐστί, καὶ τὴν κεφαλήν γε ἀποτεμοῦσα τοῦ Κύρου αὕτη ἐς ἀσκὸν ἐμβαλεῖ πλήρη αἵματος – That is Tomyris; and after she has cut off Cyrus' head she will plunge it into a wine-skin full of blood). Cf. Ambühl (2018: 29, with n. 20, 36).
75. See Muller (2017: 220) on Herodotus' interest in blood in this story.
76. Romero Mariscal (1998–9: 243) notes that Cyrus' dishonourable actions are punished by Tomyris' mirroring reactions.
77. On the significance of wine and blood in this story, see Weststeijn (2016).
78. Cf. Ambühl (2018: 32). See also Ambühl (ibid., 34–5) and Romero Mariscal (1998–9) on comparison to the bloody revenge of Euripides' *Hecuba* against the killer of her son Polydorus. Her transformation into a dog perhaps evokes the doglike Erinyes (E. *Hec.* 1259-74). Weststeijn (2016) sees Tomyris' blood revenge as inverted childbirth. On Tomyris as maternal spirit of revenge, see Amaral (1994: 27).
79. Berger (1979: 21) notes Tomyris was viewed in early centuries as a 'symbol of heroic strength and just retribution'. Cf. Ambühl (2018: 22, 30–1). Notably Herodotus does not offer his own perspective on Tomyris' actions.
80. I would like to thank Annemarie Ambühl, Simon Hornblower, Cezary Kucewicz and Lynette Mitchell for their helpful comments on this chapter.

CHAPTER 3
BODIES OF THE MILITARY COMMUNITY: A PERSPECTIVE FROM ROMAN BRITAIN

Rebecca Redfern and John Pearce

Introduction: Roman Military Bodies

Across Europe, human remains of Roman date (first to fourth centuries CE) buried with military equipment have been encountered in burial grounds and clandestine circumstances (Table 3.1). These remains are predominantly those of young adult males, some with origins distant from the place in which they were buried. Often, they are characterized by healed accidental and intentional fractures to their long bones, with musculoskeletal markers and fluctuating asymmetry suggesting intense physical activity (i.e. sidedness linked to weapon use). Perimortem weapon injuries are also frequent, indicating that these men had often met violent ends. As the likely burials of soldiers, these seem examples par excellence of the physiological impact of military service in a Roman setting. However, such cases of soldiers, buried with weapons and adequately studied from a bioarchaeological perspective, are too few to develop an aggregate understanding of the Roman military experience from the perspective of its bodily consequences, so we have turned to other evidence. The traditional approach has been to mine the thousands of surviving epitaphs for soldiers. These can provide vivid insights into individual combatants' biographies, exemplified for Britain by the cavalryman Agathopus from the Aurès uplands of Algeria (Mann 1993) or the legionary veteran Valerius from the foothills of the Alps (Sauer 2005). As will be explored below, these documents are indeed potential witnesses to key aspects of Roman military experience, age of recruitment and origin, years of service, retirement with veteran status and age at death. However, our increasing sensitivity to the potential atypicality of surviving inscriptions has introduced much greater caution in making aggregate inferences concerning typical military service (Hope 2003; Scheidel 2007; Speidel 2015). The study of human skeletal remains from an osteological and biomolecular perspective offers an opportunity to study military lives independently of some of the methodological problems associated with epitaphs. In order to illustrate their potential, and in the absence of widespread furnishing with weapons as a likely marker for identifying military burials, we turn to another means of identifying soldiers' remains, namely injury recidivism. 'Injury recidivists' comprise a clinically defined group, typically men, with an average age of twenty-six years old, who in a modern setting are repeatedly admitted to trauma centres (Redfern, Judd and DeWitte 2017a). These trauma data and the osteobiographies of several Roman injury recidivists will be explored to illuminate the combatant experience. In this chapter, we first sketch what is known of soldiers' lives,

Table 3.1 Summary of Roman military personnel excavated from across Europe

Location	*Date* CE	*Military signifiers*	*Sex and age*	*Bioarchaeological summary*	*Reference*
Resnik (Croatia)	2nd century	Socketed spearhead, Roman pottery, coin and zoomorphic fibula	Male, 35–40 years old	• Stature: 165 cm • Indicators of stress: linear enamel hypoplasia, periosteal new bone formation to the tibiae and fibulae • Healed fractures: the distal third of the left ulna and right fibula • Activity: musculoskeletal markers to arm bones and bone growth at the insertion for the Achilles tendons • Vertebrae: Schmorl's nodes	Šlaus, Domić Kunić and Pivac (2018)
Dyke Hills (Britain)	Early 5th century	Elaborate belt set and axe	Male, 30–40 years old	• Ancestry: White European • Mobility isotopes: non-local to Britain, likely to be from Northern Europe • Stature: 174.7 cm • Fluctuating asymmetry: right forearm larger than left • Congenital anomaly: dysplastic left scapula neck • Miscellaneous pathology: concha bulbosa	Booth (2014)
Belgrade (Serbia)	4th century		Male, 24–28 years old	• Stature: 169–177 cm • Indicators of stress: enamel hypoplastic defects and periosteal new bone formation to the lower limbs • Miscellaneous conditions: porotic hyperostosis to the cranial vault • Musculoskeletal markers: pelvis, leg and arm bones and scapulae • Perimortem injuries: projectile perforation of the ilium, proximal third of the right femur and a depressed fracture to the right parietal bone	Golubović, Mrdjić and Speal (2009)
Camelon (Scotland)	1st to 2nd centuries	Cist burials containing swords, shield-boss, possible belt-fitting and a spearhead	1. Indeterminate, 16–18 years old 2. Indeterminate, > 20 years old		Breeze, Close-Brooks and Ritchie (1976)

			Clandestine burials		
Housesteads (Britain) Found under a floor	4th century	Sword found mingled with the torso of the male skeleton	Adult male Fragmentary female	'From among its [male] ribs came what appears to be the point of a sword' (Birley, Charlton and Hedley 1933: 88)	Birley, Charlton and Hedley (1933)
Velsen I (Netherlands) Found in a well	1st century	Belt set, sword	Male, 20–30 years old	• Ancestry: White European • Mobility isotopes: birthplace North African Mediterranean coast, younger childhood spent in the Eastern Mediterranean • Stature: 184–191cm • Indicators of stress: dental enamel hypoplasia; periosteal new bone formation • Fluctuating asymmetry: right arm more developed than the left • Musculoskeletal markers: arm and leg bones; right arm more developed • Osteoarthritis: foot bones and vertebrae • Ante-mortem fracture: two lower thoracic body fractures; avulsion injuries to the superior aspect of the olecranon of the ulnae • Perimortem fracture: frontal and right parietal bones; healed fracture to sternal end of left second rib; healing fracture to neck of second right rib • Perimortem weapon injuries: multiple cutmarks to the pleural surface of a left rib shaft; multiple injuries to the left ilia • Miscellaneous conditions: periosteal new bone formation to the pleural surface of the ribs Osteoarthritic conditions: right foot bones	Constandse-Westermann (1982); Inskip (2020); Morel and Bosman (1989); Waters-Rist and Palmer (2016)
Canterbury Castle	1st to 2nd centuries	Two military swords	Two adult males		Bennett, Frere and Stow (1982)
			Ditch fills	Adult male with a perimortem weapon injury to skull; disarticulated male skull	

before assessing to what degree funerary evidence brings greater precision to our study of the embodied experience of service. Roman Britain is a case study, offering the opportunity to compare the evidence of inscriptions with relatively abundant data from human skeletal remains.

Roman Military Bodies

> Thus the men were carefully selected in the first place for their good physique; they were kept in excellent condition. (Webster 1998: 252)

In the Roman world, concepts of the male body and maleness were bound up with military service, with male identity being 'tightly entwined with the notion that precarious manhood was best demonstrated and won on the battlefield' (Stewart 2016: 13; Chidwick, this volume). While military *virtus* is best documented for elite males (Alston 2010), most soldiers were volunteers and mass draftees, especially from newly conquered territories, and sent elsewhere in the Empire (Watson 1983: 31; Wesch-Klein 2007: 436–9; Haynes 2013). Those who volunteered for the long duration of service, twenty-five years for both legionaries and auxiliaries from the end of the first century CE (Wesch-Klein 2007: 439), would not have been accepted without meeting certain criteria. For the legions, the first was that they had citizenship and were at least seventeen years old, although exceptions were made for illegitimate sons, and those with certain occupations were excluded from military service (Watson 1983: 40–41; Scheidel 1996).[1] It is widely accepted that potential recruits also underwent some kind of medical examination, not only to check that they were tall enough (167–77cm; Veg. *Mil.* 5), but also that they were healthy enough – at Oxyrhynchus (Egypt), for example, a weaver was excused from military service because of a cataract (Brewster 1927).

Different aspects of recruits' experience following enrolment have been foregrounded over time. Past research has traditionally emphasized the practical side of military life (see this volume's Introduction), in particular physical training, drawing on the manual of Vegetius. This frames Roman military service as an idealized response to the expectations of *labor* and *industria* as components of military discipline (Phang 2008). The benefits of access to hospitals, medical care and knowledge, bathhouses and sanitation have also been emphasized (Webster 1998).

An ontological shift in the late 1990s and early 2000s (Goldsworthy and Haynes 1999; Allison 2006; Coopey and Džino, this volume) emphasized the huge diversity in the military experience and community, not only between active and settled service, but also the plurality of military roles, including doctors and clerks, as well as fighting men (Watson 1983: 75–7; James 1999, 2006). Other military labour has also been acknowledged, of servants, *corvée* drafts and the enslaved, whether owned by soldiers or the army (Phang, Aubert and Várhelyi 2005).[2] Scholarship has also emphasized the way in which a soldier acquired a military identity through embodied performance, for example through regimes of bodily care (James 1999; Haynes 2013; Coopey and Džino; Llewellyn-Jones's chapter in this volume). Whilst mindful of literary stereotypes of soldiers as oppressive outsiders and dangerous men, James (2009, 2013) has also argued for a scholarly sanitization. He foregrounds violence as a key facet of Roman military experience, threatened and practised by soldiers in mass combat, policing and banditry suppression and in asserting personal interest. Through subjection to military discipline, soldiers too could be the victims of violence (Watson 1983: 125). Their bodies were central to their *virtus,* with their wounds and scars testament to their bravery and service, or (on their backs) as markers of dishonour, beatings or punishments, because of desertion (van Lommel 2015, 2019). Soldiers could be discharged from service if they developed a disabling physical injury or psychological condition (Watson, 1983: 121–3; van Lommel 2013).

At the completion of a partial or full period of service, veterans were still relatively young, most being in their forties or fifties (Scheidel 2007: 426–7). Depending on circumstances and unit type, they were given land or money, as well as other privileges (Mann 2002; Keppie 2003; Wesch-Klein 2007). They might remain part of a military community, remaining in settlements where they had served, but veterans often moved away, sometimes like Agathopus over long distances, from Cumbria to Algeria, 'seeking his homeland again', as his epitaph had it (CIL VIII.4800; Mann 1993; Holder 2017).

The Setting: Death on Campaign and in Camp

In principle, the burials of the military dead offer the opportunity to characterize the actual lived experience of soldiers across the empire. Discovery of the remains of soldiers might be anticipated on or by a battlefield, or in cemeteries adjacent to garrisons.

Battlefield casualties were generally buried collectively on the field of combat (Peretz 2005; Hope 2003). The practice is best attested in the case of the Varian disaster (9 CE). The pathos of the account by Tacitus of soldiers' collecting remains (*Ann.* 1.61) is somewhat undermined by the commingled scraps of human and equid bone, which is all that could be gathered six years after the slaughter (Rost and Wilbers-Rost 2017). No such burials survive from Britain, despite the battles and city sackings of the first century CE and later. Osteological testimony elsewhere to the human consequences of Roman military violence, for example for Caesar's genocidal brutality in northern Gaul (Roymans 2019), torture and mutilation in the aftermath of a civil war in Sertorian Valencia (Ribera i Lacomba 2014), and likely factional strife within the army in Skopje (ancient *Scupi*) (Jovanova 2017), hint at what might one day illuminate Roman violence directly in Britain.

Soldiers dying in service closer to camp were typically buried on the edges of garrison communities, in burial spaces lining the roads, following Roman funerary norms. Distant repatriation was rare, more likely for emperors (e.g. Septimius Severus) than ordinary soldiers (Carroll 2009). These cemeteries also accommodated veterans, household members and others whom compulsion or opportunity had drawn to garrison communities. Such cemeteries are the principal sources of the inscribed memorials surviving from Roman Britain. They comprise burial grounds serving both sites occupied by the army for a few decades during conquest (e.g. Colchester) and the long-lasting garrisons of northern England and Wales. For a variety of reasons, the inscribed memorials and burials proper from these cemeteries are best discussed separately. Their subsequent re-use means that memorials have usually not been found in association with the burials they marked. Memorials to soldiers illuminate individual biographies but are of much more limited use when exploited in aggregate for aspects of military demography, as will be explored below. In recognition of this, recent scholarship has tended to focus on their sociological and ideological interest (Hope 2003; Coulston 2007). With occasional exceptions, the garrison cemeteries of northern England and Wales have also seen very limited excavation compared to urban southern Britain, from which the majority of the well-preserved human remains derive. For this reason, and for the different methodological and interpretive challenges that epigraphic and bioarchaeological information raise, our two datasets are discussed separately.

Epitaphs and Soldiers' Stories

Occasionally individual memorials plunge us into the violent world we might anticipate on Roman frontiers. Found at Ambleside (Cumbria) in 1962, a dual epitaph reported the death of Flavius Romanus, aged thirty-five, *actuarius* or military clerk, likely 'killed in the camp by enemies', *in cas(tris) int(er)f(ectus) ab hosti(bus)* (Figure 3.1) (RIB 3218,

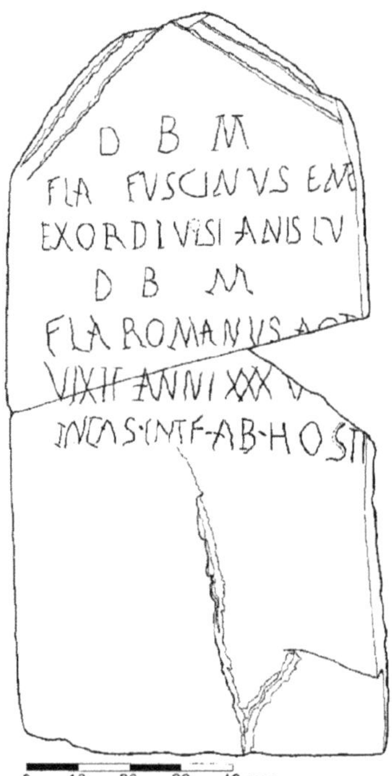

Figure 3.1 The stele of Flavius Fuscinus and Flavius Romanus, Ambleside (RIB 3218). Reproduced by permission of the Administrators of the Haverfield Bequest.

BritRom 14.13).[3] The wording evokes a beleaguered Roman outpost, but the fatality might have occurred at any time between the second and fourth centuries CE, and internecine conflict is as likely an agent of death as local resistance (Shotter 2003). However, as recorded on the 150 or so memorials commemorating Roman soldiers from Britain, such deaths are exceptionally scarce. For example a centurion commemorated at Vindolanda was killed in a war fought around the time of Hadrian's succession (117 CE) (RIB 3364, *BritRom* 5.01; Birley 1998; see also RIB 3330), and a soldier dead on an expedition in Germany (on Caracalla's expedition, by violence?), was commemorated with his mother by his sister at Caerleon (RIB 369, *BritRom* 9.28) in a presumed cenotaph, fulfilling the rite of passage in the absence of a body to bury (Carroll 2009; Hope 2003). Typical of the paucity of references in Roman epitaphs to violent death, these occasional cases remain vivid but anecdotal illustrations of the risks of military service, with most such deaths left in the shadows, perhaps because of the taint of defeat (Reuter 2009; Stoll 2019: 142–3).

An epitaph from Lincoln serves as a more typical example of a soldier's memorial, carved on a 1.5m high stele, one of a cluster commemorating legionaries from the short-lived garrison (RIB 258, *BritRom* 3.03) (Figure 3.2). The text contains many abbreviations typical of contemporary military epitaphs (Carroll 2006: 126–32), and can be straightforwardly expanded and translated:

> T(itus) Valerius T(iti) f(ilius) | Cla(udia) (tribu) Pudens Sav(aria) | mil(es) leg(ionis) II A(diutricis) P(iae) F(idelis) | (centuria) Dossenni | Proculi a(nnorum) XXX | aera [V]I h(eres) I) s(uo) p(osuit) | h(ic) s(itus) e(st)
>
> Titus Valerius Pudens, son of Titus, Claudian voting-tribe, from Savaria, soldier of the Second Legion Adiutrix *Pia Fidelis*, in the century of Dossennius Proculus, aged 30, six years' service. The heir at his own expense set this (stele) up. Here he lies.

Dated to the early 70s CE, this, like other memorials, came into being in a ritual, legal and sociological context (Mihajlović 2020). Interring (the likely cremated) remains of Pudens completed the burial process and established a place where any subsequent commemorative rites could take place ('he lies here'), the stele marking the inviolable status of the site of burial as *locus religiosus*. The adze incised on the stele symbolized the creation of the tomb in ritually proper and thus protected format (Carroll 2006: 81). In its memorial function, the stele emphasized components of Pudens' identity linked to his military service, using information likely deriving from documentation made on enrolment (Speidel 2017: 35) to bequeath to posterity an image of Pudens as the legionary par excellence. The three names, voting tribe and filiation put his citizen status beyond question. Statement of his *origo*, Savaria (Szombathely, Hungary), 2400 km from Lincoln and a cradle of Roman legionaries (Schenk, Amoroso and Blanc 2012: 233), brought to mind the long journeys into service that most soldiers would have recognized, such as Dossennius, his centurion, whose *gentilicium* (family name) is only otherwise attested in Italy (Kakoschke 2011: 94; GN 149). The reference to Pudens' legion in text and image – dolphins and trident symbolize its formation from drafting sailors (Birley 1979: 83) – marked his

Figure 3.2 The stele of Titus Valerius Pudens, Lincoln (RIB 258, British Museum). Reproduced by permission of the Administrators of the Haverfield Bequest.

affiliation to that wider, partly 'imagined community' of 5500 men (cf. James 2013: 117–18). Naming his own century recalled the comrades with whom he shared his barracks in unrelenting intimacy (Haynes 2013). Expressing his record of service as six *aera* (bronzes), an old-fashioned term by the time the monument was made (Schenk, Amoroso and Blanc 2012: 234), the stele evoked the recurring payments that rewarded the labours, privations and dangers of military life, and gave the accumulated riches expressed in the stele itself. To military passers-by, the monument advertised how the (anonymous) heir properly commemorated his likely erstwhile comrade and exceeded his obligations by paying for the monument himself. This unknown heir likely shaped the memorial according to testamentary conditions, local shared conventions and the advice of specialists – stonecutters and so on – responsible for practical execution (Edmondson 2015a).

While in principle such memorials, naming place of origin (*origo*), unit, years of service and age at death, might allow a soldierly life course to be mapped out, the dependence of memorial creation on the resources available to and circumstances of the bereaved prompts caution in drawing wider inferences from such epitaphs. In what follows, we have identified some of the patterns that may be identified and the associated problems (cf. Hope 2003). The sample's smallness, 150 epitaphs for serving soldiers and 27 for veterans, is self-evident and is not representative in several key respects.[4] Although soldiers' epitaphs are more datable than most, few can be dated with precision (e.g. Sauer 2005). Most date to the mid-imperial period, while few if any date to the fourth century CE, inhibiting study of change over time. Their geographical distribution is conditioned by stone sources (Stewart 2010), but other factors are also crucial, above all setting and unit type. There have been found 118 epitaphs found for soldiers and veterans documented at legionary fortresses and from London (7) and Bath (5), respectively a place of service on the governor's staff and a favoured sanctuary. The garrison of Legio XX Valeria Victrix at Chester accounts for almost half of these examples. Epitaphs for legionaries and auxiliary cavalry dominate the memorials to serving soldiers on which the unit name survives or can be deduced (eighty-two legionaries, seventeen of thirty-six for the auxiliary soldiers). Few auxiliary forts possess more than one or two surviving memorials (Hope 2016). For veterans, the sample is also very heavily skewed towards legionary soldiers – only three explicitly indicate that they mark the tomb of an auxiliary veteran. The paucity of surviving memorials to high-ranking officers is likely due to the shortness of their postings to Britain. The dominance of the sample otherwise by better remunerated soldiers up to the rank of centurion (i.e. legionaries, cavalry, promoted auxiliary soldiers), suggests a likely correlation between the resources available to heirs and this commemorative practice (Pearce 2015a).

Geographical Origins of Recruits

The review of epitaph evidence begins with some brief comments on soldiers' origins and service history. It is more common for the *origo* to be specified in epitaphs of first-century CE date in Britain (and beyond) than in later periods. The forty-nine documented legionary *origines* (forty-two serving, seven veterans) span the empire, from southern England (*natione Belga*, RIB 156, *BritRom* 11.16)[5] to south-east Turkey (*nat(ione) Syrus, Os[r(oenus)]*, RIB 490). Within this vast region, the predominant points of origin are more narrowly focused: the colonies of northern Italy, and nodes on the trunk routes by road and river that criss-crossed Northern Europe and the Balkans (Birley 1979: 82–4; Mann 1983). Recruitment to units in Britain conforms to a wider attested pattern. As the first century CE progressed, the places of recruitment shifted westwards and northwards from Italy, best illustrated at Chester (Carroll 2006: 215–16). The relative silence of the mid-imperial period might indicate heavy local recruitment, for example from the colonies and *canabae* of Roman Britain, a phenomenon more explicitly documented elsewhere (Mann 1983). However, names and commemorative formulae on later monuments sometimes hint at the wider zone from which the legions stationed in Britain continued to draw recruits. For example, the epitaph for a centurion at Chester (RIB 491, *BritRom* 9.21) employs the protective formula *sub ascia dedicatum*, echoing wording more common in the Rhône valley (see above), while *Dis Manibus sacrum*, a formula rare in Britain but more typical of North Africa (Tomlin 2018: 207 n. 29) opens the epitaphs of soldiers whose names also hint at origins in the same region (e.g. Irchester, Northamptonshire, RIB 233; Chester RIB 488; Birdoswald, RIB 3345). In the latter case, the connection is confirmed by a stated *origo* of Hippo Regius (Annaba, Algeria).[6]

Understanding the movement of soldiers into auxiliary units is based primarily on the geographical and ethnic names for units, and the diversity of these within Britain was as great as for any other province (Jarrett 1994). Throughout the Roman period it is likely that auxiliary units continued to draw on an empire-wide pool of recruits (Haynes 2013: 121–34). In the few instances where origins for individual auxiliary soldiers are stated, epitaphs speak to the diversity of origins through the names of the people with which soldiers were recruited, rather than a place. For example, two Treverans enrolled as cavalrymen within the *Ala Augusta* referenced their affiliation to their home communities as *Trever* / *civis Trever* (Treveran / 'Treveran citizen') (RIB 606, 3185, *BritRom* 4.1). The memorials to Sextus Valerius Genialis, originating in the Rhine delta as a *civis Frisia(v)us* (RIB 109, *BritRom* 2.07) and Nectovelius, a Brigantian (from northern England), both enrolled in Thracian auxiliary units (RIB 2142, *BritRom* 6.18), reveal the expedients adopted to redirect potential military manpower when units were short of recruits.

Neither for individual auxiliaries or legionaries do the epitaphs usually speak of the circumstances, voluntary or drafted, under which they became soldiers (Wesch-Klein 2007: 436–8). For those successfully promoted, mobility during service can be reconstructed when soldiers changed units during long careers bringing them to garrisons in Britain (e.g. RIB 203, Colchester, RIB 509, Chester) or taking them from the province (e.g. ILS 2653, *BritRom* 9, 18).

Years of Service and Life Expectancy

With the exception of these rare references to movement or (a little more frequently) the documentation of promotion, the epitaphs otherwise give little glimpse into soldiers' life courses. Portrait images, carved albeit in small numbers, represent epitomes of military virtue, revealing the theoretical hierarchies of privileges and powers more than the vicissitudes of any one life (Coulston 2007). The *vitis* (vine staff) gripped by Marcus Favonius Facilis' statuesque simulacrum, for example, signalled the centurion's right to discipline soldiers (RIB 200, *BritRom* 2.01; see also RIB 17, 9.20) (Figure 3.3), while the writing tablets suspended from the left hand of Caeciilius Avitus, *optio* of *legio XX* at Chester (RIB 492), flagged the bureaucratic control he exercised over his subordinates. Some auxiliary cavalrymen were shown in idealized confrontation with a defeated foe, the rider's *virtus* signalled by his pose and trappings, a contrast to the cowering, naked, local man trampled underfoot (Coulston 2007).

Usually, the soldier's life is abbreviated simply to years of service and age at death. It is tempting to use this information to test the number of years actually served vs normative service lengths for soldiers following their recruitment, usually between the ages of seventeen and twenty (Wesch-Klein 2007: 439–40; see above) or to establish their life expectancy. Among the fifty whose service length at the time of death is documented in epitaphs (thirty-six legionaries, fourteen auxiliaries), very few had survived fewer than five years' and few exceeded twenty-five, a similar pattern to neighbouring provinces (Figure 3.4). Age at death is preserved in a slightly larger number of memorials, mostly

Figure 3.3 The stele of Marcus Favonius Facilis, Colchester (RIB 200, Colchester Museum). Copyright: Colchester Museums; image credit: Douglas Atfield.

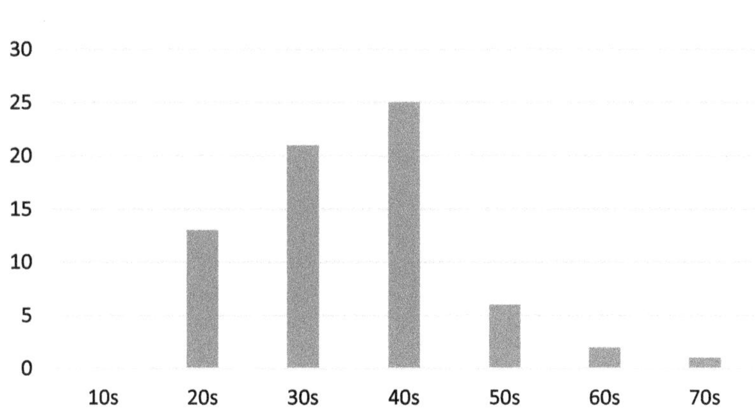

Figure 3.4 Age of death on soldiers' epitaphs, Britain (n = 68).

for legionary soldiers (forty-four legionaries; sixteen auxiliaries; nine unknown), ranging from twenty for a legionary cavalryman (RIB 403) to seventy-two for a *praefectus castrorum* for the legion at Chester (RIB 490), with no obvious distinction between legionaries and auxiliaries in terms of attested age. Among the thirteen veterans, ages at death were documented between forty-five and a hundred.

However, numerous difficulties undermine extrapolation from these figures. The confidence once expressed in drawing conclusions about age at death (e.g. Burn 1953) has been replaced by scepticism over the use of Roman epitaphs for this purpose (Edmondson 2015b: 561; Scheidel 1996). The well-known phenomenon of age-rounding applies as much to ages of soldiers in Britain (and beyond) as it does to other Romans (cf. Sauer 2005: 207, fig. 2). Of the sixty ages at death documented in Britain, forty-seven (68%) are reported as multiples of five or ten, a similar percentage to Lower Germany (Figure 3.5). Among the veterans in Britain, this phenomenon is even clearer: only one given age is not a multiple of five or ten. The contrast with the lack of rounding in years of service, also absent in the sample from Lower Germany, suggests enrolment documentation was used for the latter but not for the former (Figure 3.6). By itself, this is not a disastrous impediment, since the general impression from the British sample is one with some potential demographic credibility; a higher proportion of soldiers died in their forties than in their thirties, a situation reversed in the total sample from Roman Britain, perhaps reflecting better living conditions for the army (Figure 3.6) (cf. Sauer 2005: 208, fig. 203). However, as noted above, the statistics omit casualties on campaign (not normally recorded) and are skewed towards higher-ranking soldiers of likely longer service duration. Information on veterans is also very limited, many perhaps not being explicitly commemorated with an epitaph.

Thus, estimates of soldiers' mortality for the moment remain based on other grounds, especially model life tables, tempered by epigraphic documentation such as discharge rosters. On this basis, Scheidel (2007: 425–6) cautiously proposes that between a third and a half of recruits did not survive the full span of their service. Some privileges in military living conditions such as medical care, sanitation and diet are likely, Scheidel

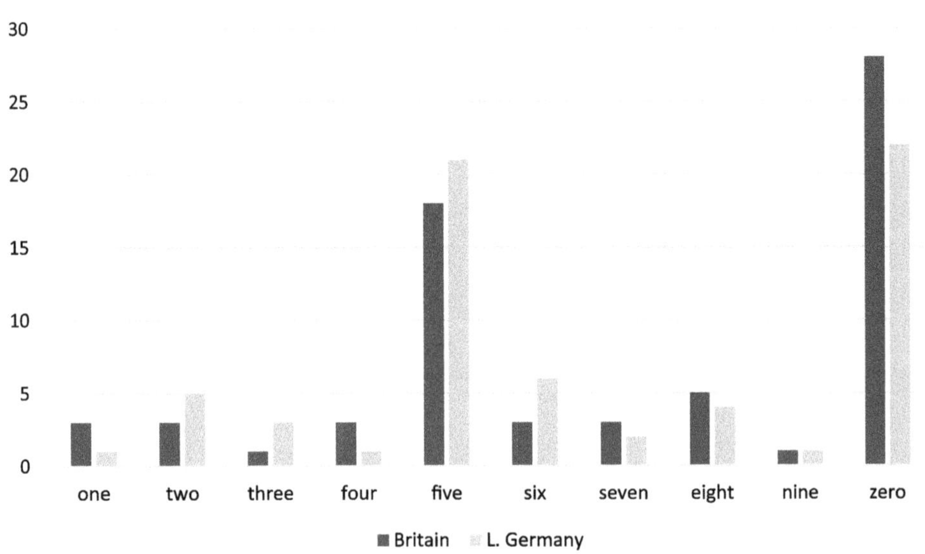

Figure 3.5 The rounding of age of death on soldiers' epitaphs, Britain (N = 68) and Lower Germany (N = 66).

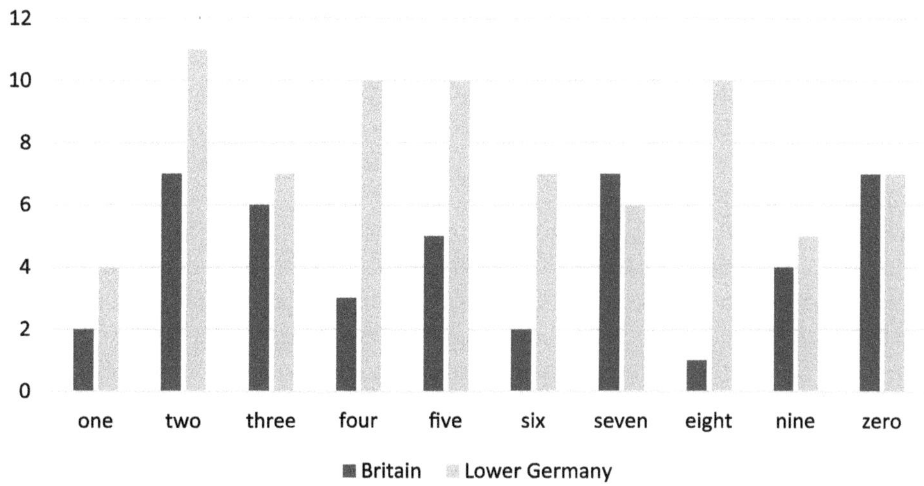

Figure 3.6 The rounding of years of service on soldiers' epitaphs, Britain (n = 44) and Lower Germany (n = 77).

estimates (ibid., 427–31), to bring military mortality closer to that of adult males in general, albeit offset by the crowded conditions of barrack life and greater exposure to epidemic and endemic disease, given garrison placement on nodal routes and in riverine settings enhancing the risk of malaria. It is hard to measure to what degree this applied in Britain. On the one hand, the long-term legionary garrisons (York, Chester, Caerleon) were not in zones traditionally susceptible to malaria (Gaardbo Kuhn et al. 2003). Anecdotal information about risk factors, for example (likely) conjunctivitis documented in the Vindolanda tablets (*lippientes, Tab. Vindol.* 154. 24) as an index of infection exacerbated by dense living conditions, or the potential presence of the Antonine 'plague', likely smallpox (Jones 2005; Hodgson and Breeze 2020), give no reason to think that substantially different conditions applied in Britain. Unless some demographic information becomes available through writing tablets, bioarchaeology data offer the best future opportunity to refine this understanding.

Bioarchaeology

Excavated Evidence

For the long-term garrison communities of northern England and Wales, it is difficult to compare memorials with the bioarchaeological data. These settings have seen limited excavation of burial areas, and the widespread cremation tradition inhibits the opportunities for osteological and biomolecular analysis, as do the acidic soils for inhumations. Even in places like York, where inhumation was widely practised and human remains are better preserved, a further methodological issue inhibits such comparison, namely the difficulty of distinguishing between the burials of soldiers (particularly veterans) and other social groups (Jones 1984; Andersen 2009). Distinctive 'objectscapes' (Pitts 2019), i.e., suites of artefacts used and deposited in funerary rituals associated with garrison communities, may over time allow the identification of more graves of Roman soldiers. However, such objectscapes are patchily represented in Britain, have 'fuzzy' edges with general burial practices, and for the moment do not supply a large sample of burials with up-to-date analyses of the human remains from which the physiological manifestations of military life can be studied. Instead, here, as a working demonstration of the potential bioarchaeology offers, we have selected individuals from Romano-British cemeteries on whom the evidence of trauma suggests a life of violent exertion of the kind characterized above as applying to (some) military biographies.

Initially, cremated human remains were included in this data collection, because most of those in cemeteries along Hadrian's Wall are of this type (Wilmott 2010; Thompson et al. 2016). However, as none of the examined reports presented evidence for 'injury recidivists' (as defined below), we focused on inhumed remains. York, with its enduring legionary presence from the late first to early fifth century CE, offers an important sample for study (Ottaway 2015). We also use the much fuller dataset of human remains published by contract archaeology for urban, small town and rural cemeteries in central and southern Britain (Pearce 2015b; Smith 2018). The Roman military presence was at its most intense in this region during the conquest decades of the first century CE but continued to be significant through the mid and late Roman periods. London is the most prominent example, where secondment to the governor's service plus the city's nodal role in communications within and beyond the province meant the presence of soldiers in large numbers (Perring 2022). Campaigns, suppression of rebellion and political violence, as well as the dispersal of soldiers for administrative, logistical and policing roles, and the return home of veterans will have generated a wider military presence in town and country;[7] for example, the presence of soldiers in late Roman Winchester has been suggested, based on an apparent reference to a workshop producing clothing for soldiers and bureaucrats (Booth et al. 2010: 527–8).

At several sites, non-normative funerary practices have also been encountered, particularly for disarticulated human remains, as well as clandestine burials.[8] Occasionally, these clearly comprise the remains of military personnel – at Canterbury Castle, for example, two males were buried in a shallow grave with their swords (Bennett, Frere and Stow 1982: 44–6). Other instances include body parts displayed for punishment purposes at military buildings. Here, however, we focus on the larger dataset published from normative burial grounds. Cemetery reports and associated publications for these sites (Table 3.2) were examined to collect the following data on injury recidivists: demography (sex and age), trauma, population affiliation and mobility. This resulted in a total of thirty-eight sites of first to fourth century CE date, dominated by data from Cirencester, Leicester, London, York and Winchester, reflecting the impact of contract archaeology (see above). Given their urban context and thus the inclusion of burials from towns and their hinterlands, we cannot exclude the possibility that these injury recidivists also included non-military personnel. In this connection we note the high levels of interpersonal violence in the Roman world (Fagan 2011), especially for those for'

Table 3.2 Number of injury recidivists in the study by location and site name

Location city and/or region and/or environs	*Site name*	*Number of injury recidivists*	*Reference*
Bristol: environs	Henbury School	1	Evans, Holbrook and McSloy (2006)
Chichester: environs	St Pancras	1	Down and Rule (1978); Down and Magilton (1993)
	The Theological College	1	
Colchester: environs	Butt Road	4	Crummy et al. (1995)
Dorset: Dorchester environs	Alington Avenue	5	Davies et al. (2002)
	Little Keep	4	McKinley and Egging Dinwiddy (2009)
	Fordington Old Vicarage	1	Startin (1982)
	Poundbury Camp	2	Farwell and Molleson (1993)
	Poundbury Farm	2	Egging Dinwiddy and Bradley (2011)
	Western Link	1	Smith et al. (1997)
Gloucestershire: Cirencester environs	Bridges Garage	7	Holbrook et al. (2017)
	North of the Fosse Way	3	McWhirr, Viner and Wells (1982)
	South of the Fosse Way	36	
Gloucestershire: Gloucester environs	Kingsholme	1	Hurst et al. (1975)
	120–122 London Road	4	Simmonds, Márquez-Grant and Loe (2007)
Leicester: environs	Newarke Street	2	Derrick (2009)
	Western Road	5	Caffell and Holst (2016); Morris (2010)
London	1–4 Giltspur Street	1	Maloney and Schofield (1998)
	America Street	4	Ives (2015)

(*Continued*)

Table 3.2 (Continued)

Location city and/or region and/or environs	*Site name*	*Number of injury recidivists*	*Reference*
	Atlantic House	3	Watson (2003)
	Chaucer House	1	Maloney and Schofield (1998)
	Cott's House	2	Maloney and Schofield (1998)
	Cutler Street	1	Maloney and Schofield (1998)
	Haydon Street	2	Maloney and Schofield (1998)
	Hooper Street	3	Maloney and Schofield (1998)
	Lant Street	15	Ridgeway, Leary and Sudds (2014)
	Mansell Street	14	Maloney and Schofield (1998)
	Ranwell East Estate II	1	Maloney and Schofield (1998)
	St Mary Spital	11	McKenzie and Thomas (2020)
	St Bartholomew's Hospital	2	Maloney and Schofield (1998)
	West Tenter Street	2	Maloney and Schofield (1998)
Somerset: Ilchester environs	Little Spittel	2	Leach (1982)
Winchester: environs	Lankhills	16	Booth et al. (2010)
	Roman cemeteries and suburbs	11	Stuckert (2017)
Yorkshire: York environs	Blossom Street	3	Newman and Holst (2017)
	Driffield Terrace	31	Caffell and Holst (2012)
	Trentholme Drive	3	Wenham (1968)

whom multiple injuries are likely to be characteristic, including performers in violent spectacle (Kanz and Grossschmidt 2009) and the enslaved (Redfern 2018).

Materials and Methods

The individuals selected for this study were from primary inhumations; it was not possible to make further selections based on preservation and completeness, because these data were not always reported or available. Individuals were included if age and sex could be established, which as a 'rule of thumb' suggests at least 35 per cent skeletal completeness. Individuals reported as being 'male' and 'probable male' based on analysis of their skull and/or pelvic morphology, and who had multiple trauma present were selected for inclusion. To account for the variations in age-classification groups, five categories were created: young adult (eighteen to twenty-five years old), middle adult (twenty-six to thirty-five years old), older adult (thirty-six to forty-five years old), old adult (> forty-six years old) and adult (> eighteen years old). 'Adult' reflects individuals for whom dental eruption and skeletal fusion show they were at least eighteen years old (Schaefer, Black and Scheuer 2009), but due to taphonomic factors a more refined age estimate could not be achieved. Males were included if they had two or more traumas (bones and teeth), with antemortem traumatic lesions including: joint ankylosis (fusion of the bones at a joint, e.g. the elbow) following trauma (i.e. dislocation/fracture), joint subluxation (an incomplete dislocation) or dislocation, dental trauma (fractures/subluxation, where the tooth is moved out of anatomical alignment), myositis ossificans (ossified tissue caused by trauma), sharp-force weapon injuries[9] and fractures (Andreasen et al. 2011; Wedel and Galloway 2014; Redfern and Roberts 2019).[10] The term 'injury recidivist' is used to describe younger adult males who are repeatedly admitted to A&E departments before dying young (Judd 2002, 2004, 2017). Although this model is not without issues (see Judd 2017), bioarchaeological and forensic studies have nonetheless demonstrated that, in varied time periods and locales, recidivists are present (Redfern, Judd and DeWitte 2017a; Tegtmeyer and Martin 2017).

For these males, we also recorded evidence for population affiliation and childhood origin. Population affiliation is estimated using various macroscopic and metric methodologies developed by forensic anthropologists (Byers 2010; Hefner 2016).[11] In order to create consistency between publications, the following groups were used, based on comparisons with the forensic analysis of modern populations: indeterminate/unknown population heritage (individuals whose results could not be matched to a population in extant datasets), diverse heritage (individuals whose results include multiple populations, and are likely to reflect individuals of diverse heritage), Black heritage (individuals whose results match those with populations from Africa or people(s) of African descent (i.e. African Americans)) and white heritage (individuals whose results match white European populations or people(s) of white European descent).[12]

In order to capture information about migration, we collated published stable isotope data for mobility based on analysis of dental enamel. Study of these chemicals can potentially reveal whether, in the case of Britain, a person was born and/or spent their earlier years local to the area where they were buried or had migrated from elsewhere in Britain or from overseas (Richards and Montgomery 2012).

Results

Demography

The majority of the study group were older adults (Figure 3.7), a result that matches the profile of a normal 'attritional' archaeological cemetery population (Chamberlain 2006) and also underscores the reality that the longer you live, the more injuries you are likely to acquire (Redfern 2016).

Population Affiliation

It was only possible to obtain data for thirty-nine males (Figure 3.8), primarily from Leicester, London and York.[13] Individuals of white European heritage were the largest group of males. More individuals were identified as having diverse heritage than Black heritage; when all non-white population affiliation groups are pooled, they make up 53.8 per cent of this sub-sample.

Mobility

Over the past decade, stable isotope analysis has demonstrated that men, women and children migrated (freely or forced) to Britain from Northern, Central and Eastern

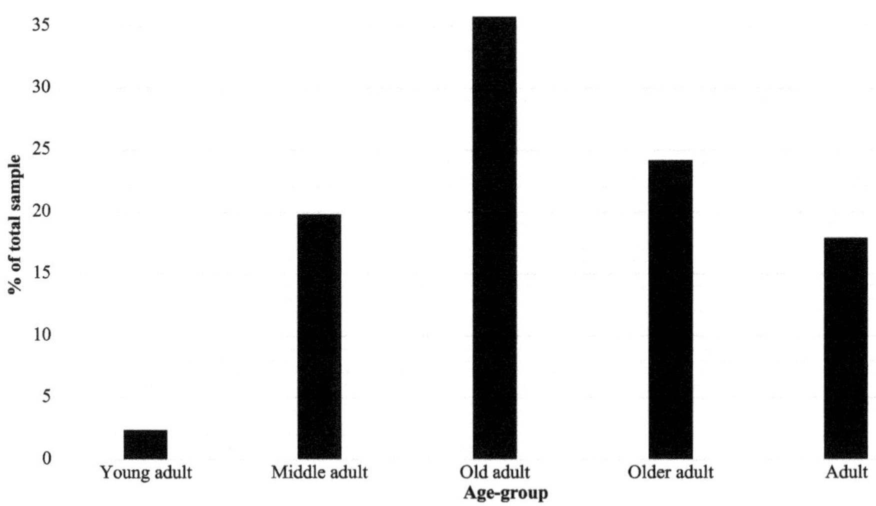

Figure 3.7 Injury recidivist demography.

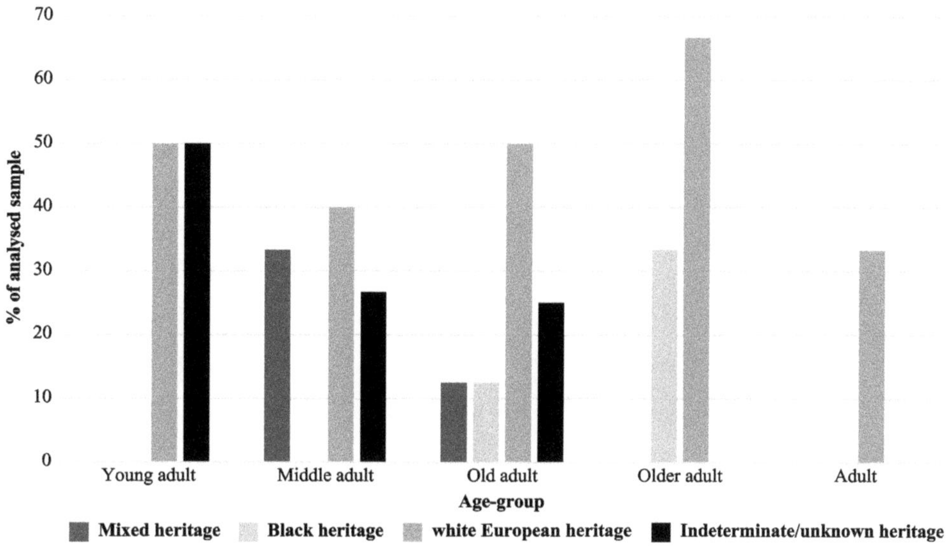

Figure 3.8 Population affiliations by age group.

Europe and the Southern and Eastern Mediterranean, complementing the limited inscription evidence (see above) (Eckardt 2010; Chenery, Eckardt and Müldner 2011; Eckardt, Müldner and Speed 2015; Redfern et al. 2016; Shaw et al. 2016). Of the eleven injury recidivists with published mobility data (Table 3.3), the majority were from Britain (54.4 per cent), 36.4 per cent were from Yorkshire and one individual (9.1 per cent) was from Central or Eastern Europe. This result reflects the populations sampled for mobility rather than a real trend, with York and sites in Yorkshire being a focus of research by the University of Reading's Diaspora project (Leach et al. 2009, 2010; Müldner, Chenery and Eckardt 2011; Eckardt, Müldner and Speed 2015).

Injury Recidivists

The highest number of injury recidivists was observed in the older adult and old adult age groups. The lowest number was observed in the young adult category (Figure 3.5). The greatest numbers of recidivists were observed at York, London east and Cirencester environs (Table 3.2).

Myositis Ossificans

Evidence for ossified soft-tissue was not observed in the younger adult age category but was reported for all other age groups (Table 3.4). All areas of the skeleton were affected, with the most frequently affected bones being the humeri.

Table 3.3 Summary of the mobility data by location, site name, individual, population affiliation and origin suggested by stable isotope analysis

Location	*Site name*	*Context*	*Sex*	*Age*	*Population affiliation*	*Origin*
York: environs	Driffield Terrace	3DT16	Male	Older adult	Unknown heritage	Yorkshire
		3DT37	Male	Middle adult	Diverse heritage	Yorkshire
		6DT2	Male	Older adult	White heritage	British
		6DT7	Male	Older adult	White heritage	Yorkshire
		6DT8	Male	Older adult	White heritage?	British
		6DT14	Male	Middle adult	Unknown heritage	British
		6DT15	Male	Young adult	White heritage	British
		6DT17	Male	Older adult	Diverse heritage	Yorkshire
		6DT21	Male	Older adult	White heritage?	British
		6DT22	Male	Middle adult	Diverse heritage?	British
Gloucester: environs	Kingsholme	B1	Male	Older adult	Not analysed	Central & Eastern Europe

Joint Ankylosis, Subluxation and Dislocation

Only large joint ankylosis was reported from three individuals, including an old adult (sk 861, Lankhills), whose right elbow was fused – he is the only male to have other trauma present. Twelve males had evidence of joint subluxation or dislocation present, and the most affected age groups were middle-aged and older adults. Nine of these individuals also had other types of trauma present (e.g. fracture), and the most frequently affected joint was the shoulder.

Sharp-Force Weapon Injuries

Injuries sustained shortly before or at the time of death (perimortem) were reported in five males and two of these also had other healed (antemortem) traumas present (Table 3.5). A young adult from London (sk 503, America Street) had cut-marks to his right temporal bone and mandible and five right ribs; he also had healed cranial fractures. The remainder of weapon injuries were observed to the upper and lower limbs. Antemortem injuries were observed in twelve males, only four having evidence for other types of traumas. Nine individuals had healed injuries to their cranium, suggestive of face-to-face violence.

Table 3.4 Injury recidivists with myositis ossificans by age group. R = right, L = left

Middle adult	*Cranium*	*R. clavicle*	*L. clavicle*	*Pelvis*	*R. humerus*	*L. humerus*	*Hand bones*	*L. femur*	*Foot bones*				
Total	1	1	3	1	2	2	3	1	5				
% of age group (5)	20	20	60	20	40	40	60	20	100				
Older adult	*Cranium*	*R. scapula*	*Pelvis*	*R. humerus*	*L. humerus*	*R. radius*	*L. radius*	*Hand bones*	*R. femur*	*L. femur*	*R. tibia*	*L. tibia*	*R. fibula*
Total	1	1	1	3	2	1	1	1	1	12	5	2	2
% of age group (50)	2	2	2	6	4	2	2	2	2	24	10	4	4
Old adult	*R. humerus*	*L. humerus*	*Forearm*	*R. tibia*	*L. tibia*	*R. fibula*	*L. fibula*	*L. leg*					
Total	1	1	1	1	2	1	1	1					
% of age group (74)	1.4	1.4	1.4	1.4	2.7	1.4	1.4	1.4					
Adult	*R. ribs*	*Thoracic*	*Pelvis*	*L. humerus*	*L. ulna*	*R. femur*	*L. femur*	*R. tibia*	*L. tibia*	*R. fibula*	*L. fibula*	*Foot bones*	
Total	1	1	1	1	1	1	1	2	5	2	2	3	
% of age group (37)	2.7	2.7	2.7	2.7	2.7	2.7	2.7	5.4	13.5	5.4	5.4	8.1	

Table 3.5 Distribution of sharp-force weapon injuries by timing (ante- or perimortem). Data given by site, individual, presence of other trauma and bone/body area. R = right, L = left

Antemortem

Site name	*Context*	*Sex*	*Age*	*Injury recidivist: other trauma*	*Frontal*	*R. parietal*	*L. parietal*	*Occipital*	*R. humerus*	*L. finger*	*R. femur*	*Present*
South of the Fosse Way	24	Male	Older adult	0	1							
South of the Fosse Way	81	Male	Older adult	0	1							
South of the Fosse Way	S	Male	Young adult	0	1							
South of the Fosse Way	101	Male	Middle adult	0		1	1					
South of the Fosse Way	305	Male	Old adult	0		1						
South of the Fosse Way	14	Male	Old adult	0					1			
Winchester	HYS G3A	Male	Adult	1								1
Winchester	LH 427	Male	Middle adult	0						1		
Driffield Terrace	3DT33	Male	Middle adult	1							1	
Driffield Terrace	6DT22	Male	Middle adult	1				1				
Little Keep	1118	Male	Older adult	1	1							
America Street	503	Male	Young adult	0			1					
			Total	4	4	2	2	1	1	1	1	1

Perimortem

Site name	*Context*	*Sex*	*Age*	*Injury recidivist: other trauma*	*R. temporal*	*Mandible*	*Ribs*	*R. scapula*	*R. clavicle*	*R. humerus*	*L. femur*	*L. tibia*
South of the Fosse Way	288	Male	Middle adult	0								1
South of the Fosse Way	21	Male	Young adult	0						1		
South of the Fosse Way	163	Male	Middle adult	1							1	
Driffield Terrace	3DT7	Male	Middle adult	1				1	1			
America Street	503	Male	Young adult	0	1	1	1					
			Total	2	1	1	1	1	1	1	1	1

The majority of injuries were sustained to the right side of the body, with most being sustained before the age of thirty-five, which corresponds to clinical and forensic injury recidivist data (Tegtmeyer and Martin 2017).

Dental Trauma

Eighteen individuals (Table 3.6) had fractures to the tooth crown and/or root; the highest frequency was observed in young adults. Teeth are often fractured by blows to the face (Redfern 2016), and two individuals also had healed fractures to their nasal bones. One also had a healed fracture to their frontal bone and another to his left parietal bone.

Skull Fractures

The left nasal bones were the most frequently fractured bones (Table 3.7). Many fractures were located above the 'hat brim line', suggesting that they had been produced by direct blunt-force blows (Kremer et al. 2008; Guyomarc'h et al. 2010). Fractures to the nasal bones and frontal bone are sustained during assaults, although they may also be fractured by an accidental fall (Wedel and Galloway 2014: 150).

Post-Cranial Fractures

The highest number of fractures was observed in un-sided ribs, and many individuals had multiple bones fractured (Table 3.8). When examined by body area, fractures to the torso were dominated by rib bones and the thoracic and lumbar vertebrae. When an individual has multiple rib fractures present, these are usually caused by direct blows or by being crushed (Wedel and Galloway 2014: 190). The majority of vertebral fractures were to the body and produced by compression or avulsion mechanisms, meaning that they are most likely to have been caused by falls and heavy lifting (ibid., 174–85). One older adult male from Cirencester (Bridges Garage, B1157) had multiple torso fractures present, two rib fractures to the angle and body and compression fractures in four thoracic vertebrae.[14]

Many of the arm and hand fractures observed on these individuals were produced by direct blows to their body, as well as by punches they themselves had thrown (Wedel and Galloway 2014: 229–43). Where information permitted, it was observed that some males had sustained Parry fractures to the distal-third of their ulna bone, a defensive injury (Judd 2008).

The leg bone data reflect fractures commonly occurring during adulthood due to accidental causes (Wedel and Galloway 2014: 257–92). A minority of males had multiple long-bone fractures present, such as 34152 from St Mary Spital, a thirty-six- to forty-five-year-old who had fractured both radii, their right ulna, left distal tibia and fibula and a rib bone.

Table 3.6 Distribution of dental trauma by age group

Age	*Dental trauma (n)*	*%*
Young adult	1	20.0
Middle adult	6	14.6
Older adult	6	12.0
Old adult	4	5.4
Adult	1	2.7
Total	18	8.7

Table 3.7 Distribution of skull fractures by bone and side. R = right, L = left

Bone/side	*No.*	*%*
Frontal	12	5.9
R. zygomatic	1	0.5
R. nasal	10	4.9
L. nasal	13	6.3
Nasals	4	2.0
L. maxilla	2	1.0
R. mandible	1	0.5
L. mandible	1	0.5
Dentition	18	8.8
R. parietal	8	3.9
L. parietal	5	2.4
Occipital	1	0.5
Cranium	2	1.0

Discussion

The bioarchaeological data show that many males from central and southern Britain (Table 3.2) have evidence of healed injuries testifying to assault and impairment: among them a large number were migrants from beyond the region or province. The extent to which the skeletal evidence reflects specific military experience or other interpersonal (often urban) violence (Redfern 2020), or a combination of both urban and military

Table 3.8 Distribution of post-cranial fractures by bone/body area and side. R = right, L = left

Bone/side	*No.*	*%*	*Bone/side*	*No.*	*%*	*Bone/side*	*No.*	*%*
R. clavicle	4	2.0	R. humerus	2	1.0	R. femur	6	2.9
L. clavicle	6	2.9	L. humerus	3	1.5	L. femur	4	2.0
R. scapula	6	2.9	R. radius	3	1.5	R. tibia	13	6.3
L. scapula	4	2.0	L. radius	7	3.4	l. tibia	11	5.4
R. ribs	28	13.7	R. ulna	8	3.9	Unsided tibia	3	1.5
L. ribs	32	15.6	L. ulna	9	4.4	R. patella	1	0.5
Ribs	39	19.0	Forearm	2	1.0	R. fibula	16	7.8
Sternum	1	0.5	Hand bones	36	17.6	L. fibula	17	8.3
Cervical	3	1.5				Unsided fibula	4	2.0
Thoracic	21	10.2				Foot bones	26	12.7
Lumbar	17	8.3						
Vertebrae	6	2.9						
Sacrum	1	0.5						
Pelvis	1	0.5						

lifestyles remains difficult to establish. Indeed, it will likely remain so until a robust way of distinguishing the inhumation burials of veterans and soldiers from those of civilians can be devised. Nevertheless, the high numbers of recidivists from York and London in particular suggest that a long-lived military presence, respectively the legion and soldiers attached to the governor's service, may influence these results and be manifested in the traumatic biographies documented (Redfern 2020).[15] The high number from Cirencester is intriguing, since the known garrison was not present beyond the mid first century CE (Holbrook 2015: 92–3).

When compared to the individuals in Table 3.1, it can be seen that many younger injury recidivists[16] share many of the characteristics observed in the skeletons of these likely military personnel, that is to say, evidence for major geographical mobility during their lifetime, healed assault fractures and (lethal) perimortem injuries, perhaps reflecting trauma and death in service. Some older injury recidivists (> forty years old) in this study might be veterans whose bodies incorporated the trials of decades-long service and movement around the empire. Many of the injuries they sustained, such as antemortem weapon injuries, would have resulted in scars, often significantly changing their physical appearance and compromising their cognitive abilities (e.g. sk 305 from Cirencester, see McWhirr, Viner and Wells 1982: 170). Males with ankylosed limb joints could represent those who had been invalided out on a medical discharge, as some traumas would have significantly compromised their mobility.

The isotope and skeletal evidence for population diversity within this group supports the epigraphic evidence for recruitment of soldiers beyond the province, including the stationing of military personnel with connections to Africa on Hadrian's Wall and elsewhere in Britain (Eckardt 2010; Tolia-Kelly 2011). It seems that both urban and rural populations contained long-distance and second-generation migrants, and people of African and diverse heritage.[17] Since soldiers had relationships and military communities contained families (Greene 2013, 2020; Ivleva and Collins 2020), it seems that the diversity within the military was mirrored in the urban centres where they were stationed.

The majority of British mobility isotope studies compare childhood and burial locations, that is to say, the start and end of a life, while the evidence for travel across their lifetimes is sadly lacking (cf. Stark 2016; Emery et al. 2018; Stark et al. 2020, 2022). Therefore, although the majority of sampled injury recidivists in this study were from Britain, we cannot assume that they had not experienced considerable mobility in their lifetime – the experience of service in Cumbria of the veteran Agathopus, who returned to his *patria* in Numidia afterwards, would not be captured by present studies (cf. Mann 1993).

Conclusions

This experimental combination of epigraphic and bioarchaeology data from cemetery settings in Britain underscores the importance of using multiple datasets to explore the Roman combatant experience, since each provides a different perspective. The inscribed memorials illuminate individual lives, but have key limitations beyond their tersely formulaic character. When aggregated they are unlikely to give a representative sample of Roman military experience (Scheidel 1996, 2007). We acknowledge that the male injury recidivist group will contain individuals who were not soldiers, and whose multiple injuries in some cases are the consequences of enslavement and structural violence. However, given the particular urban contexts of these cemeteries, it is a distinct possibility that many such individuals could have been soldiers.

The bioarchaeological data provide support for the diversity of origin and heritage seen in the epigraphic record. These data extend that diversity of geographical origin into the later Roman period, and amplify the likely attestations of individuals of Black or diverse heritage, even if the precision of individual points of origin does not yet match what is sometimes recorded in inscriptions. The human remains offer a much more nuanced impression of the reality of military life, of trauma as the price of violence, hardship and gruelling labour and its likely consequences of bodily disability, disfigurement and mental disability. In this respect, the bioarchaeological data capture the lived experience that has been elided or edited out of the formulae that summarized a soldier's life in typical military commemorative monuments. Current limitations to non-destructive methodologies for ageing adult skeletal remains mean that no straightforward comparisons can be made between epigraphically and skeletally attested ages at death. In contrast to the soldiers on active service who dominate the epigraphic record, the bioarchaeological data may well mark the presence of veterans – men whose *virtus* may be much more commonly embodied in their skeletons than can ever be visible to us in their epitaphs.

Acknowledgements

Rebecca Redfern is most grateful to her parents for providing the childcare needed to finish this work; Dr Sarah Inskip (University of Leicester) and Malin Holst (York Osteoarchaeology) who, despite the Covid-19 epidemic, generously shared their work and found time to answer her many questions; and Dr Joseph Hefner (Michigan State University) for his help and advice regarding population affiliation. John Pearce is grateful to Glynn Davis for allowing the reproduction of the Facilis tombstone from Colchester, and to Sinclair Bell for making his paper available in advance of publication.

Notes

1. The 'need' to ensure the recruits were from 'good' origins continued through into the fifth century CE in the Theodosian code (429–38 CE), which reiterates the exclusion of the enslaved, and 'that category which is debarred from military service by the ugliness of its occupation, nor shall men be dragged from houses of correction' (*Cod. Theod.* 7.13.8; Watson 1983: 42).
2. In recent years, personal relationships between military personnel and their enslaved/dependents have been re-examined, finding evidence for same-sex relationships, as well as family units (Greene 2020; Ivleva and Collins 2020).

3. Individual inscriptions are referenced by their RIB number or *Britannia* reference and Roger Tomlin's *Britannia Romana* (2018), which offers key commentary on individual discoveries, is also referenced where relevant with the prefix *BritRom*.
4. Epitaphs were collected from the key corpora, namely Roman Inscriptions of Britain I and III, supplemented by the small number of inscriptions subsequently published (from 2007) in annual reports in the journal *Britannia*. The collection of all epitaphs with a link to military service and Britain, for example including soldiers of British origin who died elsewhere, or epitaphs from other provinces of soldiers who had served in Britain, would be much more time-consuming but would not produce results which would significantly affect the patterns noted.
5. This is usually taken to refer to the *civitas* of the Belgae in southern Britain but may refer to Gallia Belgica.
6. Saturninus is a name 'notably frequent in Africa', RIB 3345, commentary.
7. Administrators, Nelis-Clément (2000); soldiers and veterans, Nicolay (2007).
8. For example, see Luff (1984); 'Appendix I' in Loe (2001); Buck et al. (2019).
9. This does not include decapitation, as that was a funerary rite applied to all age groups and sexes in Roman Britain (Crerar 2015). Many of the injury recidivists in this study were decapitated (e.g. Driffield Terrace, York), but determining whether that occurred during organized violence, a method of execution or a funerary rite remains difficult. Individuals with injuries to the skull, but without vertebral involvement, were included in this study. The position of the skull relative to the body in these cases was also checked in the site report, to verify that the evidence had not been 'removed' by taphonomic action, and to confirm that decapitation had not been present.
10. To determine whether an individual sustained a fracture by intentional violence or accident, a bioarchaeologist focuses on which is bone is affected, the location of the fracture, and fracture type. Unfortunately for the present study, such data were not consistently given and so are not reported (Redfern 2016).
11. Bioarchaeologists and forensic practitioners do not employ the term 'race' to describe human variation. Population affiliation is not interchangeable with a person's ethnicity, which is based on culture, language and is acquired through socialization (Edgar and Hunley 2009; Gannett 2014; Benn Torres 2020). Many of the individuals who form the 'known' collections used to create population affiliation studies are individuals who died within the last fifty years and those collections are dominated by the disenfranchised of the Americas, and we recognise that this perpetuates the embedded racism they suffered during their lives. Since May 2020, many forensic anthropologists have questioned whether their practice should include population affiliation, since many methods have their origin in racialized science. Many have stopped this work and others no longer teach classes on this topic (Bethard and DiGangi 2020). In this study, we recognize that the methods used in the studies collated for this project are based on racialized science, and because of embedded whiteness, their problematic history and application has not been fully engaged with (Ousley, Jantz and Freid 2009; Hefner 2016; Blakey 2020; Stull et al. 2020; Watkins 2018, 2021). For the study of the past through archaeology however, the remains of people of colour are often the only evidence for their presence, as they have been 'erased from the archive', as other primary sources are not available for certain periods, or are represented according to visual tropes (Bell 2021). Therefore, we have chosen to include this aspect of identity, in order to push back against the deeply held belief of an all-white British past and Roman Empire (Kamash 2021).
12. Indeterminate/unknown population heritage: this arises because the datasets are heavily biased towards modern populations (Redfern and Hefner 2019, 2021). None of the individuals in this study also had ancient DNA evidence for haplogroup affiliation. Multi-

disciplinary studies suggest that these different types of evidence may not always match – a result seen today between self-reported population affiliation and an individual's genomic data. This is discussed in Redfern et al. (2016, 2017b), Redfern and Hefner (2019, 2021).

13. This reflects the recording protocols undertaken for the sites in Leicester by York Osteoarchaeology, Reading University's Diaspora project which focused on sites in Yorkshire, and collaborative projects at the Museum of London.
14. This male also had a healed depressed fracture to his right parietal bone and the distal phalanx of the first left metatarsal. The male is described as having 'generalised osteopenia' (i.e. reduction in bone density) and this is given as the differential diagnosis (i.e. cause) of the vertebral compression fractures (Geber 2017). Unfortunately, the site report does not discuss how the potential diagnosis of osteopenia in this male (or other individuals) was made. As the fractures were reported under 'trauma' rather than 'osteopenia' they have been included in this study.
15. This statement should be taken 'with a pinch of salt' – all these locales have benefitted from over fifty years of archaeological excavation in response to development and infrastructure projects.
16. In particular, the Driffield Terrace (York) population who have been variously interpreted, from fallen gladiators to victims of political assassinations after the death of the Emperor Severus (145–211 CE) in York (Montgomery, Knüsel and Tucker 2011; Müldner, Chenery and Eckardt 2011; Caffell and Holst 2012).
17. Hitherto most population affiliation studies have been undertaken on urban populations (Leach et al. 2009), confirming the presence of multiple population groups and second-generation migrants (Caffell and Holst 2016; Redfern et al. 2017b). A small number of stable isotope (and/or aDNA) studies of rural cemeteries is available, and these show evidence for people born outside of Britain and/or with haplogroups suggesting familial connections outside of England (Simmonds, Márquez-Grant and Loe 2007; Wiseman, Neil and Mazzilli 2021). Our statement should not be taken to imply that people of colour could only be migrants – evidence shows that many were born in Britain.

PART II
EMBODYING SOLDIERLY IDENTITY

CHAPTER 4
CLOTHING THE SHAMI PRINCE: THE DRESS OF PARTHIAN HORSEMEN

Lloyd Llewellyn-Jones

The 'Shami Prince' (also known as the 'Shami Nobleman') is the name given to an over-life-size bronze statue that resides in the Iran-e Bastan Museum in Tehran (Museum No. 2401; Figures 4.1a–b).[1] It was discovered in the 1930s in Shami (ancient Elymais) in Khuzestan Province, alongside other fragmentary stone and bronze sculptures. The statue had been made in two separate parts – the head and the body – and joined together at the neck; it has a missing left arm and there is some damage to the waist and to the left side of the torso. It was made sometime in the Parthian period and dates for it have varied, with some scholars suggesting the early Parthian era (second century BCE) and others arguing for the late Parthian period (second century CE).[2] All in all, it seems that there is something of a consensus on a date around 50 BCE to 50 CE, which would, neatly, make the Shami Prince a rough contemporary of Augustus' Prima Porta statue.

The Shami Prince is certainly an impressive piece of bronze work, and it justifiably attracts the lingering gaze of the many who encounter it in the archaeology gallery situated on the museum's ground floor, for the statue is beautifully rendered and immensely impactful. It has a strikingly handsome face and its clothed body is well proportioned and stands straight and upright, with its legs set apart and its feet pointing outwards; much of its bodyweight is taken by the right leg. The statue's garments, carefully and intricately rendered, have been studied by Geo Widengren (1956), Vesta Sarkhosh Curtis (1993, 2017) and Gillian Vogelsang-Eastwood (2000), and all three make valuable contributions to their study. However, I am not convinced that we have completely understood the construction of the garments, nor have we yet attempted to comprehend the issue of their functionality or of their meaning in their historical context. In this chapter, I offer my thoughts on the cut and construction of the garments of the Shami Prince and I put forward an interpretation of their purpose. These types of clothes, I argue, were essential to the functioning and the identity of the Parthian horse-riding military elite who wore the original – that is to say, textile – garments.[3] This chapter explores the clothed body of the Parthian mounted warrior-archers who defined a particular, and highly successful, Iranian military system. The mounted warriors of Persia inherited their horse-techniques and dress from longstanding Eurasian nomadic traditions and their riding clothes were created for maximum physical movement and the protection of the body. This basic fact has led me to question how clothing design must follow functional needs, especially in the context of warfare. Kinesics will be employed here to explain how the form of dress of Parthian horsemen – exemplified by

Figure 4.1a–b Bronze statue from Shami, *c.* 50 BCE–CE 50, known as the 'Shami Prince'; Iran-e Bastan Museum (Museum No. 2401). With kind permission of the Iran-e Bastan Museum.

the Shami Prince – not only follows a specific function, but creates a sort of portable environment in which the clothed body of the combatant works as a complete entity. Susan Watkins and Lucy Dune, experts in the complex logistics of contemporary functional clothing design, define the 'portable environment' based on just two criteria: 'that it is (1) attached to or supported by the body and (2) that it moves with it [. . .] Much of its emphasis is on achieving both protection and portability in a functional clothing item.'[4] It is my contention that this contemporary concept can be applied to Parthian riding dress and, in turn, I suggest that the ancient Iranians were capable of thinking about the functionality of clothing design in remarkably sophisticated ways.

Central to my reading of the prince's clothing, then, is the understanding of the role of functional 'design' in the creation of those specific Parthian horseback garments. 'Design' is a fashion-related term not usually appended to studies of ancient dress, and very few scholars have considered that the construction and wearing of clothing in the ancient world needed thought and planning – in other words, that it needed 'design'. Yet dress styles did not simply appear out of a vacuum. Clothing styles evolved (often slowly) to take into account the specific availability of fibres and textiles, the technology of weaving or other forms of textile crafting, the environment in which the wearer lived, the needs of the wearer and the messages the wearer wished to convey about their culture

(Bernard 2020). Dress evolved over many centuries in ways specific to the needs and logistics of different ancient communities and the design of clothes specific to those communities developed as a series of problem-solving strategies, involving the conscious application of initiatives, intentions, decisions, actions and evaluations. I do not, however, believe in the existence of clothing 'designers' in antiquity *per se* (we have no evidence of individuals who set the mode or dictated the trends), but in global history there have been many great discoveries, both in science and in art, in which the critical link has been met by chance. It is the inquiring (individual or collective) mind that seizes on that moment and turns what was an accident into something providential. This is what I mean by 'design'. In order to create a functional clothing design, there are several important areas that require consideration: (1) Who are the users? (2) What is the activity? (3) What are the environmental conditions under which the clothing will be used? (4) What are the specific goals of the clothing, such as mobility or thermal protection? These variables influence how effective a particular design solution will be. The purpose of this study, then, is to access the functionality of the clothed body of the Parthian horseback archer, with the overall aim of becoming more aware of that in our study of war and warfare. A sharper focus on dress and kinesics will offer fresh perspectives on the battle experience, something which, in Iranian Studies, has been neglected.

Parthian Horseback Tactics

Parthian warfare operated on an *ad hoc* basis; there was no standing army. Under the command of an assertive king, nobles (*azats*) and their retainers could be summoned to join the troops of the royal household on offensive or defensive campaigns. In times of war, the Parthian elite were required to commit their private armies to the greater interest of the realm and the great aristocratic houses were expected to contribute a number of horses and riders proportionate to their status and income. When the Parthian *surena* ('general') vanquished Crassus at the Battle of Carrhae in 53 BCE, he was escorted by an army of 10,000 horsemen, formed by his bondsmen and dependants and, later, in 36 BCE, some 400 Parthian *azats* sent an army of 50,000 mounted warriors against Mark Antony (Just. *Epit.* 41.2). The Romans primarily, and correctly, thought of the Parthians as an equestrian people. 'They ride on horseback on all occasions', Justinus noted, 'on horses they go to war, and to feasts; on horses they discharge public and private duties; on horses they travel, meet together, trade, and converse' (*Epit.* 41.3, trans. Watson). Certainly, the Parthian army was made up more of horses than men, since every rider would take two or three steeds on campaign with him to ensure that he always had a fresh mount on which to gallop into battle or to retreat quickly from the field. '[T]he very word 'Parthian' (*Parθavapahlav*) acquired the meaning "knight" or "hero", while the term 'horseborne/ rider' (*asabārasavār*) came to signify a "noble" or "knight" in contrast to "foot soldier" (*payāda*), which became synonymous with "common people"' (Shahbazi 1987). It became popular to brand horses and equipment with family insignia (*nišān*) throughout the

Parthian era; the Parthian royal *nišān* – a ring on a base – can be seen adorning the horse of the last Parthian king, Ardavān (Artabanus V) (Shahbazi 1987).[5]

Parthian warhorses were specially bred on the alfalfa-rich planes of Nisea, in the foothills of the Zagros Mountains in Media (north-western Iran). Nisean horses were magnificent little steeds, universally regarded as the nimblest of beasts; they became celebrated for their bravery and tenacity.[6] An official writing from far-off China was impressed enough to note that Chinese horses could never rival those of Nisaea 'in climbing up and down mountains and crossing ravines and torrents' (Tao 2007: 96). They were the perfect animals for life in the mountains and steppes, and they were unfazed by the noise of battle or the horrors of combat.

The Parthian army usually took the field with a heavy assault cavalry at the front. These cataphract riders, carrying huge lances, were protected by body armour that rendered them mostly immune to hand-propelled missiles and arrows. The horses used for the heavy cavalry were bigger and bulkier than the Nisean types (they were more like Irish cobs) and were protected by trappers of thick felt and rawhide.[7] The psychological effect of seeing thousands of armoured warriors on the move in full battle array was terrifying, but it was the fast, light, hit-and-run mounted archers who had the greatest impact on the battle, dazzling onlookers with their moves. Most Parthians rode bareback, with little defensive armour, but they could steer their horses right into the heart of the action, striking where it mattered most. At the given signal, their assault on the enemy would begin at a trot, then a canter and finally a full gallop. The tactics of horseback writtenarchers were to relentlessly harass an enemy and goad them into a futile pursuit, stringing them out for a charge or simply exhausting them, but with large contingents and enough ammunition they could almost single-handedly win a battle – as happened at Carrhae. Parthian bowmen released their arrows at full gallop when within firing range. At a distance of around forty-five metres from the enemy, they swerved swiftly to the right and galloped off to the flank, shooting into the enemy lines. A tactic that demanded even more bravado was to bring the horse to a skidding half-turn and, as it galloped away, to shoot arrows over its rump. The use of the infamous 'Parthian shot' (hence the English term 'parting shot') meant that, if conditions were right – as they were at Carrhae – the enemy would be surrounded and the mounted archers could kill at leisure (Rostovtzeff 1943). The velocity generated by a Parthian compound bow (Persian *kāmān*) fired at close range meant that arrows could easily penetrate enemy shields, armour, flesh and bone. All the more remarkable, then, is the realization that, since a rider's hands were occupied by his bow and his body was twisted around, he relied solely on squeezing pressure from his legs to stay mounted and guide his horse. The stirrup had not yet been invented, and so the rider controlled his mount using nothing more than his knees and heels, all the time continuing to load and loose arrows. The Parthians used a saddlecloth or pad, which was held on the horse with a girth or surcingle; this meant that the archer could find some support for his weight as he steered his galloping horse with his knees. The synergy created between horse and rider was unparalleled. A horseback archer's success was very much dependent upon his kinetic response to his horse and to his external stimulus; this is why a horse-rider's clothing mattered so much.

Clothing The Shami Prince

The figure wears a long-sleeved jacket, a tunic, breeches (or under-trousers), chaps (or over-trousers) and boots (Figure 4.2).[8] If we consider the garments worn by the Shami Prince to be representative, if expensive and high-end, examples of what Parthian horseback archers wore into battle, then we have the potential to understand how clothing aided and abetted mounted warriors in their battlefield employment.[9] The design of the separate garments and the overall holistic grouping of the individual items of clothing will offer us evidence of the functionality of the body of the Parthian in combat. Let us examine the principal components of the dress in further detail.

Figure 4.2 The Shami Prince, drawing detailing the combinations of garments. Line drawing by Lloyd Llewellyn-Jones.

Tunic

Several scholars have observed, mistakenly, that the prince went bare-chested beneath his jacket, and some have identified the tiny, incised marks on his torso as evidence for a hairy chest (Curtis 1993: 63; Mathiesen 1992: 166). In fact, the incised marks indicate the piled fabric of his tunic – perhaps coarse linen or softened felt lined with cotton. A fragment of a shoulder piece from a tunic excavated at Shahr-e Qumis in northern Iran (which has been identified as the ancient city of Hecatompylos, established by the Parthians as their capital by about 200 BCE) was made of this latter combination and included a blue silk cord to secure the opening at the neck (Figure 4.3).[10] Vogelsang-Eastwood has noted that, just under the torque worn at the prince's neck, there is a thin line which, upon examination, represents the tunic's neckline.[11] It clearly did not fasten at the front and Vogelsang-Eastwood has suggested, feasibly, that it would have had a wide slit at the neck opening extending to the left shoulder, making it easy to put on and take off (2000: 34). In fact, actual tunics of this kind were discovered at the site of Halabiyeh, on the Euphrates River.[12]

No sleeves are visible, but based on the Halabiyeh finds, it is probable that the tunic had long, narrow, somewhat shaped, sleeves. As for the tunic's length, a small section of it pokes out beneath the jacket, suggesting that it was at least hip length. Vogelsang-Eastwood proposes that the tunic was rectangular in form (the very simplest type of T-shaped garment), although I believe that the use of gores and inserts, typical of Iranian and Central Asian dress, would have been used to give a more sophisticated cut with better mobility, especially at the shoulder and hips, where the archer needed the most flexibility (Figure 4.4).

Figure 4.3 Textile fragment of felt, cotton and silk, *c.* 90-50 BCE; Parthian from Shahr-i Qumis (ancient Hecatompylos), Iran. Metropolitan Museum of Art 69.24.30. Open Access.

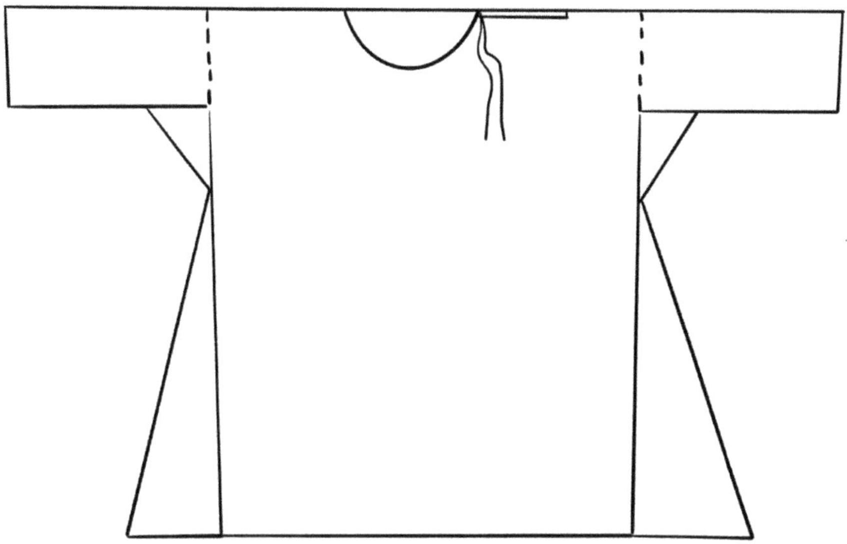

Figure 4.4 The probable shape and cut of the tunic worn by the Shami Prince, showing the neck opening at the side. Line drawing by Lloyd Llewellyn-Jones.

Jacket

Curtis describes this distinctive garment as a 'long-sleeved jacket with broad lapels, a wide hem and cuffs on the sleeves . . . [it is] tightly fitting . . . fastened with the right side overlapping the front' (1993: 63).[13] The description is, however, flawed. As Roland Barthes recognized, whilst 'human clothing is a very promising subject to research or reflect upon . . . as an object of appearance' (2006: 20), we approach the descriptive elements of dress imperfectly. He noted that in contemporary fashion magazines there is almost always an image accompanied by text. For Barthes, therefore, 'image-clothing' can only be understood when accompanied by the written-garment; how we choose to describe a garment has ramifications. Both 'image-clothing' and 'written-clothing' exist in relation to what Barthes calls 'real-clothing' (ibid.) – that is to say, the physical, worn garments themselves that are (or can be) divorced from both image and text. 'Real clothing' cannot be understood by sight (because its visual image does not reveal all its intricacies), Barthes argued, but must be known through the mechanical process of its production: its creation via seams, pleats and fastenings as they are manufactured or arranged on the body. Barthes made the verbal structures of the 'written-garment' the focus of his seminal study and although he does not deny that real clothing also works within a 'sign system', in the end Barthes felt that 'language conveys a choice and imposes it. . . [since] words determine a single certainty' (ibid., 20). How we choose to describe a garment thereby 'creates' the garment in our mind and fixes it there. How, then, do we describe Parthian garments? What images do we create of them through the process of description? Left to our own devices, we create an imperfect lexicon of dress terminology through which we read the artistic evidence. We create our own familiar terms – 'sleeves', 'lapels', 'hem', 'cuffs', 'pleats', 'drapery' – to describe artefacts that are long lost to us and are only represented in art, and, therefore, in crafting our imperfect 'written-garments', we are hostage to all the vagaries of language that Barthes identified.

With this in mind, I tentatively note that the jacket worn by the Shami Prince does not have lapels, but has a broad band of fur sewn to the edge of the garment as decoration. It is possible that the whole jacket was fur-lined, offering warmth and protection and indicating the wearer's social status. The same bands of fur decorate the back neckline and cuffs of the long sleeves, which are pushed up the arm to create multiple *ad hoc* pleats. Nor is the jacket 'tightly fitting', as Curtis interprets it. Her terminology implies a sophisticated use of pattern cutting and tailoring, which was unknown before the medieval period. In fact, the jacket was loose-fitting and constructed from a series of rectangles, squares and inserted triangles, typical of clothing production on the Iranian plateau and further east, even today. The jacket's front-right panel overlapped the left panel and was held in place by a belt. It was probably made of lined silk – a high-status textile, true, but certainly fitting for the Parthian aristocracy.[14] It is probable that the underarm seam where the sleeve joins the body of the garment was not sewn up, so that more flexibility was afforded the wearer (see Figure 4.5).

The jacket resembles Achaemenid examples and similar garments are well documented at Assur in northern Mesopotamia in the early first century CE, on sculptures from Ḵūzestān, on coins of the Arsacid kings, as well as in non-royal representations.[15] It appears frequently in Roman representations where the V-shape of the cross-over

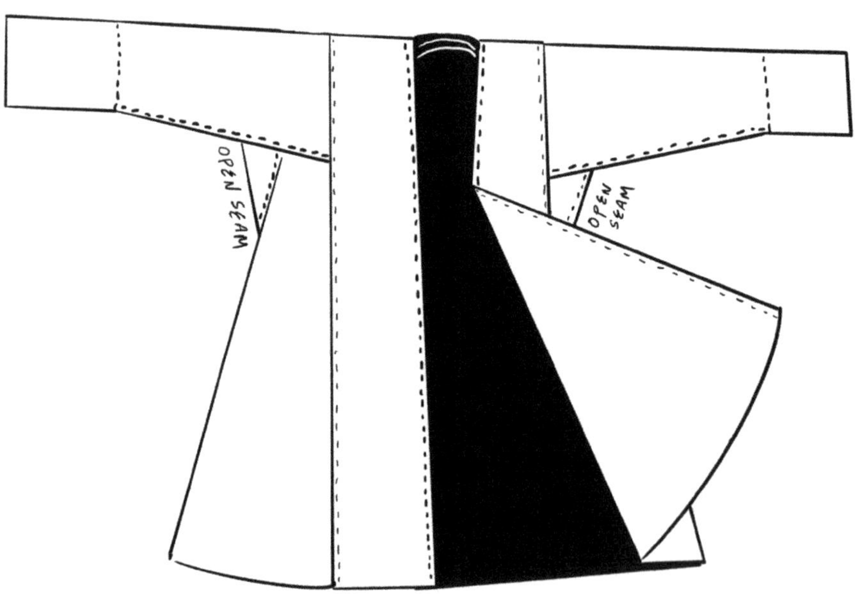

Figure 4.5 Possible cut and construction of the cross-over fur-trimmed jacket worn by the Shami Prince. Line drawing by Lloyd Llewellyn-Jones.

become the focus of special artistic attention and was used 'to portray generically the people of the East' (Schneider 2007: 56).[16]

Breeches (Under-trousers) and Chaps (Over-trousers)

It is difficult to know exactly how the breeches appeared, but a rear view of the statue gives us some answers (Figure 4.6). Most noticeable is the massed bulk of fabric that hangs from the buttocks, a style seen in other examples of Parthian stone sculpture and reliefs, including the remains of a marble statue from Susa (Figure 4.7; Curtis 1993: fig. 4.2). It is a very distinctive look. Vogelsang-Eastwood believes that the bulky rear overhang was a separate piece of cloth, separate to the trousers, and labels it a 'cushion' or a 'pad'. In this she follows Curtis, who labels it a 'cushion-type protection' (1993: 63). Vogelsang-Eastwood suggests that it was 'made from a tube of material which was gathered into folds in some manner. It was worn in order to cover the buttocks, which were left uncovered by the over-trousers or leggings' (2000: 42). This is far from a practical solution.[17] The idea that a Parthian horseman could ride for many hours on end, sometimes for days, with only a piece of cloth protecting his buttocks and thighs from being chafed raw is not feasible; the reason Iranian peoples employed trousers in the first instance was to make horse-riding a more comfortable, safe and endurable experience. Saddlesores were to be avoided at all cost.[18] Vogelsang-Eastwood is in error to speak of a separate cushioned pad. The bulky cloth at the rear of the Shami Prince's statue is part of his breeches, a garment made from heavy linen, thick wool, or maybe felt. It was cut from two large rectangles to form the trouser-legs, a smaller rectangle on the back, two

Figure 4.6 Rear view of the Shami Prince. Line drawing by Lloyd Llewellyn-Jones.

Figure 4.7 Detail of a fragmentary marble statue from Susa showing the distinctive bulky fabric at the rear of the breeches. Line-drawing by Lloyd Llewellyn-Jones.

triangular inserts and a gusset to create the crotch. When put together, these pattern pieces form trousers with a wide waist, a very full back and openings on the back of the legs level with the calf. On the upper part there were sewn channels through which a long lace passed, which was used to tie the trousers at the waist (Figure 4.8a–b). Four pairs of breeches constructed in this manner have been preserved in the archaeological record and, to all intents and purposes, they resemble the traditional *vraka* trousers still worn from Greece to Afghanistan (Figure 4.9).[19] It is likely that, like the *vraka*, Parthian breeches reached only to the knee or the top of the shins and were worn with a pair of short boots. These, too, were carefully cut and seamed: on the bronze statue, a seam line from the toe to the middle of the uppers shows that the prince's boots were shaped to fit and to be as flexible as possible.[20]

On top of the thick breeches, the Shami Prince wears a set of chaps (Figure 4.10). They have their own belt, and are fitted around the hips, resting below the thighs of the under-trousers. They are supported by suspenders, which hang from a concealed belt. His chaps are long, fitting over the foot to drape over the vamp (or upper) of the boots.[21] It is clear from the deep folds of the drapery that the chaps are not made from leather or suede (as might be expected for lower-status warriors), but from very soft fabric; I suggest they are made from a fine silk. Large amounts of good quality silk have been discovered at Hatra, Palmyra and other Parthian-related sites and Parthian art is replete with representations of the softly folding fabric (Pfister 1934: 37, 40; Hildebrandt and Gillis 2017; Callava 2019). When worn with the breeches, the chaps caused the extra fabric at the back to bunch up even more, giving that distinctive, full-bottomed look.

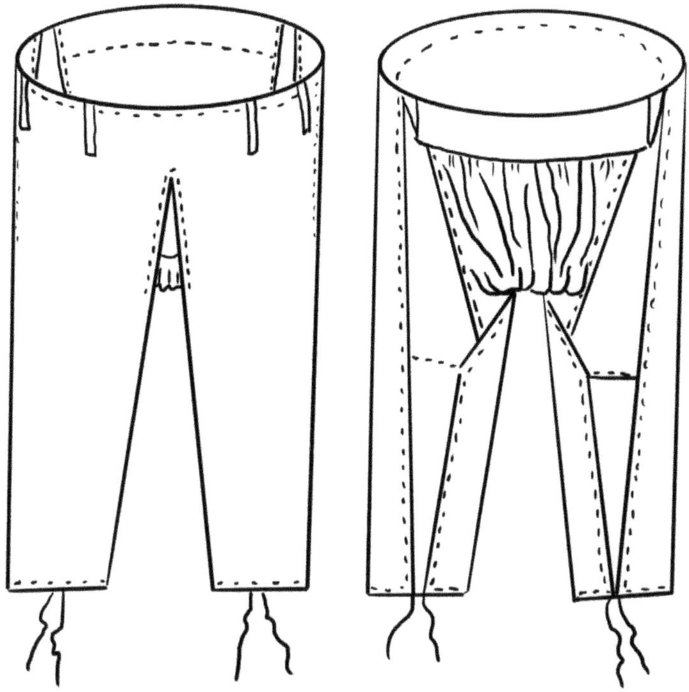

Figure 4.8a–b Possible shape of the Shami Prince's breeches, based on finds from Halabiyeh. Line drawing by Lloyd Llewellyn-Jones.

Parthian Riding Dress: Between Shape and Kinesics

In the early Bronze Age, long before the Eurasian horse peoples settled on the Iranian plateau, riding clothes would have been made from animal hides, with woven fabric only used later (Tilke 1990). This exclusive use of animal hides had a primary effect upon the initial design, composition and construction of Persian dress, which resulted in a legacy of several traditional features in later (Achaemenid, Parthian and Sasanian) times. There are two essential elements that need taking into consideration, as follows.

Shaping and Sewing

The cross-over jacket, the breeches and the chaps were constructed from straight pieces of cloth, patched together to create the desired body-fitting shape. These design-forms are often found in regions where material is primarily limited to animal hide (see Burnham 1973). Throughout the Eurasian Steppes and beyond to China, the cut of garments made from cloth reflects the fact that in ages past they were once cut from animal hides. With today's ease of manufacture, we take cloth production for granted, but in antiquity an extreme economy of material was practiced in the cutting of garments. Throughout the ancient Mediterranean, the loom was wide and capable of making large

Figure 4.9 *Vraka*, or full-seated trousers, traditionally worn by men on Cyprus; from a photograph *c.* 1950. Line drawing by Lloyd Llewellyn-Jones.

rectangles of cloth, resulting in garments that were draped and/or pinned (the Greek *peplos* and *himation*, or the Roman *toga* and *palla*), whereas in ancient Iran and Central Asia (through to China and Japan), looms were very narrow. This meant that fabric was narrow too, although length was never an issue. Garments made from such cloth were stitched together to form garments that, when compared with their Mediterranean counterparts, hung very close to the body, without drapery. In Iranian antiquity, weaving was more commonplace than cutting and sewing, and garments were made with the cut edges as straight as possible and with selvages clearly utilized to save sewing hems and to give strength where it was most needed. Garments were cut with very little wastage from the products of the narrow looms, and smaller pieces, like underarm gussets, were often cut from finer (sometimes older) fabrics. If (as was the case with ancient Iranian peoples) there was a strong tradition of cutting a garment in a certain way due to the economy of

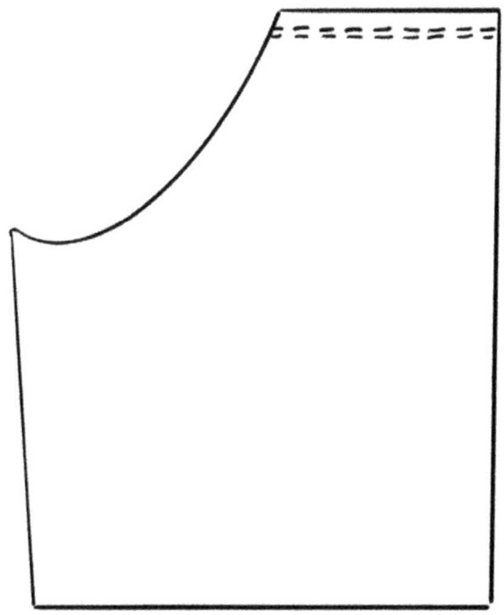

Figure 4.10 The construction of the Prince's silk chaps. Line drawing by Lloyd Llewellyn-Jones.

the width of cloth, that method of construction will hold firm long after the introduction of fabrics created in greater widths on more technologically advanced looms.[22] If, as in ancient Eurasia, animal skins were the basic clothing materials, the cut of cloth garments will retain the shapes dictated by the hides, as if the maker was subconsciously trained to work in a non-cloth tradition.

Layering

Garments were not worn alone, but in several layers. For maximum protection from the weather, from the rigours of horse-riding, and from enemy weapons, outer layers were made of strong textiles, often padded with wool, cotton, felt, or fur, while inner layers near the skin were made of fabric that could absorb sweat and moisture. The use of silk among the Parthians, which increased incrementally throughout the history of the dynasty, had a profound effect upon the wearability, mobility and efficacy of their horse-riding dress.[23] The toughness of silk fibres is comparable to that of modern nylon and Kevlar fibres and its breaking force is higher than that of cotton, Kevlar and nylon. Moreover, silk fibre is stronger than an equal thickness of steel wire (comparable to steel in terms of tensile strength, silk is the strongest natural fibre known to us) and, more importantly, it does not show the phenomenon of yielding before breaking. Silk has excellent flexibility, too; the fibres are completely elastic. Silk fabrics are warm and cosy in the winter and comfortably cool when the temperature rises. They are also highly absorbent (silk can absorb up to thirty percent of its weight in moisture without feeling damp) and silk blends well with other plant and animal fibres. All in all, when it could be acquired, silk was the must-have textile for the elite Parthian mounted warriors.[24]

All together then, the dress of the Shami Prince reveals distinctive design factors that developed out of necessity. Thus, we can observe that Parthian riding dress fulfils several key needs:

1. To cover the torso completely
2. To cover the limbs completely
3. To cover over the feet completely
4. To provide secure garment openings at sleeves, neckline and ankles
5. To provide adjustable closures for easy on/off operation
6. To provide all essential mobility
7. To protect the body
8. To work harmoniously alongside a horse and horse trappings
9. To work alongside military equipment
10. To be aesthetically pleasing
11. To convey status/rank
12. To emphasize group identity

The essence of Parthian success in mounted warfare was mobility. It was essential for horse and rider to move in harmony, as one living, organic entity. A visual framework for the way in which the horse, the rider and the rider's clothed body interact is schematically

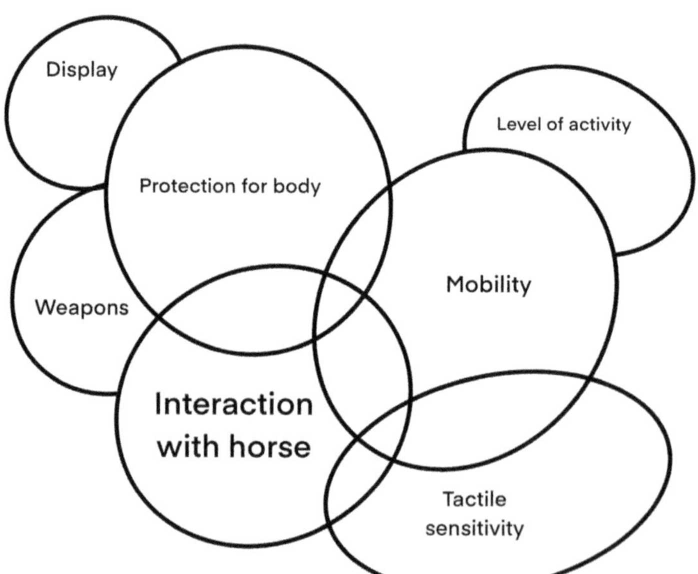

Figure 4.11 Framework of variables related to Parthian mounted combat. Line drawing by Lloyd Llewellyn-Jones.

captured in Figure 4.11, where design variables such as the activity level, mobility of the wearer, tactile sensitivity (the need to 'feel' the horse's flanks) and the need for interaction with arms (composite bow, quiver and arrows, daggers) are in operation (Loades 2016). It is the relationship between the design variables that gives cogency to the overall functionality of Parthian riding clothing. It is clear that mobility, tactile sensitivity and protection were the main concerns for the Parthians crafting and wearing clothing in the way they did.

Since clothing is intended to be a second skin, the mobility of the body in and through dress is an essential quality in design. A body moves through space without consciously controlling each movement, somehow instinctively aware of its own schema. The idea of a 'body schema' was first identified by neurologists in the early 1900s when, legend has it, Sir Henry Head and Gordon Morgan Holmes came up with the idea while watching ladies navigate their way through doorways in the very large hats with enormous plumes that were *à la mode* at the time. They noted that the women ducked just enough to clear the top of the feather or hat-pin, but no further, and they theorized that the hat had become so much a part of the wearer's implicit understanding of personal space that they instinctively incorporated it into their body schema (1911: 186–9). Head and Holmes advanced the theory that the body schema is the brain's way of understanding what is and what is not part of the body, as well as its way of recognizing where the body starts and ends. The body schema, they argued, is not always the same as the actual physical body and that held or worn items can also be incorporated into it. Head and Holmes argued that outside of the body schema is a zone that can be labelled 'peripersonal space' – a space that is within touching distance of the body and is a very sensitized area. The space beyond the body, outside the limits of touch, was labelled as 'extrapersonal space'; it is the zone within which the body schema and the peripersonal space both operate (Figure 4.12). Many subsequent studies have shown that things like handheld tools, or even cars, can become part of what the brain sees as body space; indeed, the best drivers are those with the most highly developed awareness of their body schema as incorporating the car itself.

In equine studies, it has long been recognized that whenever a rider sits on a horse, his body meets with powerful forces generated by the way the horse moves. Since the horse's natural way of moving and balancing is not conducive to carrying a rider in balance and with optimal postural strength, the rider must train his body to meet these unbalanced forces generated by the horse's movement and channel them into something different. In other words, effective riders must find synergy with the horse. Ideally, horse and rider must share the same body schema (Figure 4.13). Vicki Hearne reminds us that, '[e]very muscle twitch of the rider will be like a loud symphony to the horse' (2007: 108; also Birke and Thompson 2018) and that horses ridden by inexperienced riders experience muscle twitches like a cacophony of noise. The aim is for the rider to create harmony between their body and that of the horse, a 'conversation we call the art of horsemanship' (Hearne 2007: 108). In horseback warfare, as in equestrian sports, moving together requires joint decision-making from the horse and the rider. On a packed battlefield, and similarly on a crowded polo field, racetrack or in violent traditional Asian horse games

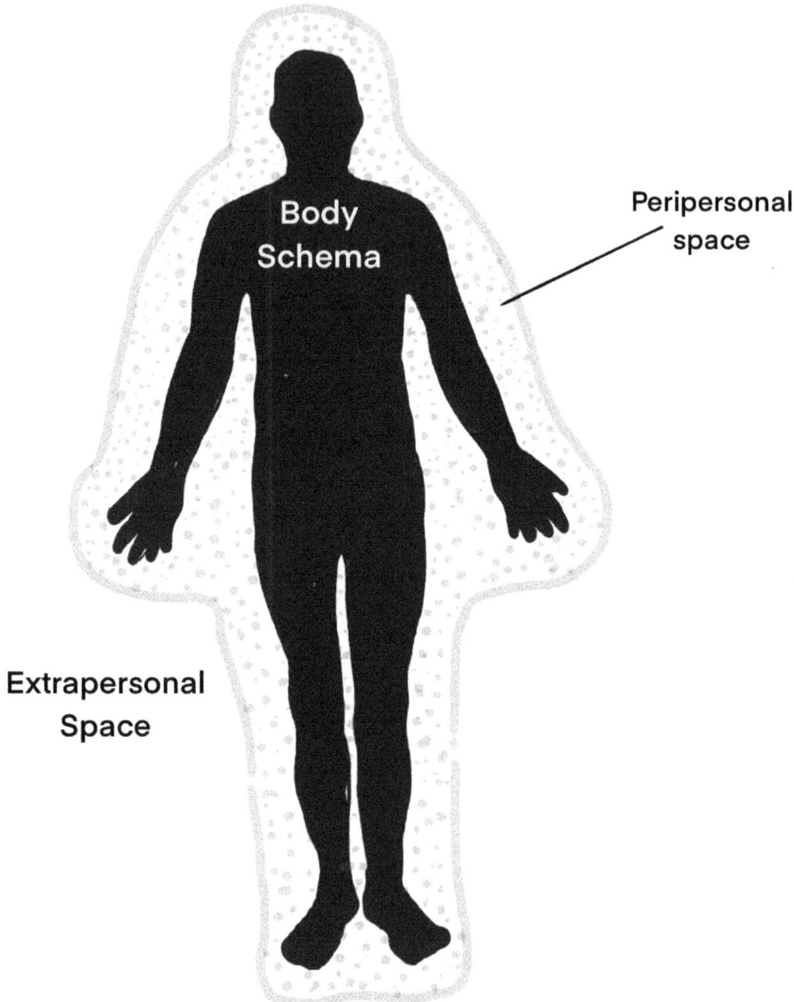

Figure 4.12 Body schema, peripersonal space and extrapersonal space. Line drawing by Lloyd Llewellyn-Jones.

such as *buzkashi* and *kokpar*, horses and riders make multiple decisions all the time, always remaining mindful of where other horses/riders are.[25] Spatial judgements are key. Human riders must work with equine cognitive skills and learn how horses respond to verbal and tactile commands (see especially Birke and Thompson 2018).

Intuitive symbiotic human-horse communication through body contact was the key to the Parthian mastery of mounted warfare. On horseback, a Parthian warrior like the Shami Prince was swift in attack and retreat; he could swivel his torso around and fire in any direction as his mount galloped on. His clothing helped fine-tune the 'conversation' his body had with the horse. His thick, padded breeches gave him much-needed protection at the thighs and rump and provided 'bounce', too. Fabric strain is always a cause of concern in the design and manufacture of today's style of trousers (Figure 4.14), but the construction of the *vraka*-style breeches worn by the Parthian riders bypassed this problem completely. The fullness at the rump and the crotch meant that there was ample flexibility built into the garment. Moreover, if, as I proposed, the breeches ended at knee-level or just above the shin, then the silk of the rider's chaps would have had direct contact with both his skin and the horse's flanks; every twitch or jerk of the muscles of the horse and rider could be felt, read and acted on. The silk gave the rider's legs tactile sensitivity to steer the horse through minute movements of his calves, ankles and booted feet. Meanwhile, his thighs did the hard work of gripping onto the horse's back. I suggest that silk chaps were adopted by the elite of the Parthian light cavalry not so much as a symbol of status (which silk no doubt was), but as a state-of-the-art technological advance in the craft of military horsemanship. The adoption of silk chaps (and later for the Sasanians, silk trousers) meant that physical and cognitive skills were enhanced for both the horse and the rider.

Beyond the horse-rider synergy created through the legs and flanks, the Parthians used riding garments that afforded ample movement and durability for the rider's torso too. The modern problem of garments 'riding-up' and wrinkling (Figure 4.15) indicates that, in modern design and manufacture, little consideration is given to free movement, but detached garment segments help determine body expansion during movement. As I suggested above, it would have been obvious to the Parthian archers that to give

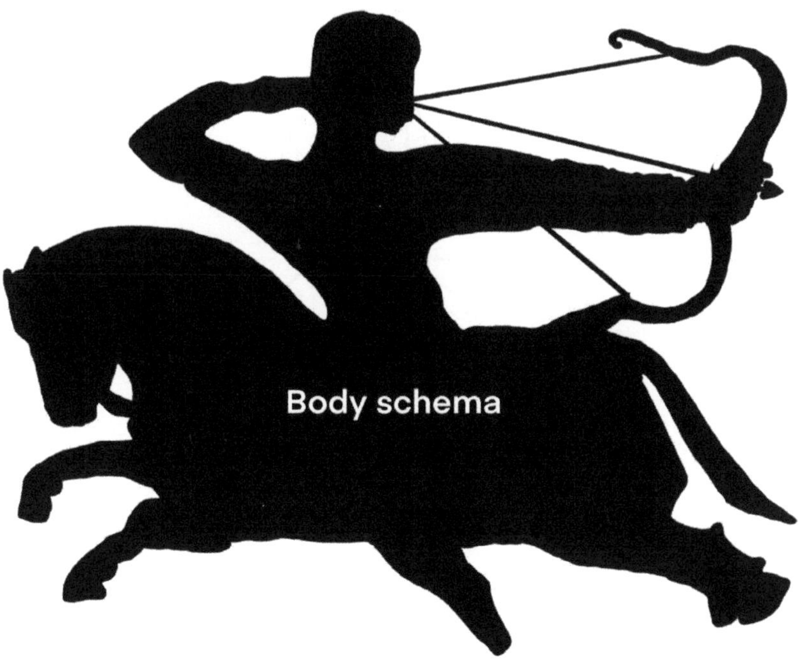

Figure 4.13 The body schema of the Parthian warrior and his horse. Line drawing by Lloyd Llewellyn-Jones.

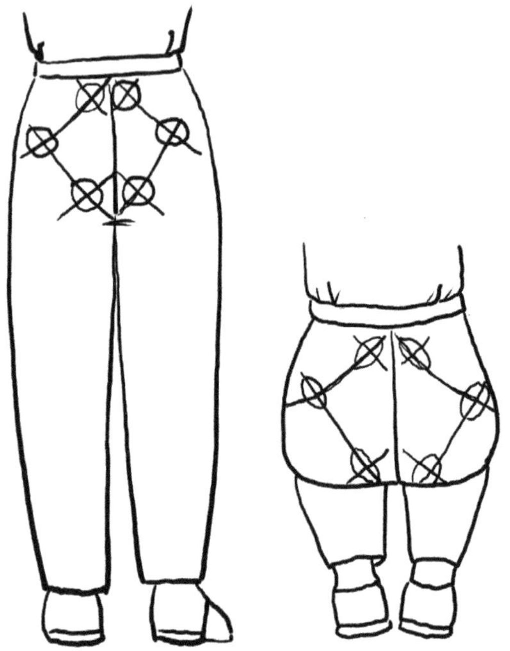

Figure 4.14 Stretch and 'give' is needed in the cut, construction and fabric choice of modern trousers and jeans. Line drawing by Lloyd Llewellyn-Jones.

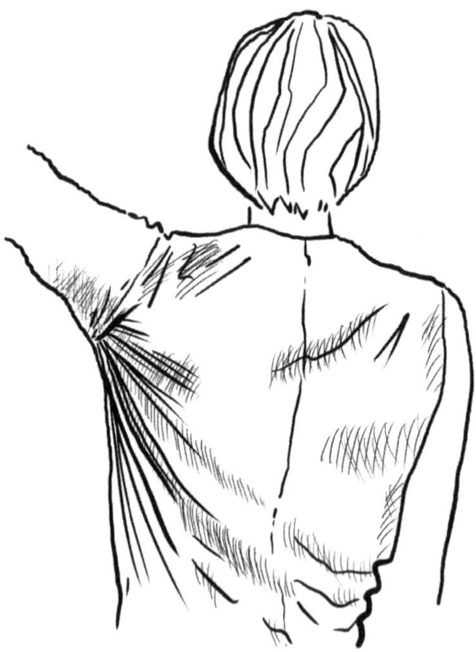

Figure 4.15 Close-set and closed underarm seams often lead to wrinkling and 'ride-up' in contemporary garments like shirts. Line drawing by Lloyd Llewellyn-Jones.

themselves the maximum flexibility of reach in their arms, they did not use joined seams beneath the sleeves. The sleeves, which were over-long and cut from ample fabric, were attached to the jacket only at the top shoulder seams (Figure 4.16). Arm movements such as bending and straightening or lifting and lowering were unproblematic (Figure 4.17). The loose cut of the tunic and the simple overlapping of the front of the loose silk jacket also gave the Parthian warrior unhindered flexibility to swivel his torso around and to shoot arrows from behind. Dexterity, after all, was key to his success (Figure 4.18).

Through many centuries of nomadic human-horse interactions, Parthian riding clothes had developed into sophisticated items of combat. Utilizing design attributes that were many centuries old, the Parthians also harnessed new technologies in manufacture and exploited the trade routes, which they controlled, to bring what was new and revolutionary into their fighting gear. Silk, the state-of-the-art textile of the moment, was employed alongside age-old forms of garment cutting and construction to repurpose the efficacy of combat dress. The clothing of the Shami Prince, the focus of this study, provides a solid case study for us to explore how the old and the new were harmonized into garments that had, we might say, design flair. The practical way in which technical knowledge of fabrics, of cut and of sewing interacted with the body of the human rider

Figures 4.16, 4.17 Open underarm seams or seam inserts allow for complete flexibility of the arms on both the sagittal plane and the frontal plane. Line drawing by Lloyd Llewellyn-Jones.

Figure 4.18 Reconstruction of the Shami Prince in battle. Line drawing by Lloyd Llewellyn-Jones.

was as sophisticated as anything that a contemporary functional clothing designer could come up with today; the fact that the garments heightened the symbiotic rapport between the warrior and his steed is only to be admired. Ancient Iranian combat dress was practical, hi-tech, cutting-edge and refined. It deserves further study.

Notes

1. I am grateful to Nasim Mohammadi of the Iran-e Bastan Museum in Tehran for providing me with photographs.
2. For a detailed description of the Shami Prince and its dating, see: Godard (1937: 285–305; 1965: 155–9); Seyrig (1939: 177–83); Stein (1940: 131–2, 146); Ghirshman (1962); Colledge (1977: 86); Herrmann (1977: 40); Kawami (1987: 58–63).
3. On dress and identity see Gleba (2008); see further Jäger (2019).

4. Watkins and Dunne (2015: xiv–xv). On clothing and the body, see DeMello (2014) and Entwistle (2000).
5. As shown in the Battle of Hormozdagān (224 CE) rock relief at Fīrūzābād, visible in Hinz (1969: pl. 52).
6. On the early history of horseback riding in Central Asia and Iran, see Drews (2004).
7. On the development of heavily armoured cavalry, see Potts (2007). For an analysis of heavy armour in the Sasanian period, see Alofs (2015).
8. The garment-types have a long history on the Iranian plateau and in Greater Iran; see Llewellyn-Jones (2020) and Kidd (2003). See also Yi-Chang (2006) and Yatsenko (2014).
9. See comments in Abbasi, Mehrafarin and Saadatiyan (2013).
10. See Nunome (1979: 341) and Kawami (1992: 14–15, 17, fig. 2, no. 1). On the use of felt as clothing, see Rudenko (1970: 83–116) and Gervers and Gervers (1974); more generally on felt in Central Asia see Bunn (2010).
11. Vogelsang-Eastwood (2000: fig. 3); more clearly seen in Ellerbrock (2021: fig. 9.12).
12. Pfister (1951: no. Hal. 119, pl. 1, no. Hal 12, pl. IIIa).
13. Schneider (2007: 56) speaks of a 'v-necked tunic', which is patently wrong.
14. While no silk dating to the Achaemenid period has been found in central Iran, Chinese silk has been found in Central Asia, an area of Achaemenid activity. It was far more commonplace in Iran by the late Seleucid period. Furthermore, Chinese silk was present in Greece in the late fifth century BCE and in the Hallstatt burials in Northern Europe in the sixth century BCE. See Wild (1984: 17–19, 22). See also Postrel (2020: 26–9, 51–4).
15. Colledge (1977: pl. 28b); Kawami (1987: pls. 7, 11); Farrokh (2005: 43); Karasulas (2004: 52); Sheppard (2020: 47); Yarshater (1983), pls. 4/1, 5/12, 6/1, 9/3.
16. For the development of the jacket into a longer riding coat in the Sasanian period, see Calament and Durand (2014).
17. Even the earliest known trousers have a gusset inserted between the legs; see Wan (2009).
18. See Beck et al. (2014) and Vogelsang (1992).
19. Kwaspen and De Moor (2013); Zink and Kwaspen (2015). These trousers show a great deal more sophistication in cut and construction than an earlier, Achaemenid-era, pair discovered in the salt mines of Chehrabad in northern Iran; see Grömer and Aali (2016) and Grömer and Bagherpour (2018).
20. Ellerbrock (2021: fig. 9.14); Vogelsang-Eastwood (2000: fig. 15); Gheibi (2009: 76–8); see also Fluck and Vogelsang-Eastwood (2004: fig. 77).
21. See Fluck and Vogelsang-Eastwood (2004: fig. 73).
22. See the layout of clothing patterns in Vollmer (2002), who transposes them onto animal-skin shapes.
23. On the appearance of silk in Iran, see Watt and Wardwell (1997). See also Hildebrandt and Gillis (2017) and Callava (2019).
24. A wondrous tradition exists between silk manufacture and horse legends; see Forrest (2016: 265–9).
25. On the Asian horse games, see Whitney Azoy (2003) and Thompson (2018).

CHAPTER 5
ROMAN *MILITES* AND DEFEATED '*BARBARI*': SOLDIERLY BODIES, DRESS AND IDENTITIES IN FIRST-CENTURY TILURIUM, DALMATIA

Ewan Shanks Coopey and Danijel Džino

Introduction

Roman military dress refers to the clothing, accessories and non-lethal equipment donned by members of Rome's many armies and their social and cultural sub-groupings. Military dress is physically preserved through archaeological finds of primarily metal/inorganic accessories but organic materials such as textiles survive in rare cases, like at *Dura Europos* (Pfister and Bellinger 1945). More common forms of evidence are public and private sculptural and painted works, and, finally, literary sources which describe, directly or indirectly, the protection, adornment and dress of Roman soldiers (*miles*, *milites*). Cassius Dio (64.14.21) and Vegetius (*Mil.* 2.18), for example, report that the equipment of Roman *milites* could be decorated with unit-specific iconography (*digmata*) or inscribed with the names of centurions (*centuriones*) and centuries (*centuria*) (MacMullen 1984: 446) – as seen with the Hagenau-type helmet found near Siscia inscribed with (*centuria*) *Lucc*(*i*) *Varronis* (the century of Lucius Varro) (Radman-Livaja 2004: 68–70).

Until recently, most research on military and provincial armaments and dress typically centred around form and function. However, scholarship has moved beyond functionality towards the socially and culturally constructive dimensions of dress and body (Coulston 2004; Hoss 2017; Gui and Petruţ 2018), with the lens provided by 'embodiment theory' proving fruitful. Emerging from a range of fields from the mid-twentieth century onwards, including phenomenology and practice theory, embodiment places the body at the forefront of experience. The idea has been adapted by archaeologists (Meskell 1996; Joyce 1998; Fisher and Loren 2003; Dornan-Fish 2012), with a unifying factor being the embodied agency and identity of both the collective and individual (Crossland 2012). The intersection of these is the focus here.

This lens facilitates considerations of the body in relation to 'the individual' *and* 'the group', understanding the body as both a symbol of 'the group' *and* as a representation of 'the individual' and their social self (Meskell 1996: 8–11). Such groups include communal, occupational and ethnic groupings. This embodiment of group and individual manifests in various ways. 'By means of dress, ornamentation, body modification, posture, gesture, and representation,' groups and individuals can establish, make claim to and negotiate their place (and others' places) within certain groupings and realities (Fisher and Loren

2003: 225–9). The remnants of these socially constructive embodying mediums in the archaeological and literary record, therefore, preserve snapshots of human action and experience (Dornan-Fish 2012: 287–9). Hence, dress and ornamentation can act as means of communication and construction, and their archaeological remains can be utilized to reconstruct the processes of belonging and identity negotiation.

This chapter draws from these theoretical premises to provide a well-defined case study of the (re)construction of identities through body, dress and sculpture in the Roman provincial landscape, examining the body of the combatant as it manifests in these different phenomena.[1] We investigate how sculptural depictions and material remains of soldierly bodies and dress communicated certain messages, fashioned identities and defined communities within the distinct geographical, cultural and historical context of the military fortress of Tilurium in recently pacified first-century CE Roman Dalmatia (Figure 5.1) – a province in which a multiplicity of social and cultural identities were negotiated and enforced on individual and group levels. We move beyond general overviews of the Roman military and provincial communities to a more specific space and time to explore how the diverse Tilurian Roman *milites* embodied new military identities to overcome cultural differences, and how similar mechanisms were deployed to impose Roman martial authority over the local Dalmatian population.

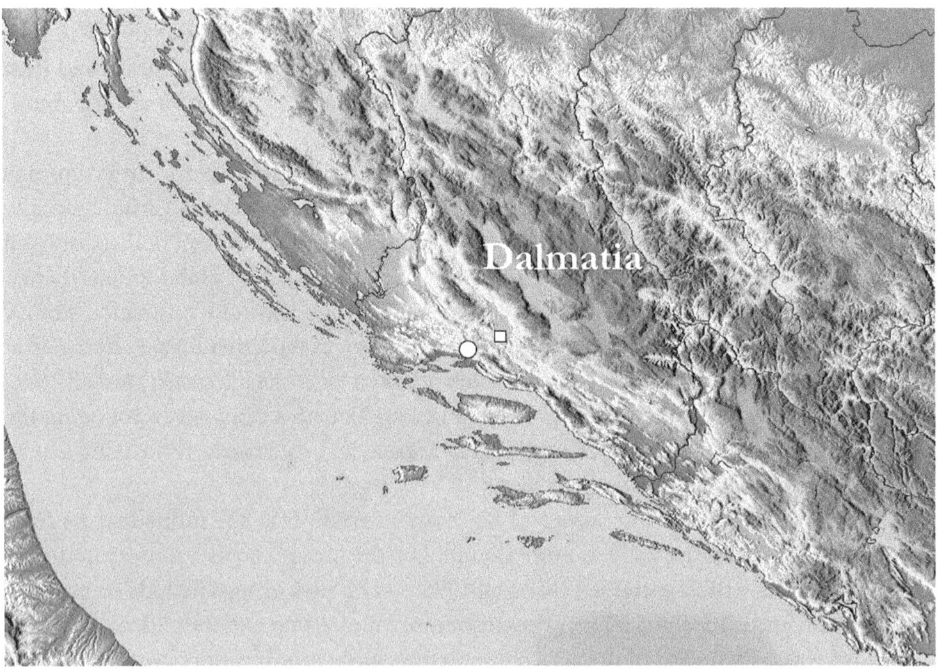

Figure 5.1 Map of Dalmatia, the military fortress of Tilurium (square) and the provincial capital Salona (circle). Copyright: Danijel Džino.

Dalmatia, The Tilurian Garrison and the Roman Military Community

First, it is important to establish the historical and local context within which our legionaries (citizen soldiers), auxiliaries (soldiers recruited from the provinces, typically without citizenship) and 'non-Roman' locals were embedded (see Wilkes 1969: 13–77; Šašel Kos 2005: 249–471; Džino 2010; Džino and Domić Kunić 2018). Roman military interactions with the future province of Dalmatia began in conflicts with the so-called Illyrian kingdom in the later third century BCE. While continuing into the second century BCE, these did not result in a permanent Roman presence. Rather, they were unconnected interventions undertaken when Roman interests were threatened. Significant imperial activity began with Octavian's Illyrian conquest of 35–33 BCE, which resulted in the establishment of Illyricum (*c.* 32–27 BCE), a province subsequently expanded northwards and eastwards between 14–8 BCE, particularly during Tiberius' Pannonian war of 12–9 BCE. The only real challenge to Roman rule was the local uprising or *bellum Batonianum* (Batonian war) of 6–9 CE, in the decades following which Illyricum was divided into Dalmatia and Pannonia, with Dalmatia covering the southern region from the eastern Adriatic coast up almost to the river Sava (Figure 5.1).

The Roman fortress of Tilurium was situated at modern Gardun in the hinterland of Split, Croatia. Archaeologically, it has been the focus of much attention from as early as the late nineteenth century, with material making its way into many museum collections, such as the AMS, MCK, MTK and FMS (Sanader and Tončinić 2013: 411–12). In antiquity, the site and the surrounding landscape transitioned from one inhabited by locals to one occupied by diverse Roman legionaries and auxiliaries. Established by the Romans in the immediate vicinity of the provincial capital Salona (Solin) in the early first century CE (Figure 5.1), it was strategically important, defending the only approach to Salona from the hinterland – the pass of Klis – and overlooking important roads and a crossing over the Cetina river (Zaninović 1996: 280–2; Sanader and Tončinić 2013: 414–16).[2] The wider region was a contested space during and following Roman conquest, and the fortress acted as part of a fortification system established within the territory of the indigenous political alliance known to the Romans as the *Delmatae* (Periša 2009: 511–12), who challenged Roman interests on several occasions from 156 BCE to the *bellum Batonianum* and inhabited the region in pre-Roman times.[3]

A permanent fortress was established shortly after the *bellum Batonianum* by *Legio* VII (later VII *Claudia pia fidelis*). This was the first garrison at the fortress, stationed alongside auxiliaries such as the *Cohors* II *Cyrrhestarum*. *Legio* VII left Dalmatia in the second half of the first century CE (Sanader and Tončinić 2010: 44–7; 2013: 416–19; Strobel 2000: 528), after which the fortress was primarily garrisoned by auxiliary units and intermittently smaller *Legio vexillationes*, such as *Ala Tungrorum* and *Legio* IIII *Flavia felix*, all of which is attested to by epigraphic evidence (e.g. *AE* 1994, 1356; 1994, 1357-58; Dodig 2007: 146). The same evidence sheds light on the backgrounds of these *milites*, revealing an ethnically and/or culturally diverse group of combatants. *Milites* of *Legio* VII, for instance, hailed from further east in Macedonia and Asia Minor or from across the Adriatic in Italy (Tončinić 2011: 201–7, T. 32) and auxiliaries are recorded as being from as far afield as *Germania Inferior* (*Cugerni*: *CIL* 3.2712/9727) and *Syria Superior* (*Berea*: *CIL* 3.14943; Kurilić and Serventi 2015).[4]

Principally, Rome's armies – including the Tilurian garrison – were forces of empire: policing, spreading, establishing and consolidating it. However, since the 1980s, there has been an ever-increasing interest in their social dynamics (MacMullen 1984), resulting in an appreciation of Rome's armies as the heart of a dynamic empire-wide social entity: the so-called 'Roman military community' (James 1999; Goldsworthy and Haynes 1999). Instead of being some modern military institution (there was no Latin word for 'the military'), it was seen as a looser collection of numerous *exercitus* (armies), supporting organizational structures and individuals – namely *milites*, *veterani* (veterans) and 'camp-followers' (James 1999: 14–15; 2011: 22–3; Coulston 2004: 140). These followers included non-servicemen, such as soldiers' wives and children, craftsmen and freedmen, who would accumulate around and within military centres (van Driel-Murray 1997; Allison 2013: 319–43) and exercise agency within the community (Greene 2016).

This amalgamation of numerous overlapping sub-communities was underpinned by unifying yet distinct identities and a strong sense of cohesion centred around shared experience and occupation (James 1999; Collins 2006: 7–9; Haynes 2014: 451). At all levels, the social grouping was relationally and symbolically constructed and rejuvenated by its members through numerous social, organizational, ritual, cultic and iconographic practices (Haynes 2014: 454–5; Walas 2015: 23–5), including one of the most significant avenues of social construction: dress. The significance of the community and this shared identity for its members was substantial, providing the diverse *milites* and their companions with something tangible to belong to and engage with – a detail Almir Marić (2019: 15–16) rightly notes should not be under-emphasized considering these soldiers were often stationed a long way from their homelands, like those at Tilurium. In a sense, the identities of Roman soldiers, at least initially, were in a state of flux, and the military community provided an anchor for new identities. It is this process that we explore in the first part of this chapter through the archaeological and sculptural evidence from Tilurium, addressing the diverse *milites*' use of dress and body to (re)construct 'Roman' and 'servicemen' identities, after which we investigate the monumentalization of Roman supremacy via projections of 'defeated' or 'subdued' local bodies.

Belts and Brooches: Embodiment of Martial Functionality and Communal Membership at Tilurium

The Military Belt

Excavations at Tilurium have been undertaken by a team from Zagreb University since 1997, revealing a great deal about the site, including some of its significant structures: defensive walls, a cistern, a warehouse, a barracks and potentially an amphitheatre. No significant trace of extra-mural settlement has been unearthed as of yet; however, much remains unexcavated (Sanader 2003; Šimić-Kanaet 2010; Sanader and Tončinić 2013:

420–2; Sanader et al. 2014, 2017, 2021). On the other hand, hundreds of various metal components of dress finds have been unearthed, with a great deal dating to the first century CE (Ivčević 2010, 2011, 2013a, 2014, 2016, 2017), and these provide some of the best insight into the soldiers' embodied identities.

Particularly significant finds are those related to sword belts (*balteus, baltei*), namely buckles, pins, fittings, military apron components, scabbards and scabbard fragments.[5] The Roman military *balteus* was worn around the waist in the first century CE, moving to the shoulder in the second century. It was made of leather and decorated with metal mounts, perhaps an apron and then closed with a buckle (Hoss 2017: 85). Numerous buckles have been found at Gardun, the most common being a D-shape with volute spirals (Figures 5.2a–c) – a typical style of military buckle limited to the first half of the first century CE.[6] Contemporaneous styles have also been found, including one elaborate scabbard fastening consisting of an ornately decorated bronze buckle and rectangular fitting, inlaid with a niello checker motif, triangles and a rosette (AMS H 4382) (Figure 5.2d).[7] Other finds include cordate buckles with a button and quadrilateral pieces, all dating to the first centuries BCE and CE.[8] While the pins are often missing, some survive.[9] Several contemporaneous rectangular apron fittings and circular rivets have also been discovered, including one 4.4-centimetre-long silver-plated incised rectangular plate (AMS H 3674) (Figure 5.2e).[10] Most of the rivets are 2–3 centimetres in diameter and decorated with small details.[11] These so-called aprons were a decorative enhancement of the *balteus*, serving as an extension of the belt strap itself or attached separately. As illustrated by archaeological finds and sculptural monuments such as Trajan's column (e.g. scenes IV, CVI-CVII), they consisted of usually three or more straps flowing down over the groin and inner thigh region, decorated with plates and studs and ending with metal pendants. They were worn by Roman foot soldiers of the first century CE, falling out of popularity by the reign of Hadrian. It has also recently been observed that wearing these aprons would have been very uncomfortable and therefore their use may have been primarily decorative and limited to specific contexts where they would have been 'on show' (Bishop 1992; Bishop and Coulston 2006: 110; Hoss 2017: 86–7), reinforcing the argument for their symbolic function outlined below.

Hanging from these belts would have been swords, and scabbard remains have been found at Tilurium and its environs. An elaborate and well-maintained 26.9-centimetre long military dagger or *pugio* of the Mainz type (late first century BCE and early first CE) was discovered near Gardun in the Cetina River in the late nineteenth or early twentieth century (MCK 5103).[12] It is enclosed within a 26.9-centimetre scabbard split into four rectangular fields, each inlaid with silver and copper and bearing floral motifs and red enamel medallions. Scabbard fittings and fragments belonging to similar elaborate pieces have also been unearthed, such as a palmette-shaped Mainz-type mount (MCK B-870) and an incised chape discovered in 2005 (MTK 971).[13]

Aprons, scabbards, belts and therefore belted tunics (*tunica*) were evidently a common piece of attire at Tilurium, comprising a key part of *miles*' dress – with the belt providing the unifying 'Roman' component. While they would have naturally been worn when soldiers were in battle gear (during campaign, training exercises, or on parade), as is

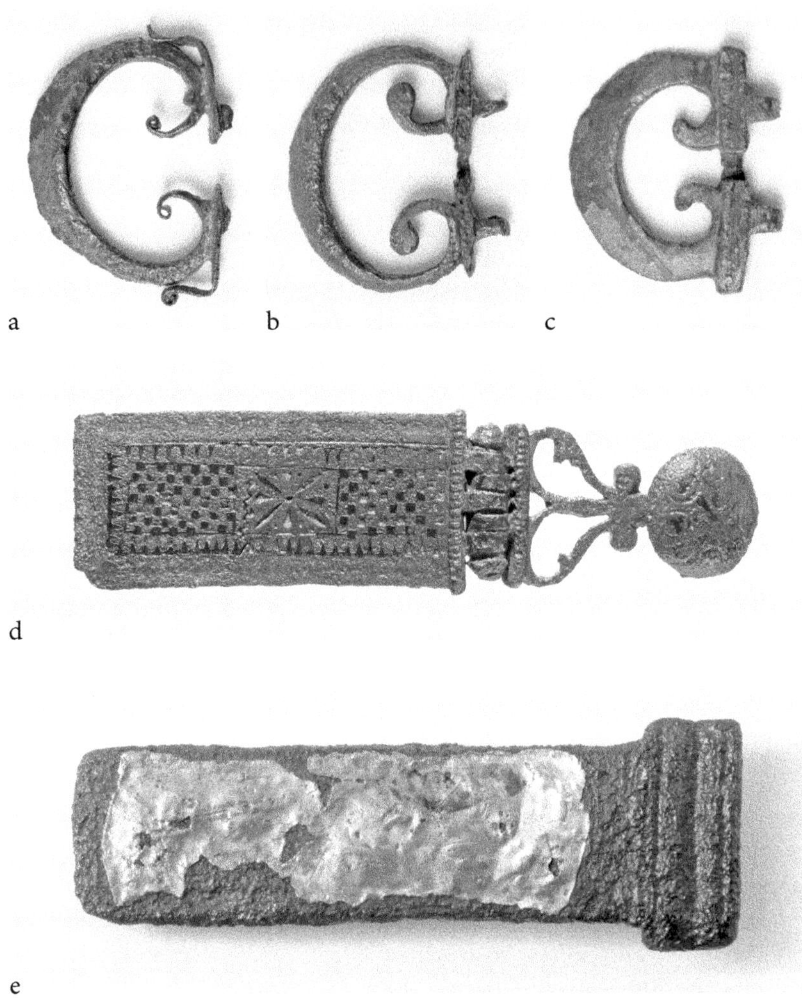

Figure 5.2a–e Small metal finds from Tilurium: (*a–c*) Belt buckles AMS H 4453, 6176, 6177; (*d*) belt buckle and fitting AMS H 4382; (*e*) apron fitting AMS H 3674. Date: First century CE. Copyright: Archaeological Museum in Split.

argued at length by Speidel (2012: 6–10), the ancient literary sources suggest that soldiers also donned belted tunics in civilian settings (Hoss 2012: 19–30). Indeed, contemporaneous military belt buckles and fastenings have been found at Salona, which is known to have been garrisoned by several military detachments (*vexillationes*), including a sizeable detachment of *Legio* VII legionaries and veterans (Matijević 2017: 132–6).[14] Members of the military community at Tilurium therefore donned these belts well beyond the fortress walls. While naturally serving practical functions, these components were 'often consciously chosen with the intent to express certain messages' (Speidel 2012: 5) – begging the question, what message did the material manifest for the Tilurian *milites*?

The belt and related pieces played a significant role in constructing military identities and a communal martial character (Hoss 2017). The *balteus* was an important part of the Roman military psyche and related to one's office, with the removal of one's belt (*discinctus*) constituting a shameful punishment (Livy 27.13; Suet. *Aug.* 24, 106). Tacitus recounts in *Histories* (2.88) that soldiers of Vitellius (69 CE) camped seven miles from Rome were made a mockery of (*ludibrium*) by civilians who cut their belts and disarmed them.[15] Hoss rightly (2017: 85, 96) refers to the Roman belt 'as the focal point and status marker of their identity', for the *balteus* and scabbard were home to the sword, the core of a Roman warrior's identity and distinguishing him from civilians. As Mihajlović (2020: 35) notes more generally, a weapon is partially what makes the soldier a soldier. In first-century CE Tilurium, then, we should understand scabbards, swords and *baltei* as signs of the Tilurian *milites*' function as state-employed enactors of violence – irrespective of whether they ever set foot on a battlefield. The *milites* were embodying their roles as 'violence specialists': a group empowered to enact necessary physical violence for the protection of society in exchange for economic benefits, but not necessarily ever needing to act in this capacity (North, Wallis and Weingast 2009: 13–20). The Tilurian *miles* was not truly a Roman *miles* without the *balteus* and sword, and the sword was not truly a Roman weapon without the *miles* and its *balteus*.

This significance of the *balteus* and its apron was evidently apparent to people outside the military community, becoming widely associated with service. Juvenal (16.48) refers to Roman *milites* as 'those men who are armed and belted' when he discusses (somewhat in spite) the supposedly bountiful benefits of soldiering.[16] Here, *milites* are metaphorically reduced to little more than their martial function via a distinct piece of dress deeply associated with this former function. This trope of soldiers as visually identifiable by their (often ornate) belts is also found in other Roman works (Pliny *HN* 33.152; Petron. *Sat.* 83; Haynes 2013: 262; Hoss 2013). As noted earlier, several of the pieces from Tilurium were indeed elaborately decorated. While the specifics of the stories in the sources may be called into question, the belt as a key component of a Roman combatant's body was evidently something other members of society were conscious of. Moreover, there was likely an audible dimension of bodily unity that complemented the visual dimension, with these various *baltei* components creating a distinct rattle and jingle, particularly when paired with the renowned hobnailed shoes of *milites* (Pliny *HN* 33.152; Coulston 2004: 142; Hoss 2017: 86–7).

The belt ensemble was therefore a medium through which the Tilurian *milites* would embody their martial identity. These accoutrements effectively (re)constructed their bodies into those of Roman soldiers by presenting them as having the capacity to enact violence via the sword.[17] By wearing the widely significant belts, aprons and scabbards found at Tilurium, combatants at the fortress – whether a long-serving legionary of *Legio* VII from Asia Minor like Lucius Ancharenus, or an *eques alaris* (auxiliary cavalryman) from Belgic Gaul like Severius Ubius – were embodying their shared role as Roman 'violence specialists' and shared identities as members of an Imperial military community.[18] In the process, the local Tilurian community of combatants was itself characterized as one comprised of military bodies distinct from those of non-combatants.

Cloaks and *Fibulae*

This bodily distinctiveness was promoted by other items of dress not *necessarily* exclusively related to martial function such as so-called 'military cloaks'. One such cloak was the *sagum*: a rectangular cloak fastened by a brooch on the right shoulder popular with *milites* during the Principate but not exclusively worn by them (Bishop and Coulston 2006: 111). While the organic material has not survived at Tilurium, sculptural depictions provide insight into their style (see below) and the metal components of this dress ensemble, *fibulae* (brooches), survive in great numbers. These *fibulae* are often of the high-bowed type found predominantly in military contexts (Allison 2013: 71–3; Ivleva 2016). One particularly popular first-century CE style is the *Aucissa fibulae*, common in *Dalmatia* (Marović 1961), the wider Adriatic and Balkans (Buora 2008; Haralambieva 2013) and across the northern Empire in Britannia, for example (Eckardt 2005). *Aucissa* consist of an arched bow, sometimes incised or ribbed, and a hinged pin and flat head plate that was often inscribed with the manufacturer's name (Eckardt 2005: 150–4; Ivčević 2011: 161–4) (Figure 5.3).

During the nineteenth and twentieth centuries, dozens of stray *Aucissa fibulae* were found and placed in AMS or MCK, with over 120 specimens held in the AMS alone (Marović 1961: 106).[19] The precise provenance of these early finds is unknown, yet their number attests to their popularity. Thankfully, recent excavations have found *fibulae* in situ, and between 2007–10 seven *Aucissa fibulae* (as well as some later styles) were discovered in several different rooms inside a complex running along the eastern wall of Tilurium.[20] While the function of these rooms is uncertain (Sanader and Tončinić 2017: 39–40), there is a particular concentration of military equipment.

Figure 5.3 *Aucissa fibula* AMS H 4372 from Gardun (Tilurium). Date: last quarter first century BCE–third quarter first century CE. Copyright: Archaeological Museum in Split.

Clearly, *Aucissa fibulae*, along with other styles popular within the military unearthed at the site, such as 'carinated' *fibulae* (Ivčević 2011: 170–1 nn. 23–7), were a common piece of attire at Tilurium alongside the military *sagum* they typically clasped. In addition to being related to these 'military cloaks', the brooches themselves held significance. Although the *Aucissa* brooch is predominantly associated with male dress, they have also been found in some instances to have been worn by women (Eckardt 2005: 152; Seldmayer 2014). Conversely, while occasionally discovered in civilian contexts, it is found primarily in military contexts – a pattern of distribution observed in Dalmatia (Ivčević 2011: 162–4). This has led scholars to conclude that these brooches acted as another marker of membership within the Roman military community (Riha 1979: 114–23; Allison 2013: 73–4), primarily, but not exclusively (Eckardt 2005: 150–4, 157–8), as a serviceman – particularly when considered in conjunction with other items of military dress. Like the sword-belt, this membership was something the wearer wished to advertise, for as Ivleva (2016: 121) so aptly states, *fibulae* had a 'highly visible positioning on military dress – at the eye-catching shoulder level' and were thus 'worn to be seen'. The diverse soldiers at Tilurium and, to a lesser extent, other non-combatant members of the community can thus again be seen using dress to overcome potential ethnic and cultural differences to construct a sense of bodily unity at Tilurium and associate with the broader community of combatants, albeit without the focus on combat function of the *balteus*. The possible use of this item of dress by the wider community at the fortress also presents a potentially worthwhile line of future enquiry.

In Death as in Life: Cloaks, Funerary Sculpture and the Military Community of Tilurium

In the first two centuries CE, funerary stelae were the commemorative monuments of choice within the military community, attested to by the vast number commemorating *milites* (Speidel 2014: 321–6). In fact, servicemen and their extended communities are often some of the earliest drivers of the 'epigraphic culture' across the provinces (e.g. Hope 1997: 255), including Dalmatia (e.g. at Salona; Matijević 2019). Naturally, then, funerary stelae have been found in high numbers around Tilurium (Gardun, Vojnić, Košute, Trilj and Dugopolje), having been reused frequently in late antiquity (Cambi 1994; Tončinić 2011: 168; Džino 2021: 64–5). While this repurposing hinders reconstructions of their exact original contexts, contemporary *necropoleis* at Salona (Miletić 1990, 1991) suggest that the monuments clustered outside the fortification walls and along the nearby roads towards Pons Tiluri (Trilj) or Salona (Sanader and Tončinić 2010: 41–5). They seem to have come from at least two local workshops: one based at Salona and another perhaps at the fortress (Cambi 2020), with the more elaborate stelae featuring portraiture.

Funerary stelae are not without their issues, however (see Redfern and Pearce's chapter in this volume). Preservation and representativeness are critical issues. It has been calculated, for instance, that over 6,000 legionaries served in Dalmatia in *Legiones* VII and XI alone (Matijević 2017: 134), rendering the epigraphic assemblage from Tilurium (and that of the whole province) a very small sample. Not only is this a matter of preservation, but also likely due to preference and cost – although there may have been military *collegia* or burial funds to offset some costs (Veg. *Mil.* 2.20; Pegler 2000: 38–9). The evidence presented here therefore preserves a snapshot of a particular group wealthy enough (individually or communally) to afford funerary monuments (Speidel 2015: 334–5). Indeed, this is why *militaria* has also been examined, so as to explore the bodies and identities of the broader community of combatants.

There is also the question of how much choice commemorators had in the design of funerary monuments: how much of the decoration on a stele is representative of the taste of the consumer as opposed to that of the stonecutter? Whilst it is unlikely every stele was cut purely on the preferences of the customer, it is very possible they had significant input. Stone-sculpting was very specialized in some regions of the Roman world (Russell 2020). Masons could sculpt monuments in advance and then add additional or personal elements at the time of purchase (Edmondson 2015a). Moreover, as mentioned above, Tilurium may have had a workshop in its close vicinity, suggesting members of the army were a significant portion of the customer base and the masons had frequent contact with them (Cambi 2020). This is not to say all stylistic choices at Tilurium were those of the customer (who could have been working in accordance with a legal will, such as those preserved upon *CIL* 3, 2710; 14932), rather, the monuments could be chosen by commemorators on account of their use of communally significant sculptural decorations; the significance of which, we would argue, the sculptors were very aware of and thus produced 'to order' or as a half-finished stylistic template. They do not need to have understood the significance of the designs, only that they *were* significant. An important implication of this stance is that the commemorator and the stonemason become key components of community (re)generation.

Following this thinking, while not meant to be lifelike representations of the deceased, portraiture at Tilurium can provide insight into the concretization of communal belonging through pieces (or depictions) of dress, such as the hooded military cloak (Coulston 2004: 151–2; Haynes 2013: 259–62). Several funerary monuments bear cloaked funerary busts, such as the stelae of *Legio* VII *miles* Gaius Longinus from Amblada (Galatia), *Claudiae Novae eques alae* Marcus Elvadius from Cugerni (*Germania Inferior*) and *Cohors* II *Chyrrestarum miles* Marcus Pytha from Berea (Cyrrhestica, Syria Superior).[21] The men are depicted wearing thick, potentially hooded cloaks, sometimes gripping them at the neck. Srđana Schönauer (2000: 256–63, 403–4 with pls T.X-XII, figs 29–33) classes these cloaks as *paenulae, lacerna* or *birri* (heavy cloaks popular with soldiers), though accurate identification is difficult. Nevertheless, they are definitely *paenula*-like, covering the body from the neck down, and they do not appear to be clasped at the shoulder like a *sagum* or *lacerna*. This cloaked portrait style seems to have held a degree of significance across first-century CE Dalmatia, likely because of their day-to-day use. Similar busts depict *cohortis* I *Bracaraugustanorum eques* Veranus and Galatian *Legio* VII *veteranus* Gaius Licinius at

Bigeste/pagus Scunasticus (Ljubuški, Hercegovina), and *Legio* VII *Claudia pia fidelis miles* Quintus Mettius Valens at Salona.[22]

As noted above, certain cloaks were associated with service in this period, with Speidel (2012: 8–9) drawing attention to the frequency of this garb on sculptural depictions of servicemen across the empire, as well as the linguistic link made between preparations for war and *saga sumere* ('taking up the cloak') in Roman literary sources (Cic. *Phil.* 8.11.32; Hor. *Epod.* 9.28; Cic. *Varr.* 5.94). Due to this popularity within the military more broadly, and the frequency of cloaked depictions amongst serving and veteran soldiers in Dalmatia (or their commemorators), it is likely that, through these representations the sculptors and commemorators at Tilurium were, consciously or not, drawing attention to and asserting the deceased's identities as soldiers (Coopey 2020: 75–6). Similar phenomena have been noted in later third-century CE Pannonia (Migotti 2013) and Dacia (Gui and Petruţ 2018: 134–5), where the stelae of military men are distinguishable by cloaks and sometimes *baltei*. This could suggest a particular affinity with cloaked busts and/or the physical cloak itself in the Illyro-Danubian region. However, this is a significant expanse of time (first–third century CE), and civilians are known to wear similar cloaks, so this association needs to be investigated further.

Still, despite their differing *domicilia* (places of origin) and, presumably, backgrounds, it is evident that Tilurian *milites* were aligned in their choice of attire in life and death – remembered as a collective of Roman combatants. Particularly when worn together with the previously discussed components of military dress, cloaks would certainly have plaid a crucial role in Tilurian *milites*' (re)construction of their social identities as combatants and members of the local (Dalmatian) and empire-wide (and perhaps regional or 'Balkan'/'Danubian') military communities – identities that would have been very significant in bringing these Tilurian combatants together.

The military identity expressed through portraiture at Tilurium was also complemented, reinforced or, on occasion, challenged through other parts of the funerary monuments, such as decorative sculptural motifs.[23] One popular funerary motif at Tilurium was the so-called 'military frieze' (Figure 5.4), through which the military occupation of the Tilurian combatants was signified by another as yet unmentioned item unique to their bodily experience: armour. While there are no literal bodies depicted (see Nováková's chapter in this volume), we see suggestions of the body via the stylized Hellenic and Roman arms and armour, namely a combination of greaves, helmets, cuirasses and shields, including the circular *clipeus*, rectangular *scutum* and three-pointed 'Amazon shield' or *pelta*. At least five Tilurian monuments are decorated with the motif, likely all belonging to *milites* of *Legio* VII – one even features a scarecrow-style *tropaeum* (trophy), a monument Romans erected on the battlefield (see below).[24] These friezes are also found across the province at sites like Salona and Bigeste/pagus Scunasticus, and further research will explore this phenomenon within the wider Romano-Dalmatian military psyche.[25] This use of sculpture to propagate embodied communal identities was also employed elsewhere at Tilurium, as seen with the Gardun *tropaeum*.

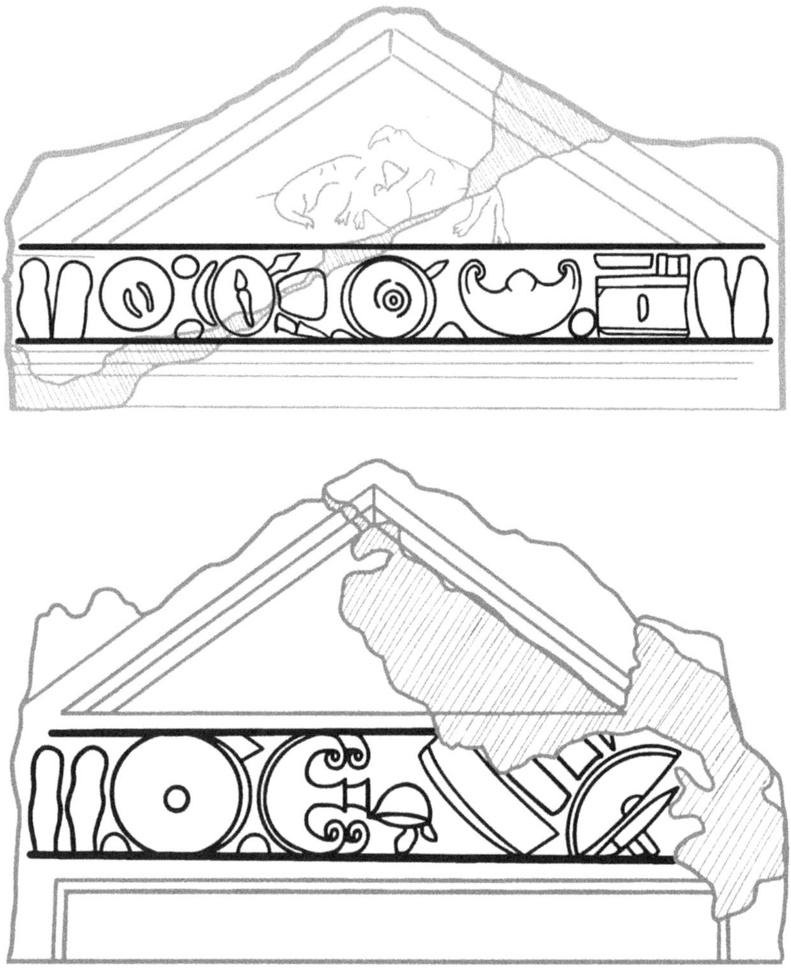

Figure 5.4 Illustration of the 'military friezes' from two stelae at Tilurium: CIL 3, 2709 and 2716. Date: *c.* 9–59 CE. Illustration: Jack Roberts. Copyright: Ewan Shanks Coopey.

Monumentalizing Bodies of the 'Defeated': The Gardun *Tropaeum*

One of the earliest monuments from the Tilurian fortress is known as the Gardun *tropaeum* (Abramić; 1937; Picard 1957: 217, 252; Cambi 1984; 2013: 12–15). Two known fragments of the Gardun *tropaeum* survive, with no known context for either find except that they were found within the fortress.[26] As such, the exact location in/around the fortress and therefore their audience(s) can only be hypothesized (discussed below), though the *milites* were of course a significant, if not primary, audience. The inscription is not preserved beyond an 'O' on the larger fragment (Figure 5.5). Both fragments belonged to the base of the larger monument, reconstructed by Cambi at an estimated 8.20–8.40 x 7.5 metres. While earlier scholars speculated that the *tropaeum* could have been erected as early as the second triumvirate (Picard 1957: 217–18, 252), Cambi (1984: 80–4) has convincingly dated it to the early Tiberian era on a stylistic basis, drawing particular attention to similarities with, and influences from, the Ara Pacis.

The *tropaeum* is a Roman monument celebrating military victory, which evolved from earlier Greek and Hellenistic models. On the battlefield after victory, the Romans usually placed a temporary trophy structure – the previously described 'armed scarecrow', essentially an anthropomorphic tree-trunk decorated with weapons and armour. They had both funerary and apotropaic values and could be regarded as belonging to the sacred domain. These 'armed scarecrows' were often memorialized as part of larger stone victory monuments, sculptures (e.g. as a decoration on the breast armour of imperial sculptures), mosaics, paintings, tombstones (as observed earlier) and coinage (Woelcke 1911; Picard 1957; Pritchett 1974: 246–75; Hölscher 2006: 29–34; Isaac 2017: 53–60). This monumental presentation of a defeated body was evidently a sign of Roman domination and supremacy, transforming military victory into a mnemonic device for the projection of political power and control, as well as the commemoration of fallen Romans (Hölscher 2006, Cass. Dio 68.8). Indeed, the erection of larger victory monuments similar to the Gardun *tropaeum*, such as the Sullan *tropaia* at *Chaeroneia*,

Figure 5.5 The larger of two fragments of the Gardun *tropaeum* (AMS D 129), depicting two seated and bound individuals below a Roman battlefield 'trophy' or 'scarecrow'. Date: Early first century CE. Copyright: Archaeological Museum in Split.

the Pompeian *tropaia* at *Panissars*, the Octavian *tropaeum* at *Nikopolis*, the Augustan *tropaeum* at La Turbie and the *Tropaeum Traiani* at Adamklissi, were devised as clear and understandable messages to the locals expressed through local visual templates (Ibarra 2009).

The smaller Tilurian fragment contains a frieze consisting of weapons and the barely visible remains of human legs (probably a prisoner's). The larger fragment contains the typical image of an anthropomorphized tree-trunk 'scarecrow' adorned with weapons and a damaged helmet (Figure 5.5). Below this sit two bound locals, or to the Romans, *barbari* (of strange speech, foreigners), distinguishable as such through their clothing, who are also defeated combatants, indicated by being surrounded by weapons. In fact, the presence of the 'scarecrow' (along with other features discussed below) may indicate they are portrayed as still being present on the battlefield. To the left of these figures is the inscription field, below which is another frieze comprising more armour and weapons, similar to those upon the smaller fragment and Tilurian funerary stelae. A recent thesis concluded that there is a mixture of weapons, some used in the first century BCE by La Tène and Germanic populations, and others being late Hellenistic (Plazibat 2017: 27–31). They therefore could have belonged to the defeated local *or* Roman troops. Some armour is also late Hellenistic (Picard 1957: 217–18, 252; Cambi 1984: 84), such as the luxurious *cuirass* depicting dolphins worn by the 'scarecrow'. Perhaps indicative of a more 'Roman' theme is the Roman *paludamentum* cloak, a Roman trumpet and the Gallic *carnyx* instrument, as well as the likelihood that the aforementioned helmet was Roman (Plazibat 2017: 29–31).

Ferris (2000: 30–62) has observed that *barbari* are usually presented in a 'state of impotence' in Julio-Claudian art, and this is exactly how the bodies of the prisoners are represented on the Gardun *tropaeum*: subdued in defeat, seated, chained and surrounded by weapons. While their faces are damaged beyond recognition, their clothing hints at their identities (Cambi 1984: 84–6; 2013: 13; Schönauer 2000: 236–9). The items of dress and bodily features are of both a local indigenous and generic 'non-Roman' flavour. The left-most prisoner has a beard and long hair, as well as a cloak and (fur?) hat – the former tied with a late La Tène *fibula* popular in Dalmatia prior to and soon after Roman occupation (by the previously mentioned Delmatae, for instance). The right-most figure wears long trousers (*braccae*) similar to that of the Dacians, Parthians, Armenians or Germans. His torso is bare, and he sits on a cloak with another La Tène *fibula*. Both captives wear sandals (potentially leather) recognized as *crepidae* by Schönauer (2000; popular contemporary Mediterranean footwear) and sit upon a pile of stones – the left-hand man upon two rows and the right-hand man upon one. Considering all these features, they are generally interpreted, following Cambi's Tiberian dating, as symbolic bodily representations of defeated Dalmatians and Pannonians following the Roman victory in the *bellum Batonianum*.

Indeed, Isaac (2017: 6–7, 45–68) has argued that *tropaea* were not simply allegorical representations of Roman victories but were rather designed to depict real events and real defeated enemies and evoke lived experiences, as well as serve as forms of sacred dedications and as foci of ritual once set in stone (Isaac 2017: 59–62).[27] While we have no direct evidence of fighting around Tilurium, some written sources hint at it. Tiberius' capture of the indigenous hillfort Andetrium in 9 CE is described in some detail by Cassius Dio (56.12.3-14.7). Andetrium is identified as Gornji Muć, *c.* twenty kilometres from Tilurium, so the trophy might commemorate this event, which signalled the end of the *bellum Batonianum* (Bekić 2011: 316). While this is the most likely explanation, Pliny (*HN* 3.142) refers to Tribulium (likely Tilurium), along with Andetrium and Burnum, as 'castella crowned by glory' (*nobilitata proeliis castella*), so it is not impossible that some kind of battle occurred at the fortress. Thus, it is very likely that the Gardun *tropaeum* commemorated a Roman victory at Tilurium, Andetrium, or another site of Roman victory during the *bellum Batonianum* that involved the Roman *milites* and the indigenous Dalmatians, with the prisoners under the armoured scarecrow representing these locals and their countrymen.

The monument also transmitted Roman military victory into lasting power over the bodies of the local combatants and their extended communities, at least to the *milites* of the fortress, but potentially to locals as well. These various functions of the images outlined above would have been enhanced by their dominating realism. As observed above, the clothes of the two defeated and chained combatants are sculpted to corresponded to the clothing of the local population *and* to be of a generic 'non-Roman' flavour. They sit subdued, stripped of weapons and armour, below an armoured 'scarecrow' adorned with luxurious weapons, enhancing their helplessness and projecting Roman military power. Their bodies are also shown to scale with the 'scarecrow', and the stone piles on which the prisoners sit are important and recognizable characteristics of the local landscape, presenting the combatants in a state of defeat in their own landscape (real or imagined). Or perhaps these stones also evoked memories of siege and defeat, with large quantities of rocks and stones hurled at the Romans during the siege of Andetrium, for example (Cass. Dio, 56.14.1-2).

While its precise location is unknown, we find it unrealistic to assume that such a piece of public imperial propaganda was not intended to be visible to locals, because two main messages – the Roman army cannot be defeated and *Roma* reigns supreme – were very relevant for them. We hypothesize that the Gardun *tropaeum* was, therefore, at least in part, intended for the indigenous population, for whom the realism would have been particularly significant – unable to read the inscription but able to identify with the images of prisoners. The bodies of the defeated locals acted as mnemonic devices, reminding them of family and ancestors who were on the receiving end of Roman military violence, either as prisoners, slaves or combatants. A potential effect was to evoke shared experiences and memories of military defeat through these bodies, representative of the indigenous onlookers. Whether the *tropaeum* had all these intended effects on the local population or produced subversion and quiet resistance, we cannot know. At any rate, the trophy had real significance for several generations of locals who remembered Roman victory and saw their defeat manifested by the subdued figures.

The Gardun *tropaeum* was of course intended for the victors as much as (if not more than) the defeated. In this regard the choice of place is particularly important (Gabričević

1984: 94) and the general location of the *tropaeum* in or near to a fortress would have elevated the Roman victors, strongly impacting their collective identities and morals. It was a perfect space for periodical and symbolic re-enactment of military victory. While this likely manifested in forms of repetitive ritual, like the annual celebrations at the *Tropaeum Traiani* at Adamklissi ordered by Trajan (Cass. Dio 68.8), the monument also served as an everyday reminder and mnemonic device. Its significance here is perhaps more direct, at least for members of units such as *Legio* VII and *Cohors* II *Cyrrhestarum*, which took part in the *bellum Batonianum* and shared personal experiences of military hardships and lasting achievement. In a mnemonically similar fashion to above, the monument and depicted bodies would have presumably reminded survivors of their fallen comrades and military victory over the indigenous locals while simultaneously transmitting these collective memories onto generations of new recruits (Ibarra 2009: 47–54), who were adopting the identity and traditions of these units. Following the departure of these units, however, the historical context in which the original *tropaeum* was placed, and therefore its message, must have been understood differently, as the new garrisons interpreted the monument in accordance with their own individual and collective experiences. Nevertheless, the primary message of the *tropaeum* – the triumph of the invincible Roman army over its enemies – certainly remained a core component within these new interpretations and the bodies of the captives played a central role in transmitting this on a symbolic level.

Conclusions

The archaeological material from the Roman military fortress of Tilurium provides an array of snapshots into the embodied identities of combatants and the ideas surrounding their profession, as well as insight into dress and sculpture as communicative media used to (re)construct the bodies of these combatants. Roman *milites* at the fortress, hailing from several territories across the Roman world and with different cultural backgrounds, would clothe themselves with particular pieces associated with the empire-wide and regional military communities to adopt and (re)construct imperial Roman military identities and fashion a sense of community in their fortress. Furthermore, the Tilurian evidence hints at the existence of a distinct Dalmatian-flavour of imperial military identity, something to be explored comparatively in future research.

Some of these items of dress, such as belts, drew significance from, and embodied the martial function of, the combatants and their distinct place in society as 'violence specialists'. Other components, such as cloaks and *fibulae*, marked these same combatants out as members of the military community and distinguished their bodies from those of civilians. While unrelated to combat function, they still carried important symbolic messages about the people who wore them and their ability to enact violence – not unlike the red cloak of the *centuriones* or the long leather coats of the Soviet secret police NKVD. The same processes of communal embodiment of identity are witnessed with funerary stelae. Portraits of cloaked combatant bodies singled the deceased out as 'military men' and were accompanied by 'military friezes', which further encapsulate their role as 'violence specialists'. Similarly, ideas of subjugation and violence are evoked by the two chained bodies of local figures upon the Gardun *tropaeum*. The two defeated indigenous combatants, along with the monument itself and the sculpted 'scarecrow', held multiple meanings, constructing the Tilurian military community (transmitting Roman military victory into power) and evoking shared experiences and memories of defeat and subjugation amongst locals, symbolizing Rome's domination over them – especially in the period immediately following the conquest.

The artefacts and monuments that adorned or depicted bodies of *milites* at Tilurium, be they *baltei*, *fibulae*, or images on stone sculptures, thus embodied the diverse combatants' status as a unified social group with a particular identity: a Romano-Dalmatian soldierly community. Likewise, these strategies of identity construction effectively implemented Roman imperialism and projected the idea of Roman supremacy over the indigenous people of Tilurium and the wider Dalmatian region via the symbolic subjugation of their persons.

Acknowledgements

This chapter is an outcome of the projects 'Among Gods and Men – The Cults and the Population of Roman Dalmatia According to the Votive Inscriptions', financed by the Croatian Science Foundation (HRZZ IP-2020-02-7214), and '*Fratres Romani* in *Dalmatia*: Community and Identity in the Roman Seventh Legion', funded by the Croatian Studies Foundation. We also thank the editor and anonymous reviewer for their constructive comments.

Notes

1. See Lloyd Llewellyn-Jones in this volume.
2. Some authors date it as early as the late 30s BCE (Zaninović 1996: 280–1; Sanader and Tončinić 2010: 41, 48; 2013: 420). Roman republican material may indicate it was first a temporary fort. A bridge and later settlement known as Pons Tiluri was later built near the same river-crossing (Milošević 2009; *CIL* 3.3202).
3. These *Delmatae* were connected with the wider Iron Age 'Central Dalmatian archaeological culture' (Čović 1987). The central eastern Adriatic hinterland was the core of the Delmataean alliance and was less visibly integrated into broader Mediterranean cultural networks prior to Roman conquest – contrasting greatly with neighbouring peoples (Džino 2006; Barnett 2014).
4. At Tilurium specifically, eight from *Legio* VII hailed from Asia Minor, three from Italy and three from Macedonia: *AE* 1995, 1232; 1999, 1230; *CIL* 3.2709-10; 3.9733-34; 3.9737; 3.9742; *ILJug* 3.1949-50 (Asia Minor); *CIL* 3.2716; 13976; 14932 (Italy); *CIL* 3.2717; 9741; 14933 (Macedonia).

5. While many of these were stray finds lacking in-depth provenance data, their general find-site is often recorded and can be supplemented with better-provenanced material.
6. Ivčević (2004: 160–1); Voirol (2000: 17). *Tilurium Cat.*: 76-8 no. 41-5; Ivčević (2017), no. 26, T.4,26.
7. *Tilurium Cat.*: 78-9 no. 46, cf. Ivčević (2010: 128–9, 134), no. 134, T.1,1. Niello was a popular first-century CE belt material (Hoss 2017: 90).
8. *Tilurium Cat.*: 80-1 no. 50-1; Ivčević (2010: 129–32, 135), T.1,2; (2016: 280–1), T.18,191-3 (cordate); (2017: 271), no. 25, T. 4,25 (square).
9. E.g. *Tilurium Cat.*: 79-80 no. 47-8; Ivčević (2016: 277–8, 189, T.17,185-9); (2017: 272), no. 27, T.4,27.
10. *Tilurium Cat.*: 81-2 no. 52-5 (plated and non-plated apron fittings).
11. *Tilurium Cat.*: 83 no. 56-7; Ivčević (2014: 178–9, 186), no. 149-50, 157-58, T.14,149-50, 14,157, 15,158; (2016: 286), no. 210, T.19,210; (2017: 272–3) no. 28-33, T.5,28-33 (apron rivets).
12. *Tilurium Cat.*: 65 no. 21; Ivčević (2013b: 436 no. 11, fig. 2).
13. E.g. *Tilurium Cat.*: 67 no. 23 (chape), 25 (fitting); Ivčević (2017: 299, 305), no. 247-8, 279-84, T.15,247-8, 17,279-84 (belt buttons and links). *Tilurium Cat.*: 66-7 no. 22, 24 (MCK B-870); Ivčević (2013b: 436), no. 10, MTK 971.
14. Ivčević (2002: 1.235-6, 2.124-32, T.I-IX); Ivčević (2013a: 303–5), no. 1-14 (Salonitan *fibulae*).
15. Herodian's (2.13.10) account of Severus' dispossessing the praetorian guard of their belts and armaments attests that this significance continued into the third century CE.
16. *Ast illis quos arma tegunt et balteus ambit.* All translations are the authors' own unless specified otherwise.
17. This explains the frequency of belts and scabbards in the funerary portraiture of *milites* throughout the first–third centuries CE (Speidel 2012: 2–5, figs 1.1–4; Hoss 2017: figs 1–5).
18. These *milites* are commemorated at Tilurium (*AE* 1999, 1230; 1994, 1356=2002, 1118).
19. E.g. Ivčević (2011: 165–8), no. 1-18 (*Aucissa*), 19 ('hinged'), 20 ('crossbow'). See also Marović (1961: 107, 109–12, 114, 116–18), no. I.2, 5, 9, III.8, IV.8, VI.2, VII.8, 12-13, 15-16, XI.3, XV.1-3, 13, inscribed *Aucissa fibulae* from Gardun.
20. Trench Z: Ivčević (2017: 300–2), no. 252-8 (*Aucissa*), 259 (later style), 260-22 (pins), T.15,252-6, 16,257-62. See also *Tilurium Cat.*: 88-90 no. 71-74.
21. *CIL* 3.9737; 2712=9727; 14934. The former dates between 6–42 CE, the latter two from mid-to-late first century CE.
22. *AE* 2000, 1174; 2000, 1179; 1991, 1291.
23. For instance, the *porta inferi* or 'panelled door' motif – a design found almost exclusively on funerary monuments commemorating *milites* in first-century CE Dalmatia (Cambi 1994: 166–71; Miletić 2015) – is depicted upon more than a dozen funerary stelae from Tilurium (*AE* 1994, 1355-8; *CIL* 3.2709-10; 9733; 9737; 13976; *ILJug* 3.1949-50).
24. *CIL* 3.2709; 2710=9726; 2716; 13976; *ILJug* 3.1949. The latter two do not *explicitly* record *Legio* VII membership.
25. *AE* 1991, 1291 (*Salona*); 2000, 1179 (*Bigeste*). These tombstones frequently, but not exclusively, commemorate individuals from Tilurian units – something to be further investigated.
26. AMS D-129; MCK no. 381, *Tilurium Cat.* 106-107 no. 112-13. AMS D-129 was accidentally discovered in 1866 and MCK no. 381 was also accidently found around 1931, some 200

metres from the first one (Cambi 2005: 24 n. 52). No other fragment has been securely identified, yet possible candidates are known (Cambi 1984: 86–7; 2013: 15; Milošević 2017: 75 pic. 97).

27. These rituals were likely similar to other forms of military and soldier commemoration in the Roman world, for instance parading and funerary commemorations (Ibarra 2009: 40–54). Indeed, it is possible *tropaea* were worked into these other rituals.

CHAPTER 6
EMBODYING MILITARY FEMININITY AND VIRGIL'S CAMILLA
Hannah-Marie Chidwick

bella viri pacemque gerent quis bella gerenda. (Virg. *Aen.* 7.444)[1]

let men carry out war and peace – men wage wars.

Men wage wars. Or, rather, the *male-sexed body* wages wars. The Roman poet Virgil begins his epic *Aeneid*, about the foundation of Rome, singing 'of arms and the man' (*Aen.* 1.1). In ancient Rome, the body wielding those arms was a normatively *masculine* one.[2] In the quotation above, the Latin warrior Turnus censures a priestess (a disguised Fury Allecto) for interfering in the confrontation between the invading Trojans, led by the epic's protagonist Aeneas, and the Latins rallying against them. As Turnus suggests, most combatants in this war are male, yet on his own side is the poem's only female military commander, Camilla.[3]

The attitude that 'war is the business of men' is hardly confined to the temporal distance of the ancient Mediterranean; Helen Snell's images foreground the male bodies and masculine expectations of twentieth-century combat.[4] There are naturally manifold differences between the armed forces of the twenty-first-century West and ancient Rome, not least that in the British armed forces, female combatants have joined frontline close combat since 2016, following the US in 2015. Yet the notion of the battlefield as a *masculine* space home to *masculine* attributes, effected by bodies sexed male, is longstanding in Western cultures and supported by statistics.[5] A 2023 report found female representation in the UK Regular Forces at 11.7 per cent (10.3 per cent in the Army; Ministry of Defence 2023). If a woman enters a military space, this can be judged a subversively political, not patriotic, act (Basham 2013; Woodward and Winter 2007). This chapter will not debate the battle-readiness of 'female' bodies, but instead explore what somatic qualities *are* desirable in close combat. Characters like Camilla trouble our understanding of how said qualities correlate with socially constructed gender ideals, typically and transhistorically *masculine*. The essentialist binary of masculine/feminine as perfect opposites underlies civilian societal structures, scaffolds forms of knowledge and expression, and most pertinently influences real-life military policies today (Dietert and Dentice 2023). Growing awareness of complex gendered identities in modern critical scholarship gives us means to push discussions of gender and war in Roman epic beyond the binary of masculine *virtus* or feminine victimhood.

This chapter will consider which bodily attributes are fit for war-making through the medium of Roman epic poetry. Contrasting the annalistic, politics-focused historiography, poetry allowed for more emotional and visceral representations of war – a popular subject for writers in the ancient Mediterranean. The epic genre played an important role in establishing the Roman elite's understanding of social hierarchies, cultural codes and, consequently, gender norms (Keith 2000: 11). Virgil's narrative occurs outside historical record, but his writing is naturally infused by his cultural values: an expanding empire owing its potency largely to war, Virgil's Rome emerged from two decades of civil conflict (49–31 BCE) into renewed patriarchal autocracy. *Romanitas* was fractious, patriotism confused. While the *Aeneid*'s characters are not yet Roman, still they embody qualities prized or denounced by Virgil's contemporaries and exemplary war-makers.

Beyond the question of how far male bodies and innately 'masculine' behaviours befit the war-space, femininity and female presences have permeated warzones throughout history (Enloe 2000: 35–48). Virgil's Camilla is one of few female soldiers singled out and named in the whole Graeco-Roman literary canon, making her a worthy case study in this volume for the way she embodies the integration of femininity and combat. As Elina Pyy argues, a female *military* leader 'appears to shake the foundations of patriarchal society' (2021: 166). Book 11 narrates the clash between Camilla's cavalry unit and the invading future Romans (11.498-915). As Camilla only exists in Virgil's gendered verse (Gransden 1998: 20), her portrayal as *virgo* (7.806), *regina* (11.499), *mulier* (11.687) makes her resoundingly 'feminine' in semantic (if not somatic) terms. Aside from Rory B. Egan's influential reading of Camilla as fundamentally an armed warrior (1983), scholarship has largely concerned itself with Camilla's gender contrariness: 'the easy binary of "male" and "female" partially disintegrates, as she proves herself superior to her male counterparts' (Gildenhard and Henderson 2018: 28).[6] Although she is further othered as an armoured Italian, her bodily comportment nevertheless seemingly epitomizes Roman military manliness, 'uniting in herself a rich diversity of apparently contradictory qualities' (Basson 1986: 65).

Yet, the question really worth exploring is how far Camilla actually 'disintegrates' these binary logics within the specific context of war, one teeming with moral, political and physical paradoxes where destruction is the *modus operandi*. It is not the ambition of this chapter, necessarily, to present Camilla as transgender, intersex, nonbinary, butch, or queer (Mackay 2021: 23–4), although all are possible framings (Xinyue 2017). Instead, this chapter explores how combat-specific traits considered masculine, feminine or nonbinary are embodied by the only warrior 'woman' in Virgil's war, sexed female by semantics. With such a study, we could be pushed into revisioning Camilla as embodying a degree of fluidity and mutability immanent in *any* soldierly body. Analysis will deal both with the interrelation of gender and the war-making body, and the physical criteria that combatants are expected to exhibit. More broadly, this chapter encourages further reflection on how gendered language underlies notions of capacity, suitability for environments and justifications for violence as an innately masculine activity.

Transitive Warriors

War makes everything, every *body*, mutable, including ideas about gender. 'Battle is an extreme situation,' write Ingo Gildenhard and John Henderson in their commentary on *Aeneid* 11, 'in which one and the same deed can be both admirable (in terms of skills or courage) and abhorrent (the casualties and the carnage, humans killing humans)' (2018: 32). Bodies involved in military and security activities are necessarily *transitive*: 'passing into another condition, changeable, changeful; passing away, transient, transitory', borrowing C. Riley Snorton's definition (2017: 5–6). Regardless of gender, combat can of course result in the permanent disabling of bodies; this has been so for millennia.[7]

Roman texts of military science convey that the body's fitness for duty was paramount (Veg. *Mil.* 1.2-6, 2.5; Front. *Str.* 4.1.1-3), but although only male recruits could enlist, not every male-sexed body was suitable. Even within the limits of a binarized system, the distinct 'masculinity' demanded on the battlefield differed hugely from that honed in the Senate house (Deminion 2020: 198). Patriarchal regulation and discipline cultivated a combat-specific maleness, primed for 'extreme situations' (Veg. *Mil.* 1.1; Frontin. *Strat.* 4.1-2; Sall. *Jug.* 45; McDonnel 2006: 10). Any bodies deemed divergent, especially bodies sexed female, were unfit for combat and therefore incompatible with military spaces.[8]

Barriers to participation for bodies not conforming to rigid masculine ideals persist in the military institutions of so-called liberal democracies today, where gender binarism still underlies policy-making and lived experiences. This is particularly pertinent for transgender bodies. Although many countries' armed forces openly welcome transgender and nonbinary combatants (Elders et al. 2015), including the UK since 2000 (Schaefer et al. 2016: 57; Ministry of Defence 2022), there are discrepancies in the health experiences of transgender versus cisgender personnel (Dietert and Dentice 2023: 996–7): stigmatizing treatment correlates with poorer physical and mental health, in an already highly stressful occupational environment (Schvey et al. 2020; Houghtaling and Osypuk 2023).

An infamous recent example of military-related gender-based discrimination is the US armed forces' controversial policy DTM 19-004, which in April 2019 implemented tight restrictions on the accession and retention of 'Transgender Persons and Persons with Gender Dysphoria'.[9] It was one of multiple policy enactments removing transgender rights across the sociopolitical spectrum, designed to appeal to President Trump's conservative base.[10] Although the ban was short-lived (revised April 2021), it was not wholly new legislation but the reinstatement of a long-running policy only rescinded in 2016. As an instance of 'emotional statecraft', as Christopher Pepin-Neff and Aaron Cohen describe (2021: 653), DTM 19-004 mobilized a widespread moral consensus that transgender bodies would never be 'fit' for military service, making the policy pertinent to any study of how gender, bodies and combat readiness interrelate. Presidential communications even constructed the admission of transgender bodies into the US military corps as a strategic, cultural and financial threat to military effectiveness (White House 2017). The policy's wording set out an explicitly *corporeal* basis for this risk.

The ideological mechanics of DTM-19004, however ludicrous and politically motivated, nevertheless prompt reflection on the way human bodies and gender ideals interact with the 'high standards' (Department of Defense 2020: 2) of physical health historically associated with war-making. According to DTM 19-004, transgender or nonbinary individuals could be denied enlistment or relegated to administrative roles unless medical evidence specified their 'stability' in one sexed identity for eighteen months minimum (DoD 2020: 2). The embodied instability of one's gender jeopardized military functionality. The notion that any body is absolutely 'stable' in one gender is the product of hyperbolic pseudo-science,[11] yet this idea was used to justify real-life policy with real-life consequences for those it policed. The policy only cites 'gender' in the singular: service personnel must identify as either 'a male or a female' (DoD 2020: 2), leaving many 'transgender troops feeling like second-class citizens', as one military psychiatrist explained (Wise 2019: 114; see also Dietern and Dentice 2023: 996). The concept of medically perceivable gender deviance naturally implies a reference point of conformity criteria that serve to self-validate when the individual identifies as 'either a male or a female', refuting the long-recognized 'full biopsychosocial complexity of sex/gender' (Rippon et al. 2021: 2).

Notably, the justification for this discrimination concerned not only mental but physical health, namely, the need to identify whether a body fits the physical 'standards associated with their biological sex' (section 2.c).[12] A body defying neat gender classification troubles this regulatory process, entailing grounds for non-enlistment or separation from service: being transgender, while 'not constituting a physical disability' could still 'interfere with assignment to or performance of duty' (DoD 2020: 10, section 2.e.3.a), like limited eyesight or missing limbs. This suggests that 'performance of duty' necessitates adherence to normative gender criteria alongside physical fitness.

The policy thereby exemplifies the role that medicine 'as knowledge, practice, and institution,' describes Ellen K. Feder, 'plays at the crossroads where gender and sexuality, disciplinary and regulatory power, meet' (2007: 60). This recent example of institutionalized discrimination reflects the way militarism and war-making can interact with, and perpetuate, conventional beliefs about gender 'instability' and transitiveness as detrimental to a culturally understood paradigm of military fitness.

Camilla: *Virgo Dura*

Judith Butler famously theorized gender as a 'free-floating artifice, with the consequence that man and masculine might just as easily signify a female body as a male one, and woman and feminine a male body as easily as a female one' (1990: 9). Our notions of gender as embodied are 'culturally specific', differing depending on the performance space,[13] including military cultures, historic and modern. To conceptualize gender in a constructivist sense as 'performative' means considering how gender manifests through bodily comportment to meet, or contradict, cultural norms. In armed combat, a violent male soldier is normal, a violent female contradictory. As such, 'combat' in modern discourse has become narrowly synonymous with the image of two armed, consenting, *male* opponents, as Cynthia Enloe notes: 'a place where men can prove their manhood' (2013: 260).

In ancient Rome, too, certain contexts demanded certain performances by Roman men (Gunderson 2003a; McDonnell 2006; Schierl 2021), and Roman (like modern) masculinity was somewhat fluid, a prime example being the self-mutilating eunuchs of the Cybele cult (masculine *galli* in Lucretius *DRN* 2.614, feminine *gallae* in Catullus 63; see Nauta 2006). A further implication of Turnus' statement '*quis bella gerenda*' is that it is the *duty* of any man (*vir*) to make war. The Latin word *vir* can translate as 'man' or 'warrior', conflating both figures (Gunderson 2003a: 7). For Lin Foxhall, in Roman society 'war was a strongly gendered activity which played a significant part in defining gender itself, founded on bodily strength and physical force. It was certainly constitutive of the principles that governed the ideals of masculine and feminine behaviours' (2013: 84; see also Chidwick 2017). Battle could be a site of Roman *virtus* (manliness, contextually courage or virtue): the masculine body was both embraced by the warzone and reinforced by war-making (Foxhall 2013: 87; Alston 2010: 205; Phang 2008: 17). The value of the masculine body *qua* fighter is evident across early Principate culture, inherited from the Republican requisite for military service before public office. Soldierly bodies symbolized national pride, while leadership was best embodied by a self-styled veteran, as the Prima Porta Augustus exemplifies.

Meanwhile, the 'female/feminine' subsisted in military spaces and discourse primarily to define desirable masculinity through opposition.[14] In an earlier battle of Virgil's epic, the Latin warrior Numanus derides his enemies by calling them *Phrygiae*, not *Phryges* (9.617; see Nauta 2006: 85–7), Trojan *females* not *males*, later jibing, '*sinite arma viris*' (9.620, 'leave weapons to men'). Accordingly, Lee Fratantuono describes Camilla as 'a true female, charging off on horseback against (effeminate) Trojans [. . .] a woman to fight women' (2007: 335). Therefore, gendering bodies in war cannot be broached solely in terms of characters' reproductive organs, but perception and performance.

Midway through Virgil's prototypically patriotic and patriarchal epic, concluding a catalogue of assembling combatants,[15] in strides 'the colourfully dramatic figure of the Volscian Camilla' (Fratantuono 2007: 322):

> Hos super advenit Volsca de gente Camilla
> agmen agens equitum et florentis aere catervas,
> bellatrix (7.803-5)

> Lastly came Camilla of the Volscian people[16]
> leading the cavalry and a company shining in bronze,
> a warrior-woman

She features in the catalogue like any other warrior but as *bellatrix*, assuredly female. *Bellatrix* conflates woman and combat, evoking a specific 'military femininity' perhaps alien to the Roman consciousness. Nevertheless, *bellatrix* connects Camilla semantically with a pedigree of war-making persons and places;[17] her name ironically evokes Roman general M. Furius Camillus, fourth-century BCE defeater of the Volsci, 'saviour and second founder of Rome' (Gildenhard and Henderson 2018: 35; Gransden 1998: 25). Camilla is later called *regina Volscorum* (11.498-9, leader of the Volsci). *Regina* also denotes Aeneas' spurned lover Queen Dido (1.303), Latinus' spouse (12.659), and extra-Virgil the Ptolemaic ruler Cleopatra VII (Cic. *Att.* 14.8.1; 14.20.1; 15.15.2; Hor. *Odes.* 1.37.7; Suet. *Aug.* 69). Camilla's epithets make her both feminine and masculine, Roman and Italian, before we even see her in action.

Consequently, scholars like Alison Keith describe Camilla as a 'particularly complex example' (2000: 27) of femininity in the *Aeneid*, compared to Dido or the virtually silent Lavinia. Keith contends that a *woman* in the catalogue of warriors 'clearly disturbs the gender code of Virgilian epic discourse, marked as masculine at the outset' (2000: 27). Despite the pervasive claim that there were 'no female rulers or generals in the history of the Roman empire' (Ramsby 2010: 13), Roman historical records do feature female war-makers, mostly opponents (famously Cleopatra VII) or 'deviant' women of the political elite (Fulvia).[18] Likewise, imperial literature derides women who deviate from exemplary feminine behaviours: namely, subservience, piety, devotion to family,[19] styling the exceptionally capable as corrupt poisoners manipulating powerful men.[20] As Ellen Oliensis writes, '[i]n narrative terms, this tends to mean that women make trouble and men restore order' (2006: 303; see also Pyy 2021).

Although Plato entertained the idea of women warriors (*Resp.* 451c-457b; see Coker 2013), the Roman historian Livy made plain that '[n]o magistracies, no priesthoods, no triumphs, no marks of power, no gifts, no spoils of war can come to [women]: elegance, adornment, clothing – these are women's marks of power' (34.7.5-9, trans. Keith 2012: 397). The fourth-century CE commentator Donatus asked, 'to what woman or man would it not be miraculous for the manliness of Mars to dwell in the female sex?' (2.108.25-7, trans. Keith 2000: 28). For the Romans, femininity typically equated to softness, passivity or unruly passion, while masculinity denoted agency, rationality and self-control. The Greek medical writings of the Hippocratic Corpus characterized women's bodies as wet and spongy, in contrast to the hard, dry bodies of men (Hippoc. *Aphorisms* 3.11, 14; *Aer.* 10, *Regimen in Health* 6), and these ideas shaped Roman thought. Although literal, somatic rather than behavioural tropes, the notion that male bodies should be 'sound', tough and impenetrable persisted (Foxhall 2013: 70; Gunderson 2003b: 59; Walters 1997; Williams 1999: 18).

Virgil's readers (ancient and modern) therefore face an immediate oxymoron when they encounter Camilla, described in the catalogue as *virgo dura* (7.806-7): not a soft, passive figure but a 'woman hardened' for battle.

> non illa colo calathisue Minervae
> femineas adsueta manus, sed proelia virgo
> dura pati cursuque pedum praevertere ventos (7.805-7)

> Not for her feminine hands were the distaff or
> wool basket of Minerva, but this woman
> was hardened to endure battle, the flight of her feet could outstrip the winds

The poet subverts expectations by noting Camilla's rejection of archetypal feminine habits.[21] She is not *adsueta* (accustomed) to spinning; the distaff is replaced by a quiver and pointed staff (816-17). What Virgil describes is a soldierly body trained for 'manly pursuits' (Donat. 2.108.8-9; trans. Keith 2000: 27). Twice, *agmen agens* follows her name (7.804, 11.433), the alliteration emphasizing her active role as leader, while her velocity is a skill shared by military figures prominent in the Roman imagination, notably Julius Caesar (Caes. *BCiv.* 1.70; Plut. *Caes.* 15.5).

Nevertheless, although her feet are not as explicitly gendered as her hands, Virgil 'feminizes' Camilla's speed as more delicate than destructive (7.808-11).[22] She is called *decus Italiae virgo* (11.508, woman, glory of Italy),[23] as she assumes a leadership role on a par with Turnus before the battle (11.519). Despite her authority, Camilla is still *virgo*, a marriageable 'maiden' more alike to untouchable Diana than Venus or militaristic Minerva.[24] Crowds are drawn by her beauty (7.812-13), a crucial 'standard' for a young woman.[25] These details force us to rethink the body signified. Fratantuono specifically calls Camilla 'no mannish woman' (2007: 338); she is not really a female-to-male 'cross-dresser' (Xinyue 2017), like famously transgressive woman-warrior Joan of Arc, a 'female-born personification of masculine virtue' (Hovey 2018: 113). Instead, for Keith, Virgil defines Camilla's exceptionality through antithesis: her feminine hands are not 'shaped' for the 'wool-basket of Minerva', but for a javelin.[26]

Moreover, her Volscian army comprises both men and *virgines Italides*: Larina, Tulla and Tarpeia, *aeratam quatiens securim* (11.656, brandishing her bronze axe).[27] Bronze is a particularly military (masculine) attribute (Basson 1986: 58). All are ascribed *decus* (11.657), often translated as 'beauty' (trans. West, Virgil 2003: 256); less passive connotations include 'honour, dignity, virtue', qualities appropriate for a soldierly body. Still, physical loveliness is not an exclusively female quality, as demonstrated two books earlier by Trojan Euryalus in his savage demise: *pulchrosque per artus / it cruor* [. . .] *purpureus veluti cum flos succisus aratro / languescit moriens* (9.434-6, the gore flows over his beautiful limbs [. . .] like a crimson bloom languishes, dying, severed by the plough). A 'purple' hue often serves to other characters as effeminate or foreign, but here and in descriptions of Camilla (7.814; 11.819) it evokes youthfulness, exuberance – traits vital in combat, shared across genders.

Constructing Masculinities

Along with her beauty and feminine (pro)nouns, Camilla embodies many of the military virtues prized in Roman males; she is even deemed *more* masculine than the Trojans fleeing her (11.732-7). A closer inspection of this embodiment reveals how military bodies and understandings of gender are constructed, and transgressed, in combat. Scholars have sought to explain Camilla's skill in close combat by pairing her ferocity with a 'paradoxical' femininity (Basson 1986: 57), diluting her prowess with patronizing epithets like 'simple' and 'naïve' (Otis 1964: 364). However, assessing whether there is anything naïve about Camilla's conduct in warfare requires knowledge of what kinds of bodies and behaviours are expected in epic battle scenes, and moreover how notions of masculinity or femininity intersect with these expectations.

The Roman army's famed machine-like discipline, wrought through rigorous training and combat experience, provides a productive metaphor for thinking about the construction and deconstruction of gendered behaviours and identities. Butler is much cited for describing gender as cultivated through a habit of repeated performance alike to drill (1990: 45):

> the repeated stylisation of the body, a set of repeated acts within a highly rigid regulatory frame that congeal over time to produce the appearance of substance, a natural sort of being.

It is habit (*habitus*) that marks a body as belonging to a specific cultural category. The 'set of repeated acts' with its 'highly rigid regulatory frame' evoke the codes of conduct and training that instil instinctive behaviours in combat. In both social and military contexts, then, with habitual execution, the body learns to conform to its gender 'standards'.

In ancient Rome, preparation began early: 'training the body from boyhood was essential for achieving the full status of manhood and was embedded in elite Roman culture' (Foxhall 2013: 77). During the *Aeneid*'s narrative, Aeneas instructs his son, Iulus Ascanius, in the ways of Latin men:

> disce, puer, virtutem ex me verumque laborem,
> fortunam ex aliis nunc te mea dextera bello
> defensum dabit et magna inter praemia ducet. (12.435-7)
>
> Learn manliness from me, boy, and real struggle,
> luck from others; now my right hand will defend you in war
> and lead you to great rewards.

Aeneas couples *virtus* with *labor*, making physical exertion central to manliness.[28] He specifies the necessity of a strong right hand (*dextera*, hand/sword) for protection, as though the true *vir* must not only train his limbs for combat, but also recognize the duality of his body as both human and aggressive technology. This is warrior (not civilian) *virtus*, as Foxhall addresses (2013: 88):

> The Latin version of 'manly courage', *virtus*, was far more cut-throat than Greek *andreia*. Romans understood *virtus* as an inherited quality that men either had or did not have. Traditionally the way to prove it was on the battlefield in single combat. [. . .] The physicality of fighting was therefore a key element for achieving masculinity in Roman society.

The notion of *virtus* as a quality 'inherited' by Roman males is supported by the prefix *vir-* (man), yet it must be outwardly *proven* through bodily performance. Ascanius should observe and regularly recall examples of other male warriors' struggles (*exempla tuorum*, 11.439), especially his father, or uncle Hector. While *virtus* is strongly gendered, it is more flexible and multifaceted than any classification of male sex based in biology (Chidwick 2017). Attributes similar to *virtus*, like *fortitudo* (both feminine nouns), are ascribed in rare instances to women, without the same physical connotations.[29] *Virtus* can be learnt, earnt and activated by entering the combat zone – a notion that persists today: 'the idea of becoming a soldier is intimately tied up with the idea of becoming a man,' write Rachel Woodward and Trish Winter (2007: 61), an argument echoed in Joshua Goldstein's *War and Gender*: 'men are made, not born' (2006: 264). We could even consider *virtus* a transitive quality: 'passing into another condition, changeable, changeful' (Riley Snorton 2017: 5–6), as in this context *virtus* is produced in the transition from civilian to combatant. *Disce, puer* – learn, boy, make yourself into a warrior.

Strangely, the Camilla episode offers us insight into the technicalities of how men are 'made' for battle. Camilla is made, not born, *bellatrix*. We see this happen in the narrative of her upbringing, recounted by Diana (11.532-96).[30] Violent women in myth and popular culture often have their aptitude for aggression 'explained' by misfortunes that forced them to adopt lifestyles atypical for young, Western women.[31] When Camilla's father Metabus was exiled from the city of Privernum, he and his infant daughter escaped to a rural life:

hic natam in dumis interque horrentia lustra
armentalis equae mammis et lacte ferino
nutribat teneris immulgens ubera labris.
utque pedum primis infans vestigia plantis
institerat, iaculo palmas armavit acuto
spiculaque ex umero parvae suspendit et arcum.
[. . .]
tela manu iam tum tenera puerilia torsit
et fundam tereti circum caput egit habena
Strymoniamque gruem aut album deiecit olorem. (11.570-80)

Here, in the bushes, amongst the bristling dens of beasts,
he fed her with the teats and milk of a herd of wild horses,
milking the udders into her soft lips.
So, as soon as the infant had planted her first steps
on the soles of her feet, he armed her palms with a keen javelin,
and from the little girl's shoulder hung a bow and arrows.
[. . .]
Then, while her hand was soft, she twirled childish spears
and whirled the sling on its smooth thong around her head
and razed the Strymonian crane and white swan.

The language describing Camilla is particularly striking: it is clear that her developing body is not innate, not chosen by her, but *shaped*. The gendered nature of Latin establishes her femininity, but her reproductive health, the basis by which sex is typically determined, goes unmentioned. *Infans* (573) can signify masculine or feminine, as though gender remains in flux at a young age; then *parva* (575) indicates she is female. Meanwhile, her otherness is emphasized as she is suckled by wild animals, and her cloak is a tiger skin (11.576; see Horsfall 2016: 59).

'As soon as the infant had planted her first steps' she is armed: preparing to fire becomes as habitual as walking. 'While her hand was still soft', before she became *virgo dura*, she was spinning javelins and hurling the sling (practising to face the Trojans, who are likened to Strymonian cranes, 10.265). As above (7.803-17), her bodily features are isolated for scrutiny, especially her hands, evoking the importance of the *manus* for a Roman combatant (see the Introduction). In Nicholas Horsfall's analysis, her *tela puerilia* (578) are 'toys' (2016: 59) of the sort that entertained young Roman boys while simultaneously readying them for military service. Camilla receives the same education. Barely walking, she nevertheless trains with warrior's *arma*, conducting formative play with war technology and learning to kill (*deicio*).

Camilla's *bellatrix* persona was fashioned – 'congealed', to borrow Butler's term – into a 'hardened' body that subverts the typically squishy (*mollitia*, see Keith 2000: 27) feminine form. Her soft, infant flesh was *moulded* around her weapons in a process evocative of that reportedly endured by Roman male infants, physically shaped at birth to ensure their development as masculine.[32] Foregoing this physical shaping process risked boys growing to be malformed – or, worse, women. Through this medical practice, the *construction* of masculinity is made physical, manifesting Butler's contention that the gendered body 'itself is a construction' (1990: 12). Still today, as Feder reports, comparable 'gender training' and physical (surgical) shaping processes are conducted on intersex children (2007), coercing them into embodying a single-sex identity. Such practices, Feder argues, leave a 'mark of power' that the child bears, bodily, as the product of corrective enforcement of gender difference (ibid., 63). Virgil shows us that Camilla's infancy, too, involved reconstruction without her explicit consent (see Fratantuono 2007: 337).

Camilla's narrative reveals not merely a woman adopting typically masculine habits, but specifically *the making of a combatant*, an identity forcibly embodied. Camilla's weapons literally, physically, define her, as her body is shaped to wield them, much as a soldier moulds themselves for war. Military anthropologist Kenneth MacLeish writes, 'although a weapon is an extension of the human body (as is acknowledged in their collective designation as "arms"), it is instead the human body that becomes in this vocabulary an extension of the weapon' (2016: 67). Shared physical habits make bodies fit (in its dual meaning) into their environment; habits including physical exercises act as 'a chief ritual through which individuals are integrated into the ranks of military organizations' (Bousquet, Grove and Shah 2020: 8–9; see also MacLeish 2016; McSorley 2014: 115). In ancient Rome, these rituals correspond to the daily *armatura* (weapons training) and *ambulatura* (military step) conducted by Roman military units.[33] Like any green recruit or aged veteran, Camilla embodies a history of activities that have shaped her for bellicosity.

Camilla's experience in some ways recalls that of adult women entering military environments today and consciously aiming to 'reshape' themselves, sloughing off their feminine traits to better suit the military sphere. Frank J. Barrett's description of a twenty-first-century US Navy recruit beautifully expresses this transitive process (2002: 64):

> She strives to shape her body just as a man would, standing straight, squaring shoulders, deepening the voice [. . .] the men get respect, so I wanted to look like one.

Accounts like these exhibit the physical unworking of archetypal lived and embodied femininity, to fit a space that otherwise rejects the female. During WW1, women were scorned for adopting 'mannish attitudes' and donning men's apparel like trench coats (Marcus 1989: 158). Wearing men's uniform is, unfortunately, still a necessary inconvenience for many serving female soldiers, as personal equipment designed for female bodies is sometimes scarce or non-existent (House of Commons Defence Committee 2021: 31–2).

Yet, Camilla shows no signs of striving for a specifically masculine identity, only pugnacity. The juvenile Camilla trains with *arma Dianae*, the javelin and bow of Diana.[34] As Diana's handmaiden, Camilla's drill – there is no better word for it – has been partly justified by her dedication to the deity of hunting, wildness and women.[35] This justification is overturned once she flouts Diana's will and enters battle (11.535-7), removing herself from both divine protection and the cultural limitations of femininity. As a Volscian general her only prey are Trojan men who also employ the skills of the chase: her eventual slayer, the skulking Arruns (11.765), sports a spear and matches Camilla's Lycian quiver, connecting him to Diana, too (11.773; Gildenhard and Henderson 2018: 24). Even Aeneas is compared to a hound (12.751). Camilla's prowess as a hunter therefore does not confine her to femininity but exhibits a capacity for battle shared by her male opponents.

Furthermore, Camilla shows herself to be just as competent with a soldier's 'service weapons', the sword and spear (Watson 1983: 59). When the hunter Ornytus (11.678), who, like Camilla, wears animal-skins, seemingly considers her mere 'game' (11.686), she jeers:

> advenit qui vestra dies muliebribus armis
> verba redargueret. nomen tamen haud leve partum
> manibus hoc referes, telo cecidisse Camillae. (11.687-9)
>
> The day is come when a woman's weapons
> refute your words. Still, this name, by no means insignificant,
> you will take to your ancestors, felled by the spear of Camilla.

Here, Camilla unequivocally identifies herself as *mulier* (woman, female), but unites the word with *arma* and altogether changes the meaning of both terms. 'Woman's weapons' refers not to Diana's arrows, nor 'elegance, adornment, clothing' (Livy 34.7.5-9), but an expertly wielded *telum*, a military missile not exclusive to hunters, certainly not to women.[36] Virgil makes the point of saying she faces her enemies *paribus armis* (11.710, with equal arms), bearing *ensis nudus* and *pura parma* (11.711, a naked sword and unsullied shield). In battle, she fights independently with an energetic style (*indefessa*, 11.651) typical of centurions (Phang 2008: 51), showing aptitude for that same fearless *virtus* that brings home the *corona aurea*.

> sequiturque sequentem;
> tum validam perque arma viro perque ossa securim
> altior exsurgens oranti et multa precanti
> congeminat; vulnus calido rigat ora cerebro. (11.695-8)

> she pursues her pursuer,
> then her powerful axe through the man's armour and through bones,
> through begging and pleading, rising above him
> she struck repeatedly; the wound wetted his face with hot brains.

This is one of the most grisly deaths in the epic (Fratantuono 2007: 342); hardly an instance of 'simple' or 'naïve' (Otis 1964: 364) combat. Virgil conveys syntactically how her movements mirror that of her opponent (she 'pursues her pursuer'). Her line of attack evokes other epic warriors, including within the *Aeneid*: for instance, *congemino* is repeated during the final fight between Aeneas and Turnus (12.714). Scholars have read this contest (11.648ff) as Camilla's Homeric *aristeia*, an epic convention usually reserved for male warriors (Gransden 1998: 124):

> at medias inter caedes exsultat Amazon
> unum exserta latus pugnae, phaeretrata Camilla
> et nunc lenta manu spargens hastilia denset,
> nunc validam dextra rapit indefessa bipennem (11.648-51)

> And in the middle of the slaughter, exulting like an Amazon,
> exposed on one side to the fight, with a quiver, Camilla,
> sometimes with one hand throwing pliant javelins to thicken [the air],
> sometimes her tireless right hand seizes the powerful double-axe

The participle *spargens*, along with *denso* and the repeated *nunc... nunc...*, convey the velocity promised by her first appearance. But note the language of excess here and in the passage above: *ex-surgo*, *congemino*, *et... et...*, *exsulto*, *exserta*, the prefix *ex-* meaning 'outside', extreme, extraordinary, not confined to one category.

Although Virgil never portrays Camilla as weaker than her opponents in combat, her own assertion to wield *muliebra arma* echoes the poet's habit of appending feminine modifiers that mitigate her resemblance to her male fellows or foe. He apostrophizes to her as *aspera virgo* (11.664), juxtaposing her juvenile epithet with antithesis – rough, wild, savage. She is not awarded the simple designation of 'combatant' without a reminder of her 'paradoxical' nature, leading Virgil to call her *Amazon* (11.648), a (sub)version of femaleness that his Roman audience might accept. The mythical Amazons reflect not only Camilla's militaristic femininity,[37] but also her otherness as an Italian: as foreigners (Foxhall 2013: 85; Pind. *Ol.* 13.87-90), the Amazons appeared regularly in Greek and Roman imagery, symbolizing a monstrous threat to civilization (duBois 1991: 49–77). This same point of reference persisted into twentieth-century conflicts, where women wearing men's attire, especially khaki, were dubbed 'Amazons who aped men' by the press (Marcus 1989: 158).

Camilla's bodily otherness and gender deviance are further heightened when she retreats, Parthian-style, firing arrows behind her (11.654; Gransden 1998: 125; Virg. *Georg.* 3.31), connecting her with another 'effeminate' enemy force.[38] Her opponents are shamed ('One woman has routed this whole army!' 11.734), with a repudiation typical of generals' speeches emphasizing Roman superiority over foreign peoples. In appearance, she combines exoticizing purple, an Amazon's bared side, Diana's quiver and Parthian arrows, plus she is *cingitur* (11.536, belted) like a good Roman soldier (see Džino and Coopey's chapter in this volume). Altogether, she amalgamates the qualities of a worthy enemy.

The persistent feminizing allusions tell us more about Roman expectations of warring bodies than they do about Camilla's gender. Camilla not only disproves the necessity of a medically recognizable, single-gender (masculine) identity to meet the high physical standards for combat: she also forces us to recognize the transitiveness, the fluidity, of gender in warfare, a context in which normative categories are inexorably breached.

Still, scholarly consensus claims that one aspect of Camilla's character wholly divorces her from the norms of femininity or military masculinity: exuberance in combat. Even Virgil cannot count how many she slays (11.665; Basson 1986: 61 has at least eleven). Ancient military writers admonished excessive *impetus* (Frontin. *Strat.* 1.10; McDonnell 2006: 64–71), as Sara Elise Phang notes: '*animus* (confidence, morale) and *impetus* (onslaught, energy)' were prized soldierly traits, but when excessive were 'hard to distinguish from *furor* (madness) and *ira* (rage)' (2008: 37). Pride makes Camilla *furens acrique accensa dolore* (11.709-11, raging, enflamed with piercing anger), closely resembling the moment Allecto instils in Turnus *furor* for battle (7.456-7). Anger and rage may be undesirable, but they are hardly uncommon fuel for real or literary warriors (μῆνις, *Il.* 1.1; see Shay 2003: 77–99), especially in epic, where *furor* is often attributed to depraved characters but is not always divorced from *virtus* (Behr 2007: 42; Hershkowitz 1998: 207; see also McHardy's chapter in this volume).

What is more, Camilla's bloodthirstiness appears to *physically* transform her: her victims freeze in terror at the sight (11.700-1). Even in Turnus' eyes, she is *horrenda* (11.507), visually unrecognizable as human female, leading scholars to call her 'an androgynous monstrosity [. . .] unclassifiable' (Gildenhard and Henderson 2018: 30). For Thomas Rosenmeyer, 'she has something of the monster about her. [. . .] that ranges her with such figures as Mezentius and Charon and the Sibylla in the moment of her divine madness' (1960: 160), once again invoking feminine and masculine counterparts. When death comes for Camilla, Arruns' spear spills *cruor virgineus* (11.804, maiden blood), pointedly reminding us that her body cells are female even if her actions are not. Virgil explains that she is defeated only because she is consumed by a 'feminine desire for spoils' (*femineo praedae et spoliorum ardebat amore*, 11.782): the glamorous gold and purple clothing worn by Trojan Chloreus (*Phrygiis fulgebat in armis*, 11.769, glittering in Phrygian arms), colours particularly associated with non-Italians (Gransden 1998: 134), renders his appearance equally as 'exotic' and effeminate (Nauta 2006: 89–92) as Camilla's 'feminine' hunger for it. Scholars have asserted that her death is not due to womanly failings, but 'a fatal moment of lost concentration' (Sharrock 2015: 159), owing to 'a weakness to which many warriors succumb' (Ramsby 2010: 17). Her intentions are unclear – to display the coveted gold and purple as victor's *praeda* or to wear it herself (11.778-80), like the golden torques sported by Iulus Ascanius and his fellow aspiring warriors (5.558-9, 10.137-8). Camilla is more like a soldier slaying for plunder than a stereotypical Roman female (Morello 2008).

Conclusions

It befits the false logic of gender stereotype pseudo-science (Mackay 2021: 25–6) that so much attention has been devoted to debating Camilla's gender. Virgil's repetition of feminine adjectives creates a semantic picture of a female body with feminine counterparts, but one equipped and trained for male-dominated battle. However, while it is certainly the case that the *Aeneid*'s prosody and narrative complicate our understanding of her embodied identity, if Camilla is paradoxical it is not because she performs a kind of 'female masculinity' or vice versa, but because she exposes the expectations devastatingly imposed on bodies to meet military 'standards'. War can be *horrendous*, so it follows that the bodies involved might have 'something of the monster' about them. Camilla's story showcases the lifelong effects of war-making and preparations, which can transform the body into an extension of the combat zone, reshaped by weapons and wrath. Through Virgil, she portrays how bodies make war and *are made* in war. Gender is both entwined with cultural ideals of the soldierly corpus and also detached from them: a high standard of military conduct, in practical terms, by no means depends on having a single-sex body.

We could revere Camilla as an *exemplum* of female empowerment and pride, like the matrons inspired by her death to defend their city (11.891-5; Pyy 2021: 132–3; DeBoer 2019: 132–3), but we should also acknowledge that her need to fight is the product of patriarchal systems predicated on power purchased with violence. Like many others, Camilla embodies the identity she is forced to perform. If 'womanly' signifies weakness, then its juxtaposition with 'weapons' highlights through antithesis the brutality of battle, the bodily de- and reconstruction that occurs in any warring environment. As Swati Parashar writes, still today, 'in an age of revolutionized and advanced warfare technology (aimed at minimizing civilian casualties), we witness more injured bodies and the increasing use of the human form as a war tool' (2013: 621).

War narratives allow for the exploration of issues in an inherently unsettling, destabilized situation. War-making has long been intimately connected with the male/masculine and this *is* a problem, for reasons beyond the issue of inclusivity. Attitudes to the feminine and the treatment of female bodies in war spaces are indicative of wider societal trends (DeBoer 2019). In a recent discussion of the 'enduring resistance to women joining combat roles' today, B. K. Greener writes that this resistance, 'ostensibly to protect those women, is paradoxical when juxtaposed against the dangers faced by women daily' (2023: 306). We are once again prompted to reflect on our conceptualization of 'combat' and the bodies that act as its agents, even when they are unrecognizable. As *horrenda*, Camilla seems a fitting participant in Virgil's *horrida bella* (7.41), a foundation myth for an empire forged in bloodshed.

Acknowledgements

This research was first presented at the *(De)Constructing Masculinity* conference, King's College London, 2018; thanks are due to conference delegates and organizers Hannah Burke-Tomlinson and Grace Emmett, for informing the writing of this chapter, and to Annemarie Ambühl, Vanda Zajko and the anonymous reviewer for their valuable editorial notes.

Notes

1. Latin text from Virgil's *Aeneid* comes from Horsfall (2000); Gransden (1998). Translations are my own.
2. The central importance of 'the man' resounds in the proems to Latin epics and in descriptions of epic. See Keith (2000: 23–4), who cites Virg. *Ecl.* 6.3; *Aen.* 1.1., Ov. *Am.* 1.1.1; Hor. *AP* 73; Luc. 1.1-4; VF. 1.11-12; Stat. *Theb.* 1.41; Stat. *Ach.* 1.3; Sil. 1.1, 5-6.
3. See Sorscher, this volume: characters like Creusa, Lavinia and Dido are all combatants of a sort, given their involvement in conflicts, and DeBoer (2019: 142–6) argues for Dido's 'militancy'. Camilla is exceptional because she serves as a leader in close combat.
4. See also Redfern and Pearce, and Sorscher, this volume; Rousseau (2015); DeBoer (2019) on ancient and modern discourses regarding war as 'men's business'.
5. 'Biologically sexed as male' (Mackay 2021: 23–4). The UK and US military institutions constitute this author's cultural frame of reference. See Barrett (2002); Higate (2003: 203); Millar and Tidy (2017: 146): 'the male/masculine (as interchangeable) nature of combat is apoliticized and naturalized into the empirical description of an objective social phenomenon'. Also, the Atherton report revealing the UK Armed Forces as 'still a man's world' (House of Commons Defence Committee 2021: 5), including evidence of sexual harassment and structural sex-based discrimination (33). Coincidentally, the first British woman to fight was transgender (BBC 2016). See Campanile, Carlà-Uhink and Facella (2017); Surtees and Dyer (2020); Wyke (1998), for approaches to gender diversity in the ancient Mediterranean.
6. Also, Xinyue (2017); Pyy (2010); Sharrock (2015); Gildenhard and Henderson (2018: 28): 'one epic oddball, who adds to the book's battle scenes a spectacularly wrongfooting feminine

touch'. Morello (2008) counters centuries of receptions to argue for Camilla's competence in battle despite gender expectations.

7. See other chapters in this volume, especially Draycott. For statistics, see Keeling et al. (2021). See also, Scarry (1985: 112): 'the same arms and legs lent out to the state for seconds or minutes and then reclaimed may in war be permanently loaned in injured and amputated limbs'.
8. See Sharrock (2015: 157) on the various roles played by women in Roman epic warfare.
9. 'A history of gender dysphoria is disqualifying, unless, as certified by a licensed mental health provider, the applicant has been stable without clinically significant distress or impairment in social, occupational, or other important areas of functioning for 18 months' (DoD 2019: 7, section 1.2.b.1) Challenged by Wise (2019: 115). See Ford and Schnitzlein (2017: 2) for a definition of terms.
10. Pepin-Neff and Cohen (2021: 654), e.g. the removal of federal protection for employment discrimination.
11. See Fine (2018); Butler (1990); Rippon et al. (2021); Mackay (2021): most physical and behavioural norms demarcating male/female exist only as constructs in the cultural imagination (23–44), reaching far wider than a 'cursory glance at genitalia at birth' (23). 'Transgender' refers to someone experiencing a disconnect between the 'sexed' body and perceived gender identity.
12. Standards include: 'e.g., medical fitness for duty; physical fitness and body fat standards; berthing, bathroom, and shower facilities; and uniform and grooming standards' (DoD 2020: 2). The UK Armed Forces currently operate 'gender-free' fitness testing (HoC Defence Committee 2021: 14).
13. See Mackay (2021: 30–1); Atherton (2016: 243); Foxhall and Neher (2011: 493). 'Bodily performances are negotiated, carried out and perceived differently depending on the space. Space here refers to factors that can affect the positionality of the body' (Nasser-Edin 2017: 147).
14. See Portillo, Doan and Mog (2022: 574–5) for an overview of scholarship on these issues.
15. On Camilla's placement in the catalogue rendering her 'other' and comparison with Herodotus' Artemisia and Virgil's Penthesilea, see Gildenhard and Henderson (2018: 24–5); Boyd (1992).
16. Old enemies of Rome occupying southern Latium, defeated and colonized in the late fourth century BCE, who became Roman citizens in 90 BCE.
17. Egan (1983: 20–1) uses etymology to connect Camilla with 'armour'. *bellatrix gleba* (VF. 7.612, 'the warlike soil'); *ista bellatrix iracundia* (Cic. Tusc. 4.24.54, 'this warlike rage'); to describe Minerva (Ov. *Met.* 8.264) and Rome itself (Ov. *Trist.* 2.1.321); cf. *rebellatrix Germania* (*Trist.* 3.12.47). See Xinyue on *bellatrix* as 'diametrically opposed to *matrona*' (2017: 169).
18. See Hallett (2015). Cleopatra and Antony (Plut. *Ant.* 25.5-28.1, 29); also, Artemisia (Hdt. 7.99, 8.87-8). Sempronia (Sall. *Cat.* 24.3-25); after Virgil, Boudica (Tac. *Ann.* 14.31-7; *Agr.* 14-16).
19. Keith (2012: 395–7); Pyy (2021: 130–1). The Republic's foundation myth features noblewoman, Lucretia, who suffered rape then died by suicide to salvage her womanly honour, spare her husband's shame and provide a precedent for future 'unchaste' women (Livy 1.58.10). See Sen. *Controv.* on women's behavioural virtues and vices (1.6, 6.6, 7.3, etc.).

20. See Lomas (2021: 995); Wood (1999: 84); Foxhall (2013: 70). Livia, 'first lady' of Rome during Virgil's writing life (Wood 1999: 77) is painted as a conniving trouble-maker in Tacitus' *Annals*, e.g. 1.10.23: *Livia gravis in rem publicam mater*, translated by Grant, 'Livia was a real catastrophe, to the nation, as a mother' (1996: 39).
21. See DeBoer (2019: 147–50) on Camilla's militancy as 'anti-maternity'.
22. Many thanks to Annemarie Ambühl for contributing this point.
23. On Camilla as *dia* (*Aen.* 11.657, godlike), a word rarely used by Virgil, see Gransden: 'the heroic and spiritual incarnation of archaic Italy' (1998: 126). Her death is often deemed symbolic of old Italy's defeat and colonization (Pyy 2010: 88, in Gildenhard and Henderson 2018: 26).
24. See Surtees and Dyer (2020: 8–10) on Athena/Minerva's 'hyper-masculinity'.
25. 'Camilla [. . .] is attractive and beautiful, but utterly devoted to the virginity that is the *sine qua non* for any devotee of Diana' (Fratantuono 2007: 338).
26. Keith (2000: 28): 'Virgil signals her transgression of gender norms by drawing attention to her sex (11.507, 664, 705, 734, 808), characterizing her weapons as "womanly" (11.687) and her behaviour on the battlefield as "feminine" (11.782)'.
27. Tulla and Tarpeia are overtly Roman names, from history and myth; see Gransden (1998: 125); Sharrock (2015: 164).
28. On the association of Roman military service with physical exertion, see the Introduction.
29. Williams (1999: 127); McDonnell (2006: 161) on L. Afranius, *togatae* 65. 'A woman could not, by definition, possess *virtus*, which derives from *vir* and implies specifically manly courage' (Wood 1999: 79). See also Hovey (2018) on historical 'virtuous transmasculinity'. Gransden describes Camilla as 'the incarnation of Italian *virtus*' (1998: 21), though Virgil does not connect this term with her.
30. See Pyy (2010: 183 n. 8) for key bibliography on Diana's narrative.
31. For instance, Callisto in Ov. *Met.* 2.415. Camilla is 'an ancient prototype of such contemporary action heroines as Wonder Woman, Jennifer Lawrence' Katniss Everdeen (*The Hunger Games*), or Lucy Lawless' *Xena Warrior Princess*' (Gildenhard and Henderson 2018: 22).
32. According to Galen, *Hygiene* 1-3; Soranus, *Gynaecology* 2.16.32-34. See Chidwick (2017: 6). In the *Aeneid*, Numanus proclaims his own people *durum genus* (9.603, a hardy race), and lists practices for strengthening young boys and men (603-13).
33. Polyb. 10.20; Livy 29.22.1-3; Veg. *Mil.* 1.9; Cic. *Tusc.* 2.16; Chidwick (2018); Phang (2008): 41; Watson (1983: 54–70). On poor training, see Sall. *Jug.* 44; Tac. *Hist.* 2.99; *Ann.* 13.35.
34. Being a devotee of Diana is not exclusive to females, cf. Hippolytus (Horsfall 2016: 56). On devotion to Diana/Artemis, see Horsfall (2016: 56). 'The *arma Dianae* borne by Camilla may be seen as a misuse or perversion of the weapons of the chase, in contrast to the arms given to Aeneas in book VIII by Venus' (Gransden 1998: 21).
35. Aside from female participants in the *venatio* (artificial hunt) or as gladiatorial combatants, Roman women were not normally seen to be violent (Brunet 2014: 551–3). Keith (2000: 30–1) sees Camilla as a prototype for Ovid's Atalanta (*Met.* 8.317-23) and Statius' huntress (*Theb.* 4.267-8, 6.563-7).
36. *Aen.* 5.520, 12.284; Caes. *BGall.* 1.26, 1.46-7, 3.4, etc. See Egan (1983: 21): 'No other individual in the *Aeneid* uses anything like the variety of weaponry ascribed to Camilla over the course of about 350 lines,' rendering her 'the embodiment and personification of a well-stocked ancient arsenal'.

37. On Penthesilea in the Trojan cycle, see Gransden (1998: 22–3). On comparisons with Penthesilea and Harpalyce, see Xinyue (2017: 170); Sharrock (2015: 162–4); Basson (1986: 57–8).

38. See Xinyue (2017: 171) on symbols of 'eastern effeminacy'; also, Llewellyn-Jones in this volume.

PART III
THE POST-COMBAT BODY

CHAPTER 7
RETHINKING ANCIENT GREEK WAR MONUMENTS: BODIES, VIOLENCE AND NATURE

Lucia Nováková

Introduction

Looking at ancient Greek military monuments, what resonates most is the many fighting bodies. Their perfect likeness tempts us to a simplistic understanding of these monuments in the context of the celebration of victory first, but on a closer look, the view of ancient warfare broadens. The typified naked bodies emphasized the role of the Greeks as heroic combatants and othered them from their opponents. However, the constant glorification of war is rethought from the perspective of scenes that show suffering on both sides of the fighting. The death of the enemy was an ever-recurring motif in battle scenes, where empathy for the opponents' experience or mercy had no place. A certain change occurred during the fratricidal Peloponnesian battles: the attention of artists was increasingly directed towards the internal experience of the warrior, emphasizing emotions such as anger, pain and fear. Gradually, images of combatants began to show their sensory experience and affliction, surrounded by figures of dead and dying warriors lying on the ground or contorted by pain. The torment that war brings was also depicted through the gestures of those involved, especially the civilians. The depiction of the warriors' bodies had an even broader significance: reflecting the anthropocentrism of Greek artists, it stood in for images of the natural environment. The various wounds reflect a landscape scarred by war and the body can be likened to a battlefield. The dead bodies become the terrain, the place where the fighting continues, which is trampled upon and through which the living wade in blood. Even nature herself becomes a warrior: her elements are unbound, strong and defy the greatest Greek heroes.

War, Combatants and Monuments

Military monuments had a diverse character and were combined with images of struggle and violence. These were mostly reliefs on public buildings (temples, treasuries, altars) or free-standing statues, often preserved in Roman copies. For a comprehensive understanding it is necessary to supplement these images with descriptions of the battle scenes and the warriors' bodies in written sources. The following lines present the combatants' bodies depicted on the war memorials. In the foreground was a figure of a warrior in the scrum of battle. This archetype appears in every culture, whether in texts or images, and its Greek form has undergone a certain evolution. The figure of the semi-divine combatant (*heros*) who was born at the turn of the Bronze and Iron Ages (twelfth to eleventh century BCE), was, in the classical period (fifth to fourth century BCE), gradually identified with an anonymous citizen. There was an obvious attempt to depict a citizen with the body of a warrior. Later, individuals emerged in iconography to inspire contemporaries and future generations. This is best seen in the figure of Alexander III of Macedon (356–323 BCE), who styled himself as a Greek demigod and Near Eastern ruler (Figure 7.1).

Figure 7.1 Bronze statue of a warrior reflecting a portrait tradition of Alexander established during his lifetime. Hellenistic period. The Walters Art Museum, Baltimore.

Works referring to the military achievements of the Greeks served to celebrate victory, justify violence and help communities honour the memory of the fallen. Often, they featured remembrance rituals and festivities. Combat has always been depicted as an expression of heroism, but while in Homer's works we feel the raw side of killing with pathos or anger (*Il.* 4.500, 11.235, 20.445) and the emotion breaks through outwardly, in later works this emotion is often wiped away, also reflected in the visual arts. In the archaic period (700–480 BCE), a negative view of the afterlife resonates along with the related awareness of the tragic aspects of war (Hom. *Od.* 11.486; Hes. *Sh.* 264-270). The period of Athenian democracy (508–322 BCE) meant a totalitarian approach to citizens and the glorification of military action. During this remarkable period, the Athenians were almost constantly at war (Pritchard 2013: 174–5). Sacrifice for the homeland was glorified, the 'bodies of the citizens' became the defence and power of the state and it was necessary to constantly 'produce' and celebrate them. They were the very walls of the city (Pl. *Alc.* fr.28, *Laws* 6 .778d; Arist. *Pol.* 7.1330b). The more citizens, the more ways the polis had to protect herself.

Classical art developed two basic concepts of the warrior: an older man and a young man who had reached an age suitable for combat. The two examples are connected by two men, Harmodius and Aristogeiton, depicting the assassination of the tyrant Hipparchus who became a symbol of democracy (Thuc. 6.56; Lattimore 2006: 456–58). While the art of the late archaic and late classical periods provides many examples of heavily dressed and armed soldiers, high classical art (450–400 BCE) emphasized, above all, the idealized naked figure of a warrior, possibly wearing only a *chlamys* and a helmet. It was a clear depiction of the body of a standardized Greek combatant without part of the armament. This nudity corresponded to the longstanding artistic tradition of depicting *kouroi* and athletes, blurred the differences between the individual citizens and emphasized the difference between Greeks and others (Hdt. 1.10). The Athenians fighting the soldiers in Persian clothing in the south frieze of the Temple of Athena Nike on the Acropolis are naked, depicted in a furious but clearly victorious battle. The heroized naked body, with no paraphernalia, stressed the man's role as a warrior. *Heros*, the offspring of god and man, a naked male figure in his artistic expression, was depicted in a state of strength and perfection. It is in the fight with the enemy that his fearlessness and invincibility come to the fore, passing the test from which he emerges as the clear winner, without harm to his body. In fine art, heroes often fought with their bare hands as if they did not need weapons or protective equipment. This was in stark contrast to the othered opponents, who were decked in armour.

In the classical period, it is possible to see a kind of fusion of the citizen and the *heros*: posthumously in the case of the tombstones of specific individuals or groups as evidenced by celebratory speeches (Thuc. 2.42.2), epitaphs of the fallen, or anonymously, as is the case of idealized figures of citizens on public monuments. The monuments did not depict 'real' bodies or battles. The appearance of civic sculptures did not capture the individual features of a person, but rather idealized or generalized traits typical for certain groups of people, reflecting the ideas of contemporaries and subsequent generations (Cohen 2012: 472–6). After all, a realistic depiction was not as important as the presentation of a versatile and fit individual. The battlefield realities were not necessarily ideal, nor were the deaths necessarily heroic, so embellishing them was necessary to maintain a shared civic awareness and sense of unity. The physical parameters of important generals are also usually recorded in written sources. For example, we know that Agesilaus II was 'crooked' but otherwise beautiful. In his case, positive leadership abilities prevailed over bodily deficiencies (Plut. *Ages.* 2). Regarding Eumenes of Cardia, Plutarch noted that his figure was beautiful and slender, not emaciated but symmetrical, like a work of art (Plut. *Eum.* 11).

Physical indisposition disqualified a person from being a capable soldier.[1] The Spartans, in addition to other groups, relied on physical fitness, so any deviation and imperfection raised doubts.[2] Bodily appearance played an important role in the characterization of a person and could evoke, in the case of perfect parameters, a positive relationship. The most common characteristics of the Achaeans according to Homer were, in addition to beautiful shins, long or beautiful hair (*Il.* 2.11, 65, 323, 7.442, 448, 459, 8.53, 9.45, 18.6).[3] Long hair characterized archaic *kouroi*, and the Spartans, according to Xenophon, let long hair grow to make them appear taller, more sublime and more frightening (*Lac.* 11). It is perhaps no coincidence that a similar type of hairstyle, with *anastole*, later also appears in Alexander's portraits.

The manner of depiction on military monuments changed from strong schematization and idealization in the archaic and classical periods to naturalism and hyperrealism in the Hellenistic age. The change in the depiction of military scenes can be seen during the Peloponnesian Wars (431–404 BCE) and the period of civic unrest in Athens, which is perhaps related to the transformation of the polis and the drama of civil wars. During these conflicts, the Greeks no longer faced a foreign enemy, but each other. Prolonged wars, the destruction of settlements and their rural background, not to mention the huge loss of life and large numbers of enslaved or expelled people, had a significant impact on society at the time and perhaps led to greater sensitivity to loss, death and pain in art (Lattimore 2006: 456–79). This opens up the possibility to show the inner experience of the warrior, the victor as well as the vanquished. The depicted emotions, no longer conceived in the context of the pathos of combat, but rather of contemplation, offer an opportunity to reconsider the representations of battle scenes and the bodies of warriors. The significance of victory was underlined by the high price at which it was won and what had to be sacrificed for it. The wounded bodies depicted were those of the whole of Greek society: that is, both sides. During the civil wars, the link between the citizen and the *heros* became more pronounced.

At the same time, the number of public images and monuments dedicated to particular combatants increased in political art. The military general Conon was to have his likeness placed on the Athenian agora (Dem. 20.70). For the Monument of the Eponymous Heroes, in the same place at the end of the fourth century BCE, Athenians created statues of mythical figures as well as mortals. These included Antigonus I Monophthalmus and his son Demetrius Poliorcetes, who were later replaced by statues of Ptolemy III Euergetes and Pergamese king Attalus I. Artemisia II had a celebratory monument built in Rhodes consisting of a group of bronze statues representing her brandishing a sword and personifying the conquered city (Vitr. *De arch.* 2.8.15). According to Pausanias, such honours were not unusual since in those times those who had power were celebrated in different ways (6.3.16).

Most of the war-glorifying monuments from the late classical and Hellenistic period (fourth to the first century BCE) are not preserved in the Greek environment. The lack of these finds is not only balanced by the number of written sources, but also by finds outside the Greek world.[4] The images on the state monuments became more detailed and contained a number of characters. Scenes of attacking cities, which had a parallel in Near Eastern historical reliefs (Figure 7.2), reflected the way in which the war was conducted and the popularity of historical figures as conquerors of cities (Serrati 2013: 179–94). Attalid rulers stylized themselves as protectors of the Greek world and bearers of Athens'

Figure 7.2 Siege scenes in Greek art reflect the tradition of large-scale siege scenes covering the palace walls of Near Eastern kings. Relief from King Ashurbanipal's Palace at Nineveh. Neo-Assyrian period. The Walters Art Museum, Baltimore.

cultural heritage, portrayed their military opponents with dignity, and highlighted the magnitude of their own victory (the Ludovisi Gaul). The pride of the enemy at the moment of defeat forms a contrast to the dynamic battle scenes of earlier periods, yet it seems to underline the triumph.

The Body, Armour and Weapons on Monuments

The Greeks used the spoils of war to finance the construction or reconstruction of public buildings and to build statues and other works that celebrated and commemorated victory. They also expressed their success by displaying enemy arms and armour in public places or in shrines (Figure 7.3). Armour was extremely valuable as an integral part of the warrior's personal property. Literally, it was an extension of his body. Descriptions of arms and armour completing the overall appearance of the warrior can be found in written sources. The *strategos* of the Achaean League and the general Philopoemen recommended that the soldier's armour should shine, which was to intimidate the enemy (Polyb. 11.9.1-7).

Armour personified the owner itself, and its removal after death was proof of victory and absolute superiority, while a loss was a disgrace to the defeated party.[5] In the Battle of Tanagra with the Spartans in 457 BCE, the Athenians did not allow their ostracized general Cimon to fight. His friends took his armour in the middle of their unit and all, about a hundred men, fell (Plut. *Cim.* 17). It was even worse if the armour or part of it fell into the hands of the enemy during the life of the warrior.[6] In the *Iliad* we find descriptions of several fights in which the winner, in an effort to immediately remove the armour from the body of the fallen, ceases to be vigilant and becomes a victim of another attacker; in fact, it might be impossible to obtain the armour because the fallen were taken to safety by their companions (Hom. *Il.* 4.464-72, 4.517-35).[7] The battle of Achilles and Memnon over the dead body of Antilochus is depicted on the eastern frieze of the Siphnian Treasury in Delphi. The western frieze of the Temple of Athena Nike on the Acropolis, on the bastion of which were probably hung the captured Spartan shields, was decorated with soldiers fighting for the bodies of the fallen comrades together with a shield hung on a tree (Lippman, Scahill and Schultz 2006: 551–63). The Amazons on the Bassae frieze use Greek shields in combat (Sporleder 2015: 23–33), which they also take away as booty.

The focus on the human figure created from the enemy's armour can be seen in *tropaia*, which served as a symbol of collective victory (De Vivo 2014: 163–84; Kinnee 2018: 46–60).[8] They were built on the site of the first military encounter or where the defeated army turned to flee. Immediately after the end of the conflict, arms and armour were simply hung or attached to a tree trunk or a wooden wheel as a kind of personification of the defeated enemy (Janssen 1957: 245; Woelcke 1911: 25–8). During the Graeco-Persian Wars (499–449 BCE), monuments made of durable materials like stone and metal were erected to commemorate the Greek victory over the Persians, often taking the form of statues on pedestals, or a combination of a small structure and

Figure 7.3 The Corinthian helmet was one of the most common forms of helmets among hoplites. Its distortion (the cheekpieces bent outward) suggests that it was dedicated in a sanctuary. Archaic period. The Walters Art Museum, Baltimore.

a statue (Vanderpool 1966: 105–6; West 1969: 7–19). Thus, their appearance was distinct from the earlier representation of a 'warrior', symbolized by piled-up armour. Ancient historians have highlighted the rare moments of military clashes when the highest military commanders met face to face in a battle and a life-and-death struggle took place between them. For example, King Pyrrhus of Epirus, who at the end of his life not only killed the Spartan commander Eualcides in a cavalry battle, but also killed soldiers protecting the corpse of his enemy after jumping from a horse (Plut. *Pyrrh.* 30).[9]

The attention of visual artists, however, has long been focused on the idealized representation of a male body rather than on arms and armour.[10] Descriptions of armour are also lacking in the works of historians. An exception in this regard is the chain of information that comes down to us via Herodotus, the father of history.[11] The perfect bodies of warring heroes and gods seem to have no need of equipment: in artistic and literary expression the weapon could be practically anything. In real combat, in addition to boulders, table utensils, stones or bare hands could be used in the battle, such as in the Centauromachy depicted on the west tympanum of the Temple of Zeus in Olympia or on the metopes on the south side of the Parthenon. The frieze adorning the Hephaisteion in Athens was decorated with scenes of naked figures fighting with stones (Theseus against the Pallantides?) and the Temple of Apollo Epicurius in Bassae depicts a scene of a Lapith holding a stone being attacked by a centaur with a branch (Dinsmoor 1956: 401–52; Morgan 1962: 221–35).

The use of stones as weapons emphasized that pure physical strength caused death and devastating injuries, just as traditional weapons did. The first reference Homer makes to fighting with a rock was Thracian Peiros. Other examples could be found with the Achaean heroes Diomedes, Ajax and Patroclus, but Hector also distinguished himself in this style several times (*Il.* 4.517-35, 5. 297-310, 7.264-74, 8.326-9, 12.384-6, 14.409-19, 16.402-13, 577-87, 731-64).[12] However, fighting with a stone was not only reserved in artistic expression for heroes. It often appears as a weapon of choice for centaurs and giants. Such heroic struggles may be beyond our understanding of fighting, and we would rather displace these to archaic and mythical times, but Phidias still depicted himself on the shield of the Athena statue in a scene of struggle with the Amazons, holding a stone over his head about to strike (Plut. *Per.* 31; Harrison 1966: 107–33). The fight with the rocks evokes superhuman strength, which could only be possessed by gods, primordial powers (giants) or heroes.

The focus of attention in these scenes was primarily on perfect male bodies in a variety of positions, with weapons and equipment playing only a secondary role. A certain change can be seen in the formation of mercenary armies, when the diversity of weapons was also reflected in the artistic expression. The bodies depicted make a visual distinction between standing and mercenary armies. The shield and sword, and more rarely the spear and helmet, were relatively common in the repertoire of weapons in high classical fine art, while armour or greaves were usually missing. In the fourth century BCE, we encounter a larger number of armed figures who are not only equipped with a helmet, a round shield and a shorter spear, but with a cuirass, a short sword and *knemides* (greaves). At that time, Greek hoplite armaments were already widespread throughout the Mediterranean and therefore cannot be used to distinguish the ethnicity of the characters in the scenes (Lee 2006: 480–508; Lee 2013: 147–50). The reliefs of the Heroon in Limyra depict a Lycian phalanx of soldiers with various types of helmets from Attic to Thracian and Macedonian, carrying a shield, a spear or a sword in their hands (Borchhardt 1976: 72–8). Perhaps this was an attempt to emphasize the ethnic diversity and size of the troops, following the example of the army of the Great King, in which Greeks also fought.[13]

At the same time, written references to the armaments of specific historical figures were recorded. The Syracusian tyrant Dionysus was saved by a cuirass (thorax). On the contrary, the Spartan tyrant Machanidas was betrayed by his purple cloak and horse equipment, which cost him his life in combat (Diod. Sic. 14.88.3-4; Polyb. 11.18.1-4). Iron armour, specially made for Demetrius Poliorcetes, was to weigh 40 *mines*, a little over 17 kilograms (Plut. *Demetr.* 21). Alexander III of Macedon apparently did not have such 'heavy' protection when he jumped from the walls of an Indian city and an arrow was shot through his armour and wounded him in the chest (Arr. *An.* 6. 9-10). The Macedonian ruler, who was later imitated by many Hellenistic monarchs, used the fusion of the figure of the *heros* and a specific person to legitimize his power. One of the many forms of the ideal ruler became a victorious warrior on horseback, leading the army, nude or dressed in a cloak as a figure of an athletically conceived combatant with a sword, or even more often, with a spear.

In both cases changes in the tactics of the Macedonian phalanx are expressed, in which hoplites armed with the *sarissa* and cavalry played an important role (Lee 2006: 480–508; Serrati 2013: 182–93).[14] When the *strategos* of the Achaean League Philopoemen stabbed the Spartan tyrant Machanidas in a duel fought with a spear, it became the subject of his equestrian statue (Plut. *Phil.* 10). The number of similar statues grew from the second half of the fourth century BCE. The Athenians allegedly had more than three hundred bronze statues erected in honour of Demetrius of Phalerum (Diog. Laert. 5.5.75). Epigraphic finds (*IG II*^{2} 450b) document the presence of honorific equestrian statues on the Agora in Athens. Lysippos' bronze monument commemorating the victory at Granicus, consisting of twenty-six life-size figures, became the prototype for later equestrian monuments depicting the ruler as the victorious commander (Stewart 1994: 127).

War, Combatants and Trauma

Herodotus says through Croesus that no-one is fool enough to choose war instead of peace (1.87). War always brings pain, regardless of the period, but the suffering of its participants was sometimes emphasized more in texts and pictures, either for political reasons or because of the author's style and dramatization. The visual rendering as well as the lyrics reflect the same eulogistic tendency: to see one's own pain and suffering through a prism of pride and determination that will not submit to the enemy even after death and defeat. Victory is measured in the suffering of bodies.[15] However, there are also exceptions. For Hesiod, the image of war is relieved of heroic embellishments. One can see this in the archaic epic poem *The Shield of Heracles*, which parallels a similar description of Achilles' shield in Homer.[16]

Archaic and classical art replaced emotional expressions with tranquillity, or the grandeur of warriors, who did not show pain or other emotions (Figure 7.4). However, at the decisive moment, the eyes of the warriors met and the winner killed the defeated while looking into his eyes. Wounded Amazons, lying on the ground and leaning on one handd, the other stretching towards the Greeks as if begging for mercy, decorated several classical buildings. The well-known story of Achilles and Penthesilea is evidence of this clash of perspectives, as otherwise impassive faces could have been full of the most intense emotions and loss. Emotions were more clearly displayed on Zeus' altar in Pergamon: the eastern frieze of the Hellenistic victory monument of the Attalid dynasty captures the moment of the death of the giant Alcyoneus. The pain is clearly reflected not only on his face at the moment of death, but also on the face of his mother Gaia, who stretches out her hand in vain asking Athena for mercy. These depictions parallel the cruelty depicted in written sources. Their emotions, however, are nothing compared to the atrocities described by Homer. He displays the cruelty to defeated enemies perhaps too far. Achaean heroes kill their enemies in ways that are not heroic, but, e.g., when the wounded and helpless hug the knees of their killers and beg for their lives.[17]

The violence in the *Iliad* was rendered in archaic war scenes, where victorious warriors were usually depicted spread out with their hands up, holding a spear piercing the enemy. The defeated, on the other hand, were depicted in various positions: running trying to escape, kneeling or lying on the ground (Hölscher 2003: 1–17). The classical period brought much greater variability in the positions of the bodies. Greek warriors usually pulled Amazons' hair, tore them off their horses or stabbed them while the Amazons knelt on the ground. A similarly brutal scheme depicts a centaur jumping over a dead companion, biting the neck of a Lapith, who is simultaneously piercing him with a sword (Temple of Apollo in Bassae). Right next to this picture, another Lapith is kneeling on a prostrate centaur (Dinsmoor 1956: 401–52; Jenkins 2006: 130–50). Similar dynamics were reflected in the metopes of the Parthenon. The torsion of the warrior's body was characteristic for the friezes on the Temple of Athena Nike and Hephaestion, while the chiastic position of the figures is more common on the friezes on the Mausoleum at Halicarnassus.

In individual cases, whether from the archaic or the classical period, the weapons of the victorious warriors did not seem to harm their opponents: there are no visible devastating injuries on the body; defeat is indicated by the position of the body. Even in written sources, we only rarely find a description of the direct impact of the fight on the warriors themselves and thus the forms of injuries or cause of death. This is mainly related to the focus of historical works on the general presentation of events. Alexander III of Macedon is one of the few historical figures for whom we have a detailed account of a real fight (Plut. *Mor.* 341b).[18] Arrian portrayed him like Homer's *heros* in the Battle of the Granicus (334 BCE), where he threw the Persian Mithridates off his horse with a spear to his face and subsequently fought off another Persian by thrusting a spear into his chest through the armour (*An.* 1.15). This is reminiscent of the scenes from the *Iliad*, where we often see an almost surgical description of a cut or stab wound and a forensic description of the method of killing. With these 'successes' of such heroes, we most often see a piercing of the breast with a spear, and in the case of Idomeneus' victory over Alkathous, we are even 'enriched' with the detail of the heart in which the spearhead stuck and which moved its handle for a while (Hom. *Il.* 4.480-2, 5.144-7, 13.435-44).

Wounds to the body were a measure of success in battle. The more wounds on the opponent's body, the greater the hero's strength. Cutting off or injuring the hand, or more curious injuries to the buttocks or groin, were just additions to the most common targe

Figure 7.4 Wounded Amazon. The depiction of these female warriors fighting against the Greeks was a widespread motif on military monuments. Roman copy after Kresilas. The Walters Art Museum, Baltimore.

of attacks, which was clearly the head. Homer notices here, in addition to cutting off the head or attacking the neck, wounds to the ears and the upper part of the skull, the most numerous were wounds, especially with a spear, to the face. In this area, mouth wounds with the details of the impact of a spear are clearly the most common (Hom. *Il.* 5.291-4, 16.345-9).[19] In the idealizing art of the classical period, the experience of emotions from the depiction of the Greeks on the outside was removed. After all, in archaic art emotion was usually expressed through gestures, not grimaces. The rough, wrinkled faces of the centaurs were in contrast to the peaceful grandeur of the Greeks (the western tympanum of the Temple of Zeus in Olympia, the southern metopes of the Parthenon, the Ionian frieze of the Temple of Apollo in Bassae). The Greek himself never asks for mercy, even if he is in a hopeless situation and lying on the ground, defending himself with a shield.

This description operates with the pathos of tranquillity and heroism, as if free from the bloody chaos of a real fight, obscuring the reality of defeat and showing a possible defeat as a 'moral' victory. Yet, it was a little different for women. Greek women, abducted by centaurs, have peaceful faces, but gestures reveal their despair. Outstretched hands asking for help appear in scenes in vase paintings as well as in public reliefs. For scenes in the Centauromachy in Bassae, women embrace in horror, and Apollo himself saves a woman with a small child in her hands who is being attacked by a centaur (Jenkins 2006: 130–50; Sporleder 2015: 23–33). The half-draped bodies of Greek women molested by centaurs adorned the southern metopes of the Parthenon or the western tympanum of the Temple of Zeus at Olympia. The groped Deidomeia pushes away her attacker, pulling back from his outstretched hand and elbowing him in the face. The scene of the smaller podium frieze decorating the Nereid Monument depicts a woman in a besieged city pulling at her hair, and her lament foreshadows her fate: rape, enslavement and possibly death (Sidebottom 2004: 25–8). Historians' works, like these scenes, are only imperfect mediators of the suffering that war brought to the warriors, not to mention the suffering of civilians, especially when it became extreme.

The rape of women and girls was a common part of war as a form of revenge (and lineage breakdown) in front of their menfolk,[20] as evidenced by Diodorus' description of the conquest of Selinunte (13.58.1-2) or the massacre of the oligarchs in Syracuse (19.8.1-5). The entire *Iliad* is interwoven with the Trojans' fear, both men and women, of defeat and enslavement (Seaman 2020: 29–38). Euripides' *Trojan Women* is one long lament for these captives (41-4, 116-20, 726-35, 876-84, 1089-95, 1305-10). Here the contrast between written sources and visual art is clear. Although similar scenes (violence against women, torture and the death of children) are described in texts, they rarely appear in visual art (with the exceptions such as the rape of Cassandra and the death of Astyanax on the Greek vases). The monumental art refers to the rapes of women and girls in the scenes of Centauromachy, but the focus is more on the heroic struggle between the Greeks and their opponents. Nor are scenes of infanticide common. Artists concentrated on more 'noble' combat activities than on depicting atrocities against civilians, which the audience clearly did not want to see. The play *The Capture of Miletus* by the dramatist Phrynichus was banned by the Athenian authorities because the depicted scenes of destruction and violence against civilians were too disturbing and had the potential to negatively affect the morale of the Athenian citizens (Hdt. 6.21).

Emphasizing the details of atrocities was vehemently criticized by ancient historians, especially Polybius and Diodorus, placing history and truth in opposition to the works of poets and performances in tragedies (Polyb. 2.56.7-8, 10-16; Diod. Sic. 8.4). Even in the case of historians of the classical period, there is no special effort to create sensation. Rather, they adopt a factual tone, more in line with the scenes on the monuments glorifying war, and usually did not give details of horror unless they were directly involved somehow (Bosworth 2009: 168–87). However, the war scenes in the visual arts of the late fifth century BCE began reflecting, perhaps under the influence of the events of the fratricidal Greek wars, the pain and suffering on both sides. At the same time, there was an unwritten rule, according to which, after the victory of one polis over another, only *tropaia* of short duration and not permanent monuments could be built so they did not ignite new internal conflicts (Cic. *Inv. rhet.* 2.69; Duffy 2018: 31–2; Kinnee 2018: 40–5). Furious battle scenes without a clear victor decorated Hephaestion, the Temple of Apollo in Bassae and the Mausoleum at Halicarnassus.

Sometimes there are even ceasefire scenes where the Amazon supports the wounded Greek and the Greek carries the wounded Amazon (Bassae). On the smaller stage frieze of the Nereid Monument, we see a wounded soldier dressed in a chiton and a cuirass, who is supported by his companion in battle as he walks. Epigraphic evidence shows that public, or rather civil, monuments were built more often as the society was exhausted by internal fights and disunity. The increasing number of public documents, honorific decrees and celebratory statues testifies to an intense effort to encourage and highlight the merits of citizens for the homeland, becoming a kind of mass official production. The typified portrayal of the warrior remains unchanged, but the variety of characters depicted expanded. In addition to statesmen, commanders and orators, depictions of artists and philosophers, and finally also ordinary people, came to the fore (Chaniotis 2005: 18–37). The art of the fourth century, perhaps under the influence of Lysippos and Scopas, foreshadows interest in the emotional experience of the main protagonists, later fully manifested in the pathos of the Hellenistic 'baroque'.

Mutilation was a cruel and permanent sign of defeat for survivors. Greeks showed their spirit of invention here as well. The people of Samos, on the other hand, were not satisfied with just any mutilation, but branded an owl (a sign of Athens) onto the foreheads of the defeated Athenians, which was essentially revenge for the time the Athenians burned a ship onto the foreheads of their defeated ones (Plut. *Per.* 26). More often, however, we see mutilation in the case of punishment or torture before execution. This is the case of Bessus, who Alexander III of Macedon ordered to be lead in wearing an iron collar, and whose nose and ears were later cut off before he was hanged (Arr. *An.* 3.30, 7.7). Similarly, but with greater swiftness, Achaeus, the opponent of Antiochus III, was executed.[21] We lack the details in the fine arts, but in the case of military reliefs, emotions are conveyed in the gestures of civilians. Conquering the city, which is depicted on the smaller podium frieze of the Nereid Monument, shows the fighting undertaken by soldiers on the walls as well as the besieged population, who watch the fighting soldiers in front of the city gate or throw stones from the walls (Jenkins 2006: 186–202; see also Hornblower 2007: 43–7). The despair of civilians is not surprising. Diodorus mentions their punishment, torture or murder (Diod. Sic. 13.57.1-6; 13.111, 4, 6). Thucydides sometimes also goes into detail, especially in the case of Thracian mercenaries plundering the town of Mycalessus in Boeotia (Thuc. 7.29).

War, Combatants and Nature

Greek war memorials were primarily depictions of fighting bodies, but the concept of the warrior has varied considerably in art. All the various wounds on the combatant's body reflect a landscape scarred by war and his body can be compared to a battlefield, where a life-and-death struggle takes place. This is alluded to by the tangle of fighting bodies filling the entire space of the late classical and Hellenistic reliefs. In other cases, nature itself becomes the combatant. In the *Iliad*, the 'revolt of nature' against its destruction is the attack of the angry god Skamandros on Achilles due to pollution and, in particular, the desecration of his waters with corpses, accompanied by arrogance towards the deity (Hom. *Il.* 21.214-382). The anthropocentric concepts of victorious monuments, flowery and landscape-related motifs were captured through various figures (personification) or artistic abbreviations (branches, rocks or indications of terrain). The southern metope (1) of the Parthenon is sometimes interpreted as a scene in which a centaur is attacking a Greek with a branch, which could indicate that the fight took place in the open (Brommer 1967: 71–6; Schwab 2005: 175–6). The same applies for scenes with the immortal Caeneus, who is being buried in the ground by centaurs: we can only see them holding rocks in their hands (Hephaisteion, Temple of Apollo at Bassae).

Similarly, we can only sense the nature or the terrain itself, rather than 'see' it in the descriptions of paintings that have been lost. In the scenes from the Athenian Stoa Poikile depicting the Battle of Oinoe, the fight between the Athenians and Theseus with the Amazons, or the fall of Troy, the 'Tourist guide to Greece' does not provide any details about the surrounding country. The painting of the Battle of Marathon depicts the Persians on the run 'pushing each other into the swamps', and when Phoenician ships are mentioned, we get a sense of the seashore and its proximity. At the end of the scene, the perspective of the landscape is sketched by Theseus, who seems to have emerged from the earth (Paus. 1.15). This is similar to descriptions of Polygnotus' paintings of the Fall of Troy from Delphi. Although these are extensive passages, Pausanias notices the group of characters in various poses on the coast in front of the conquered Troy. He also mentions the coast only incidentally, except for the character of Nestor, whose horse looks as if he wants to roll in the sand on the beach (Paus. 10.25).

In the case of peaceful scenes decorating sacral buildings, the geographical scope of the event was represented by figures of local (usually river) deities, placed in the corners of tympana. In dynamic battle scenes, we encounter, from the late archaic period, prostrate wounded, dead or dying warrior figures who defined the battlefield itself (the

Temple of Athena Aphaea in Aegina). In the mixture of figures of sculptural art of the high and late classical periods, the bodies of fallen and wounded warriors created the terrain on which the battle took place. In the intense battle passages of the *Iliad*, we find identical behaviour in both Achilles and Hector. They were both stepping on the dead and their shields, and the axles of their chariots were covered with blood, as well as the edges of the frame, on which drops were falling as they splashed from the horses' hooves (11.534-7, 20.498-504).

This colourful picture recalls a description of Polygnotus' painting of Iliupersis from the Leche of the Cnidians in Delphi. It forms only a part of the painting, and even though it looks schematic, in the details we can imagine individual victims, whether that is the naked Pelis, turned on his back, or those lying underneath him such as Eioneus and Admetus, who are still wearing their shields (Paus. 10.27). The fighting figures on the friezes in Bassae fill the entire space, but the terrain of the battlefield also consists of stones and horizontal figures. Both infantrymen and riders step on the bodies of the dead and wounded. Many twisted, dead bodies are also found on the eastern Ionian frieze of Hephaisteion (Morgan 1962: 221–35).

In artistic expression, we encounter tendentious imagery suggesting a different treatment of the bodies of the fallen by Greeks and non-Greeks. In most cases, Greek soldiers respect dead bodies (Figure 7.5).[22] This expands the possibilities of interpreting battle scenes where the bodies of warriors, dead or alive, play a key role.[23] Herodotus says through Pausanias that insulting the dead is a disgraceful act, unworthy of Greeks (Hdt. 9.78-9; Muller 2014: 41–72). The southern metope (28) of the Parthenon depicts a centaur triumphantly trampling on a fallen Lapith (Brommer 1967: 120–1). The western metopes of the same temple (11, 13) show similar scenes of dead or wounded figures found under the hooves of horses saddled by Amazons. An Amazon sitting on a horse, stepping on the prostrate, twisted body of a Greek dressed in an *exomis* (tunic) is also in a frieze decorating the Mausoleum at Halicarnassus (Cook, Ashmole and Strong 2005: 42–64). On the south frieze of the Temple of Athena Nike, the Greeks, unlike enemies, do not step on dead bodies; they are captured trying to protect wounded fellow fighters, and one of the dead is even partially hidden behind a rock (Blümel 1923: 11–19).

However, there are exceptions: the northern frieze of the same monument depicts a battle scene among the Greeks themselves, interpreted as Eurystheus' death, where we clearly see a Greek warrior stepping on his enemy's severed head (Schultz 2009: 128–67). Also, the scenes from Bassae show solidarity between the traditional opponents of the Greeks: one Amazon pulls away the body of her dead companion, while the other protects another that is lying on the ground but still alive and holding a Greek shield (Jenkins 2006: 130–50). In the written sources, we occasionally see the mutilation of dead bodies in the case of the Greeks. Probably the most famous example is Achilles dragging Hector's body around, despite his previous pleas (Hom. *Il.* 20.335-71, 24.15-20). After retreating from Cunaxa, Greeks mutilated the fallen attackers to give the enemies the most terrible view possible (Xen. *Anab.* 3.4). Plutarch, according to other sources, states that the Massalians built walls around the vineyards from the bones of the many fallen enemies from the Battle of Aquae Sextiae, not to mention that the land, thanks

Figure 7.5 Black-figured amphora with Ajax carrying the dead body of Achilles. Respect for the dead bodies of fellow soldiers seems to have been a distinguishing feature of Greek soldiers. Archaic period. The Walters Art Museum, Baltimore.

to dead bodies (and winter rains), had greatly improved in terms of its fertility (*Mar.* 21).[24]

One of the few exceptions in artistic expression in which stepping on the dead or trampling on the wounded is shown by all the parties from the archaic period is the traditional motif of the Gigantomachy. Participants in the fights on the northern frieze of the Siphnian Treasury in Delphi are also stepping on the dead (Moore 1972: 305–35). This embodiment of primordial forces connected to the Earth is exceptional in the way it depicts the bodies of warriors who draw strength from the ground, from their mother. The giants, the sons of the goddess of earth, Gaia, who were firmly attached to her, were an inseparable part of the terrain in a mixture of bodies on the stage frieze of the Pergamon Altar. The mother appears to help her sons, depicted with half of her body in the ground (Thimme 1946: 345–57).

It is interesting to mention the fallen at Thermopylae, where King Xerxes allegedly tried to raise the morale of his army by 'exhibiting' the bodies of dead Greeks. There were supposed to be 4,000 of them, and they were put in one place so that his troops could see them (Hdt. 8.24-5). This 'enthusiasm' for piled-up corpses could be seen as a kind of tangible proof of victory, but mostly propaganda aimed at the troops themselves. It seems that this was not just the behaviour of non-Greeks. In describing the Marathon battlefield, Herodotus states that the Spartans, who 'missed the battle', wanted to at least see the corpses of fallen enemies (Hdt. 6.120). The dead bodies were supposed to show the father of history, who came as a tourist, the place of the battle between the Persian and the Egyptians at Pelusium (Hdt. 3.12). After one victory of the Spartans at Corinth in the civil strife in 392 BCE, Xenophon amazed his contemporaries with the piles of corpses, which he compared to piles of grain, wood or stones (*Hell.* 4.4.12).[25]

In some ways, nature is more than just a passive object: it is both victim and combatant, just like humans. Nature suffers the same; her wounds bleed like human bodies. Whether it be trees, rivers, meadows or fields, their pain is also alive and devastating, and they can feel and, in their own way, die. We see destruction during the conflicts of King Philip V of Macedonia with the Aetolians and Pergamon (Polyb. 5.11.3-6). The Syracusian tyrant Dionysius I, frustrated with a failed attack on Rhegium in 394–393 BCE, attacked the surrounding country, which was burned and orchards cut down (Diod. Sic. 14.90.7, 100.2). Similarly, the Spartan Agesilaus II, during the civil strife in Corinth (390 BCE), expressed his anger and frustration after the destruction of one Spartan division at the Corinthian port of Lechaio (Xen. *Hell.* 4.5).

Xenophon describes the battlefield after the Battle of Coronea in 394 BCE. The land was soaked with blood, and according to him, in a single pile lay all the fallen, comrades and enemies. There were pierced shields lying on the ground, broken spears, unsheathed swords without scabbards, some on the ground, others stuck in dead bodies or still in the hands of warriors (*Ages.* 2.14). In this case, nature appears to be the material on which the bodies of the fallen and their blood rested. The confusion of the fight suddenly ceased, and it was left with only lifeless matter intertwined within the armour. Hesiod's words from Heracles' fight with Ares' son, Cycnus, in which the earth groaned, come to mind (*Sh.* 371-2). He compared the collision of both warriors to an avalanche of boulders from the mountains, which took everything with them (*Sh.* 373-83).

Nature and the elements have an irreplaceable role in artwork, whether literary (Pliny *HN* 34.61) or artistic, which did not contradict the focus of Greek culture on the human body. The reference to nature or battlefield terrain in these images can be found in the form of artistic abbreviations and symbols, and often also in the form of the human figure. Perhaps no other culture has placed the human body on a pedestal to the same extent as that of the ancient Greeks. If we disregard the aesthetic categories, the naked male figure could symbolize a whole range of meanings and references. In the first place, it was a warrior representing the best in the polis, as the elite of the state. In this sense, he pointed to the mythical past of Homeric heroes, connecting human society with the world of the gods, with something beyond the ordinary and human in man. On the other hand, he emphasized the heroic deeds of that time, which put at least an imaginary sign of equality between Achilles and the former *heros* – a citizen. In the artistic expression of the classical period, we meet the character of a warrior in an idealized form, sometimes out of the context of combat chaos. The character fundamentally represents the winner, who not only defeats the enemy, whether represented by mythical or real beings, but also symbolizes victory over destruction and death. This depersonalization, without obvious emotions, representing heroism in battle such as Athenians facing 'the whole world,' the mythical and the real, also showed a form of artistically perfect propaganda. The glorification of war did not always correspond to historical records or literature, where we find a much more threatening picture. It was suffering in the endless conflicts among the Greeks themselves that probably brought about a change in the depiction of battle. Gradually, more and more emphasis was put on emotions and the pain of the depicted characters, in which they replaced their previous seriousness and nobility. War and death were also freed from ornaments by depicting the suffering of victims, injuries and atrocities against civilians, which already fill the pages of the *Iliad*, not to mention historical works and ancient tragedies. In the literature of the late classical and especially Hellenistic periods, there is a noticeable shift in the portrayal of heroism, moving away from the anonymous collective or mythological figures to focus on individual, historically identifiable figures, such as specific commanders or rulers. This transition reflects a broader cultural shift from a communal to a more individualistic representation of heroism.

Notes

1. See Jane Draycott's chapter in this volume about prostheses and injury in battle.
2. We do not know how the Spartan Androclidas ended up, but when he was rejected as a soldier because of his crippled leg, he argued for his ability to fight (Plut. *Apoph.* 13.1).
3. This aesthetic stands out even more when baldness, in combination with the bending and deformation of the body, signifies the anger of the troublemaker and the ugliest of the Achaeans, Thersites (Hom. *Il.* 2.216-19).
4. The walls bordering the Heroon in Trysa were decorated with scenes from the conquest of cities and were combined with Greek mythological stories. Battle scenes, long reserved for the celebration of the Greek world, appeared on the tomb of the Carian satrap Mausolus.
5. In Hesiod's account, Ares was threatened with rebuke by the other gods because he let his armour fall into human hands (*Sh.* 358-66).
6. The poet Alkaios coped with such a situation in his own way after his escape from the battle for Sigeion. His verses, making use of a striking symbol of cowardice derived from throwing a shield on ground while running, are perhaps the most original example of self-irony in history (Hdt. 5.95).
7. Ajax was prevented from removing Amphius' armour by a hail of Trojan arrows, and Diomedes was also unable to gain any booty because he was struck by Paris' arrow (Hom. *Il.* 11.368-80). Achilles killed the grandson of the river god Skamander Asteropaeus: he jumped on his chest, stripped off his weapons and disrespected not only the dead, but also his grandfather in a bragging speech (Hom. *Il.* 21.161-98). He similarly sinned in dragging Hector's body about, even though it had already been deprived of its armour (22.376, 396-409).
8. Cf. Danijel Džino and Ewan Shanks Coopey's chapter in this volume about Roman *tropaia*.
9. A brutal duel full of hostility and hatred, approaching the reality of battle, led by Neoptolemus and Eumenes of Cardia. Eumenes eventually removed the armour as booty from the corpse of his enemy, but the details supplied to us by Plutarch deprive this battle of any pathos or heroism and turn it into an 'ordinary battle' (Plut. *Eum.* 7).
10. Magnificent armour or foreign clothing could, as in the case of the proud Spartan general Pausanias and his Median clothing, be the subject of criticism (Thuc. 1.132).
11. In a military clash in front of the city of Eretria, Herodotus describes the fight between the Persian cavalry commander Masistius and the Athenians, during which he fell off his horse.

Even his armour made of gold scales and purple clothing did not save him because they killed him by stabbing him in the eye through the visor of the helmet (Hdt. 9.22).

12. Heracles injured Hera with a rock, but Homer offers a much more surprising picture in Book 20 of the *Iliad*, where Athena attacked Ares in the same way (20.403-9, 5.392-5).
13. We also encounter a similarly varied mixture of armaments and equipment on the battle scenes on the smaller podium frieze of the Nereid Monument.
14. A detailed description of equestrian armour, from the helmet to the horse's equipment, can be found in Xen. *Horse*. 12.
15. Cf. Fiona McHardy's chapter in this volume on head-hunting in Greek warfare.
16. The horrible pictures on the personal armour of one of the most popular heroes show the goddesses of violent death (Keres), who ruthlessly devour the bodies of the fallen and wounded, their souls forever languishing in the darkness of the Underworld. Similar is the scene in which the author confronts his listeners with the image of grief (Hes. *Sh*. 264-70).
17. Diomedes and Achilles showed themselves in this way multiple times (Hom. *Il*. 10.454-7, 11.143-8, 20.455-94, 21.64-75, 115-20). In addition, when it seemed that someone, specifically Menelaus, had mercy for his rival, Agamemnon convinced his brother otherwise (Hom. *Il*. 6.38-46, 62-5).
18. Alexander certainly did not avoid fights, as shown by his numerous injuries such as a strike from a catapult through his shield and armour into his shoulder near Gaza (Arr. *An*. 2.27).
19. Such details, whether it is a mouth, nose, jaw, eye injury or in the case of Achilles' rage, even an ear injury, we find a large number in the fight scenes of Homer's *Iliad* (13.615-19, 671-2, 14.493-9, 16.404-5, 17.617-18, 20.472-6).
20. See Hannah Sorscher's chapter about 'second-phase' warfare in the *Iliad*.
21. His ears and nose were also cut off and his head was immediately cut off, put in donkey skin and his remaining body was crucified (Polyb. 8.21.3).
22. These visual depictions seem to be in line with the generally accepted respect for the fallen among the Greeks and the effort to provide the last honours so that religious and social conventions are observed (Thuc. 4.42-5; Plut. *Nic*. 6). Macedonians also put special emphasis on the burial of the fallen (Curt. 2.9, 3.12, 5.4). See also McHardy, this volume.
23. Unlike the Greeks, the Persians did not bury their dead, at least not in this sense, which can be explained by the religious customs of Zoroastrianism (Arrington 2015: 23–4).
24. The distance between the place of the battle and Massalia already excludes the use of the bones of the fallen, but even such records, more than two centuries after the event, testify to the legends with which various battlefields were glorified.
25. We can see a similar scene in the case of the conquest of Selinunte by the Carthaginians in 409 BCE: the dwellings were burned or razed to the ground and the whole territory was filled with blood and corpses (Diod. Sic. 13.57.1-6).

CHAPTER 8
ASTYANAX'S FATE AND SECOND-PHASE WARFARE IN THE *ILIAD*

Hannah Sorscher

The meeting of Hector and Andromache at *Il.* 6.390-502 has been called 'the most famous of all Homeric scenes' (Kirk 1990: ad. 6.440-65). This encounter between husband and wife, hero and household, exudes a powerful emotional force. Prominently featured is the baby Astyanax, whose adoring parents frequently speak of him, and his future, here and elsewhere (22.477-514, 24.725-45), but their speculations regarding his future are neither coherent nor realistic. I here analyse the inconsistencies and improbabilities in Hector and Andromache's speeches to show that these variations reflect the consequences for the bodies of women and children in the aftermath of war. I argue that the multiple depictions of Astyanax's future maximize the poem's portrayals of second-phase warfare – the systematic rape, murder and enslavement of conquered women and children – for an audience that likely included people of all genders and ages who had experienced it themselves, directly or indirectly. Other chapters in this volume are historical and archaeological in focus; my contribution shows that literature too involves bodily realities and gives us insight into the strong emotions that attach to them.

This chapter relies on scholarship about children (see Ingalls 1998; Pratt 2007; Le Meur-Weissman 2009), lament (Alexiou 1974; Murnaghan 1999; Dué 2002; Tsagalis [2004] 2012; Perkell 2008), misdirection (Morrison 1992), performance context[1] and second-phase warfare in the *Iliad*,[2] and contributes to the growing body of work on the fates of conquered women and children in the ancient world, who are more marginalized in modern scholarship on ancient warfare than in Homer's portrayal thereof.[3] More importantly, examining second-phase warfare in the ancient world 'puts back in' the bodies of women and children.[4] The traumatic and deadly effects of war on women and children challenge our understanding of what defines a 'combatant's body' and show that the killing and violence of war spreads far beyond the battlefield.[5]

'Second-phase warfare' denotes the rape, murder and enslavement of the conquered during and after the ravaging of a settlement, ordered and carried out by the conquerors. I borrow the term from Kathy Gaca,[6] whose work on second-phase warfare in the ancient world is particularly applicable for my own study of ancient texts. Similar gender-based violence perpetrated upon victims by invading soldiers (and other military and even peacekeeping forces) is often discussed using different terminology in modern conflict studies, but the parallels are astounding.[7] In the ancient world, populace-ravaging warfare was a common technique of martial aggression, in which murder, rape and enslavement occur in patterns based on the sex and age of the victims. In general, fighting-age men are slaughtered in combat, the first phase of warfare. Then girls and fertile women are raped, both as a method of demoting them to enslavement and in order to compel the procreation of enslaved people (or those lowest in the social order) for the benefit of the conquerors. Young boys, if enslaved rather than killed, can be castrated. Mature women are often fatally gang-raped and older men are killed. In all cases, the bodies of the conquered are targeted. One strategic, political and economic endpoint of ancient warfare was the acquisition of young, child-bearing bodies as resources for the conquering community. Study of second-phase warfare ties together aspects of ancient warfare that are commoner in scholarship (strategic, political, economic and scientific research into state-sanctioned violence) with the innovative perspectives of this volume, namely a focus on the body and its centrality in war-making.[8]

In Homeric warfare, the bodies of women and children were never absent: they were highly valued as enslaved property and, through childbirth, as a means of reproducing that property.[9] Indeed, the *Iliad* is one of the earliest sources for the practice of second-phase warfare.[10] Although the epic focuses on the first phase of warfare, men killing men, and gives less space to the violence suffered by women and girls (Gaca 2008: 164), its many references to second-phase warfare, from brief mentions to similes and more extended descriptions, show that the conquest and exploitation of the bodies of women and children, sanctioned and ordered by military leadership,[11] was understood as the inevitable outcome of war throughout the world of Homeric epic.[12] The killing of fighting-age males was not the only goal of warfare, but was committed in part to achieve the enslavement and social debasement of female bodies,[13] bodies as integral to war-making as those of combatant soldiers.[14] The facts of second-phase warfare challenge the perception of 'soldierly' bodies: women and children suffer transformative bodily violence just as fighting-age men do.[15] Such bodies in the *Iliad* can be viewed as combatants in their own right, and their vulnerability to second-phase warfare would have resonated across the audience of Homeric poetry.

The dismissal of second-phase warfare as less integral to martial aggression than fighting between men is a modern rather than an ancient phenomenon. Euripides' *Trojan Women, Andromache* and *Hecuba*, for example, speak to the centrality of women's bodies and their grief in the Trojan myth.[16] The popularity of such retellings shows the ancient audience's investment in these aspects of the story, largely disregarded by scholars of war.[17] The consequences of ravaging warfare were wide-ranging, and people of all ages and genders who had experienced it, at various removes, would have populated the audience of Homeric epic. Women and men who had participated in war or experienced captivity could relate images of second-phase warfare in epic to their direct experience (Gaca 2008: 168); further, not only those who had personal experience, but also those whose parents and grandparents might have been soldiers or captives (Gaca 2016a: 1047). In particular, enslaved persons in the epic's audience (whether in the archaic or later periods) would be affected by hearing accounts of their trauma, and indeed, non-enslaved persons might listen, fearing this very trauma.[18] As Jose González writes on epic performance, '[c]hoices of subjects and their particular variants, thematic emphases and de-emphases, are carefully tuned to the breadth of audience reception' (2013: 17).[19]

I argue that the *Iliad* responds to the certainty and, in particular, the multiplicity of the epic audience's physical and emotional experience of second-phase warfare.[20] To make this point, I consider the three passages from the poem about Astyanax and his family, beginning with a celebrated passage from Book 6. Hector, as an experienced fighter, is familiar with second-phase warfare, as is shown when he outlines what will happen to Andromache if he does not fight to save Troy:[21]

ὅτε κέν τις Ἀχαιῶν χαλκοχιτώνων
δακρυόεσσαν ἄγηται, ἐλεύθερον ἦμαρ ἀπούρας·
καί κεν ἐν Ἄργει ἐοῦσα πρὸς ἄλλης ἱστὸν ὑφαίνοις,
καί κεν ὕδωρ φορέοις Μεσσηΐδος ἢ Ὑπερείης
πόλλ᾽ ἀεκαζομένη, κρατερὴ δ᾽ ἐπικείσετ᾽ ἀνάγκη·
καί ποτέ τις εἴπῃσιν ἰδὼν κατὰ δάκρυ χέουσαν·
'Ἕκτορος ἥδε γυνή, ὃς ἀριστεύεσκε μάχεσθαι
Τρώων ἱπποδάμων, ὅτε Ἴλιον ἀμφιμάχοντο.'
ὥς ποτέ τις ἐρέει· σοὶ δ᾽ αὖ νέον ἔσσεται ἄλγος
χήτεϊ τοιοῦδ᾽ ἀνδρὸς ἀμύνειν δούλιον ἦμαρ.
ἀλλά με τεθνηῶτα χυτὴ κατὰ γαῖα καλύπτοι,
πρίν γέ τι σῆς τε βοῆς σοῦ θ᾽ ἑλκηθμοῖο πυθέσθαι. (6.454-65)

when one of the Achaeans dressed in bronze leads you away in tears, having deprived you of your day of freedom. And in Argos you might weave at the loom for another woman, and you might bring water from Messeis or Hyperia, although very much unwilling, and strong force will be laid upon you. And someone, sometime, seeing you pouring down tears, might say: 'This one is the wife of Hector, who was the best at fighting among the horse-taming Trojans, when he was fighting around Ilium.' Thus someone, sometime will speak: and there will be a fresh pain for you from the lack of some such man to ward off your day of slavery. But may the heaped earth cover over me, dead, before I at all know of your cry and your carrying-off.

Andromache might weave, and she might carry water from Messeis or Hyperia. G. S. Kirk points out the difficulty of identifying these two places, concluding that they 'are generic and descriptive names that could be given to many springs in many different places, and were chosen for precisely that reason by the poet' (1990: ad. 6.456-7).[22] Hector uses the optative in his descriptions of Andromache's hypothetical tasks (Kirk 1990: ad. 6.455), but he uses the future indicative to say that she will have 'strong force laid upon' her, and be 'unwilling' (458): the circumstances for women might vary, but violent capture and rape are certainties.[23] At the same time, the alternatives, generality and verb tenses contribute to a portrayal of a range of consequences that might have resonated with different audience members during performance.[24]

Hector is certain that his wife, and the city, will be captured, remarking: 'There will be day at some point when holy Ilium is destroyed' (6.448, ἔσσεται ἦμαρ ὅτ᾽ ἄν ποτ᾽

ὀλώλῃ Ἴλιος ἱρή).[25] The word ἦμαρ, referring first to the day of Troy's destruction, then to Andromache's last 'day of freedom' and 'day of slavery', links these occurrences with one another.[26] But Hector does not mention what might happen to his son, Astyanax, who will also be brutalized by wartime violence. He is likely to be either enslaved or killed,[27] given the norms of second-phase warfare, with which his father is familiar. Hector suppresses these likely scenarios in favor of an opposite outcome. Just after picking up the child, famously frightening him with his helmet,[28] he prays to Zeus and the other gods for his son's future success in battle:

> Ζεῦ ἄλλοι τε θεοί, δότε δὴ καὶ τόνδε γενέσθαι
> παῖδ᾽ ἐμόν, ὡς καὶ ἐγώ περ, ἀριπρεπέα Τρώεσσιν,
> ὧδε βίην τ᾽ ἀγαθόν, καὶ Ἰλίου ἶφι ἀνάσσειν·
> καί ποτέ τις εἴποι 'πατρός γ᾽ ὅδε πολλὸν ἀμείνων'
> ἐκ πολέμου ἀνιόντα· φέροι δ᾽ ἔναρα βροτόεντα
> κτείνας δήϊον ἄνδρα, χαρείη δὲ φρένα μήτηρ. (6.476-81)
>
> Zeus and the other gods, indeed grant that this child of mine also, just like I myself, become distinguished among the Trojans, with just as good strength, and grant that he will rule Ilium by force: and sometime someone might say of him returning from war, 'This one is far better than his father.' And may he bear bloody spoils having killed a warlike man, and may his mother delight in her *phren*.

In Hector's vision, Astyanax has grown up and returns from victory in war to Troy and to Andromache. Hector is not present at his return, but his mother is delighted by his success. The vision of Astyanax-as-warrior evokes his future soldierly body: strong, forceful, committing violence against the bodies of his enemies (6.478, βίην, ἶφι ἀνάσσειν; 6.480, ἔναρα βροτόεντα; 6.481, κτείνας δήϊον ἄνδρα).[29]

This picture is at odds with Hector's previous idea that Troy will certainly fall, and Andromache will be captured. It is also strikingly discordant with the mythic tradition, in which Astyanax suffers a violent death during the fall of Troy. This outcome is not mentioned directly in the *Iliad*, but it is found in other literary sources, including the Epic Cycle,[30] and visual depictions, common from the archaic period on (Richardson 1993: ad. 24.734-9).[31] Astyanax's fate is confirmed by the near-unanimous testimony of the later literary tradition, from Euripides to Tryphiodorus.[32] Much of the epic audience would have been familiar with this tradition, which, indeed, the *Iliad* includes throughout.[33]

It is puzzling, then, to see Hector hypothesize a scenario so at odds with the myth, especially given his established familiarity with the practices of second-phase warfare. Scholars have read the scene many ways: as evoking Hector's denial (Briggs 2003: 28; Schein 1984: 177; Pratt 2007: 30), as producing tragic irony (Le Meur-Weissman 2009: 33–4), as showing Hector's fixation on *kleos* (Murnaghan 1999: 212–4; also, Nappi 2015: 43–4) and as creating suspense in the epic (Morrison 1992: 70). Most, if not all, of these ideas are likely at play, but the inconsistency remains, and invites reflection on Astyanax's future throughout the epic.[34]

Hector's prayer for Astyanax is also at odds with Andromache's idea of the consequences of Hector's death. She asks him to stay in the city 'so that you do not make your child an orphan and your wife a widow' (6.432, μὴ παῖδ' ὀρφανικὸν θήῃς χήρην τε γυναῖκα).[35] Hector's death will indeed orphan Astyanax and make Andromache a widow, but it will also enslave them, unless only Andromache is enslaved while Astyanax is killed. To Andromache's mind, it is perhaps possible that Hector could die, but the Trojans could still win the war, leaving her and Astyanax as widow and orphan. Hector's fate, however, is widely understood as representing the fate of Troy: his death will bring on the fall of the city, as Priam predicts (22.38-76), Andromache herself predicts (24.725-39), and Zeus implies (15.68-71). The audience would know, since the Trojan War was to end in a year's time, that even if Troy did not fall immediately at Hector's death, Astyanax's orphanhood would not last long if it were to occur.

Andromache's omission of enslavement and death is especially notable because her family has experienced these consequences of second-phase warfare. In an earlier speech to Hector, she describes how Achilles killed her father and brothers and captured her mother (6.413-26).[36] Kirk notes that she details her family history – with which Hector would already be familiar – as a rhetorical device to emphasize the corporeal consequences for her at the fall of Troy (1990: ad. 6.414-28). Andromache knows better, because of her family's tragedy, than to think Astyanax will merely be orphaned in the event of Hector's death.[37]

The multiple depictions of second-phase warfare in this conversation would have been especially affecting to enslaved members of the audience because of the presence of Astyanax's wet nurse.[38] Her body, and her role of holding and comforting the child during the interaction between his parents, comes into focus when Andromache meets Hector: 'with her went a maid holding the tender child to her bosom, just an infant' (6.399-400, ἅμα δ' ἀμφίπολος κίεν αὐτῇ / παῖδ' ἐπὶ κόλπῳ ἔχουσ' ἀταλάφρονα, νήπιον αὔτως), and when Astyanax cowers away from his father's helmet: 'The child leaned back, crying, into the bosom of his well-belted nurse' (6.467-8, ἂψ δ' ὁ πάϊς πρὸς κόλπον ἐϋζώνοιο τιθήνης / ἐκλίνθη ἰάχων).[39]

Astyanax's babyhood is highlighted with the phrase 'just an infant' (6.400, νήπιον αὔτως),[40] and the word αὔτως is often translated as emphasizing the tininess of his body and 'underlining the helplessness of the little child'.[41] Just a few lines later, Andromache pleads with Hector, using the diminutive substantive νηπίαχον (6.408, 'infant child') to refer to their son, again emphasizing his vulnerability.[42] The focus on Astyanax's babyhood cultivated by both Andromache and the narrator sharply contrasts with Hector's vision of Astyanax as a future soldier, discussed above.

The presence of the nurse cues us to consider the reaction of enslaved listeners in the audience, especially those whose bodies have been appropriated to provide nourishment for their captors' offspring. As Daphne Briggs points out, Astyanax's nurse is most likely a war captive who has experienced great trauma and loss herself. 'Listening to the talk of doom between Astyanax's parents, we can only imagine what his nurse must have been reliving of traumatic loss and displacement earlier in her life: at that moment she was perhaps cradling the baby as much for her own comfort as for his' (Briggs 2003: 28).

Enslaved women in the audience would have listened to the prospect of Andromache's capture from the perspective of those whose bodies had already been violated, damaged, used and transformed by the ravaging process of second-phase warfare;[43] and indeed, they would likely be experiencing this narrative and reacting physically through the body, the site of stored trauma (see Rothschild 2000; van der Kolk 2014). Free women might have listened from a perspective more akin to Andromache's, knowing that their status could change as a result of capture in war at any time.

Despite her knowledge of second-phase warfare, whose spectre looms over both the bodies and the conversation in Book 6, Andromache expands on the vision of herself as a widow and Astyanax as an orphan in Book 22. She has just learned that Hector has been killed, and she knows the likely consequences of his death for her and her son. Regardless, she gives an account of Astyanax's future in which she excludes the inevitability of bodily injury or death:[44]

ἤν περ γὰρ πόλεμόν γε φύγῃ πολύδακρυν Ἀχαιῶν,
αἰεί τοι τούτῳ γε πόνος καὶ κήδε᾽ ὀπίσσω
ἔσσοντ᾽· ἄλλοι γάρ οἱ ἀπουρίσσουσιν ἀρούρας.
ἦμαρ δ᾽ ὀρφανικὸν παναφήλικα παῖδα τίθησι·
πάντα δ᾽ ὑπομνήμυκε, δεδάκρυνται δὲ παρειαί,
δευόμενος δέ τ᾽ ἄνεισι πάϊς ἐς πατρὸς ἑταίρους,
ἄλλον μὲν χλαίνης ἐρύων, ἄλλον δὲ χιτῶνος·
τῶν δ᾽ ἐλεησάντων κοτύλην τις τυτθὸν ἐπέσχε,
χείλεα μέν τ᾽ ἐδίην᾽, ὑπερῴην δ᾽ οὐκ ἐδίηνε.
τὸν δὲ καὶ ἀμφιθαλὴς ἐκ δαιτύος ἐστυφέλιξε,
χερσὶν πεπληγὼς καὶ ὀνειδείοισιν ἐνίσσων·
'ἔρρ᾽ οὕτως· οὐ σός γε πατὴρ μεταδαίνυται ἡμῖν.'
δακρυόεις δέ τ᾽ ἄνεισι πάϊς ἐς μητέρα χήρην. (22.487-99)

For even if he flees the very tearful war with the Achaeans, always for him there will be toil and grief in the future: others will take away his lands. The day of orphanhood cuts a child off from his peers: he lowers his head entirely, his cheeks are covered in tears, in need the child goes up to the companions of his father, pulling at one by his cloak, another by his chiton: and of the ones pitying him, one holds out a small cup, he moistens his lips, but he does not moisten his palate. And one with both parents alive forces him from the feast, having struck him with his hands and attacking him with reproaches: 'Get out now: your father does not feast with us.' The child in tears goes to his mother, a widow.

The irreconcilable futures of Astyanax, as portrayed by Hector earlier and Andromache here, are linked by verbal parallels: Hector prayed that Astyanax would return (6.480, ἀνιόντα) to his mother from war; Andromache imagines his returning (22.499, ἄνεισι) as an orphan rejected from a shared meal. Astyanax's 'day of orphanhood' (22.490, ἦμαρ δ᾽ ὀρφανικὸν) recalls Hector's repeated ἦμαρ to describe the day of Troy's destruction and Andromache's enslavement, and Andromache's phrase 'deprive him of his lands' (22.489, ἀπουρίσσουσιν ἀρούρας) echoes Hector's ἀπούρας (6.455, 'deprive').[45] Such parallels draw attention to the inconsistency of the parents' visions for their son.

Andromache goes on to describe Astyanax's heretofore happy childhood, as a contrast to the orphanhood she predicts after Hector's death:

> Ἀστυάναξ, ὃς πρὶν μὲν ἑοῦ ἐπὶ γούνασι πατρὸς
> μυελὸν οἶον ἔδεσκε καὶ οἰῶν πίονα δημόν·
> αὐτὰρ ὅθ' ὕπνος ἕλοι, παύσαιτό τε νηπιαχεύων,
> εὕδεσκ' ἐν λέκτροισιν, ἐν ἀγκαλίδεσσι τιθήνης,
> εὐνῇ ἔνι μαλακῇ, θαλέων ἐμπλησάμενος κῆρ. (22.500-504)

> Astyanax, who before, on his father's knees, ate only marrow and the rich fat of sheep: then when sleep took him, he stopped playing like a child, he slept in his bed, in the arms of his nurse, in soft bedding, his heart filled with happy thoughts.[46]

Astyanax's childishness is emphasized with the rare word νηπιαχεύων (de Jong 2012: ad. 22.502). Andromache also describes him again as νήπιον αὔτως just before these lines (22.484); both uses recall his vulnerability in Book 6 (6.400, νήπιον αὔτως; 6.408, νηπίαχον). Andromache again depicts Astyanax as small and protected in the arms of his nurse. His little body and bodily functions – eating and sleeping well – as well as his unmarred happiness, highlight the pathos of his true fate for anyone who knows his story.[47] But Andromache stops describing her son and his fate at this point, and moves to a description of her husband's dead, naked and mangled body (22.505-10). She considers the brutal end for his body, but not her son's.

Comparison of the scenes in Book 6 and 22, prompted by verbal and thematic parallels, inevitably brings up their contradictions. Will Andromache be a war captive, dragged to Greece as in Book 6, or a widow living in disgrace in Hector's 'halls', as in Book 22? Will Astyanax be a warrior prince or an orphan, and does he fit in at all to the scenario of his mother's capture, perhaps as a captive himself? Most of all, why do neither of his parents, in either scene, account for the possibility, which they surely understand, that he might be captured or killed? Astyanax's grandfather Priam mentions this outcome of second-phase warfare in his description of Troy's hypothetical capture earlier in the same Book (22), when he urges Hector to save the Trojans by coming inside.[48] He says that if Hector does not, 'small children will be thrown to the earth in the din of battle' (22.63-4, καὶ νήπια τέκνα / βαλλόμενα προτὶ γαίῃ ἐν αἰνῇ δηϊοτῆτι), while the women in his family will be taken captive (22.62-5). Priam's line about the death of infants (22.63, νήπια τέκνα) hints at the fatal consequences of a Trojan defeat for Astyanax, identified as a νήπιος throughout the poem.[49]

In the moment the king describes, the bodies of both women and children suffer the violence of combat, and his impassioned plea evokes the emotional and physical pain of the young and old who do not fight on the front lines but wait behind the city gates for destruction. Priam mentions the consequences of war for children in general. But the image of a small body meeting death is perhaps too raw an outcome, and too harsh a defeat, for Astyanax's parents to articulate or face yet.[50] As Rabinowitz observes of

Astyanax's death in Euripides' *Trojan Women*, 'The death of Astyanax *embodies* the horror of war and the cruelty of the Greeks' (2014: 199, italics mine).[51]

But Astyanax's fate comes up once more, in Andromache's lament for Hector in Book 24. Here, his mother's speech is consistent with the practices of second-phase warfare:[52]

> ἆνερ, ἀπ᾽ αἰῶνος νέος ὤλεο, κὰδ δέ με χήρην
> λείπεις ἐν μεγάροισι· πάϊς δ᾽ ἔτι νήπιος αὔτως,
> ὃν τέκομεν σύ τ᾽ ἐγώ τε δυσάμμοροι, οὐδέ μιν οἴω
> ἥβην ἵξεσθαι· πρὶν γὰρ πόλις ἥδε κατ᾽ ἄκρης
> πέρσεται· ἦ γὰρ ὄλωλας ἐπίσκοπος, ὅς τέ μιν αὐτὴν
> ῥύσκευ, ἔχες δ᾽ ἀλόχους κεδνὰς καὶ νήπια τέκνα·
> αἳ δή τοι τάχα νηυσὶν ὀχήσονται γλαφυρῇσι,
> καὶ μὲν ἐγὼ μετὰ τῇσι· σὺ δ᾽ αὖ, τέκος, ἢ ἐμοὶ αὐτῇ
> ἕψεαι, ἔνθα κεν ἔργα ἀεικέα ἐργάζοιο,
> ἀθλεύων πρὸ ἄνακτος ἀμειλίχου, ἤ τις Ἀχαιῶν
> ῥίψει χειρὸς ἑλὼν ἀπὸ πύργου, λυγρὸν ὄλεθρον,
> χωόμενος, ᾧ δή που ἀδελφεὸν ἔκτανεν Ἕκτωρ
> ἢ πατέρ᾽, ἠὲ καὶ υἱόν, ἐπεὶ μάλα πολλοὶ Ἀχαιῶν
> Ἕκτορος ἐν παλάμῃσιν ὀδὰξ ἕλον ἄσπετον οὖδας.
> οὐ γὰρ μείλιχος ἔσκε πατὴρ τεὸς ἐν δαῒ λυγρῇ. (24.725-39)

> Husband, you have perished from life young, and you leave me a widow in your halls: and your child, just an infant, whom you and I, unfortunate, produced. Nor do I think he will reach youth: for first this city will be destroyed entirely: for indeed you, its guard, have died, who protected it, and you held fast the dutiful wives and infant children: who indeed soon will be carried in the hollow ships, and I myself with them: you in turn, child, either you will follow me, to where you will labour at shameful tasks, toiling for an ungentle lord, or one of the Achaeans will throw you taking you by the hand from the wall, a mournful death, angered, whose brother, perhaps, Hector killed, or father, or even son, since very many of the Achaeans bit the great earth with their teeth at the hands of Hector. For your father was not gentle in mournful war.

Here perhaps Andromache has realized the visceral consequences of Hector's death for both her body and her son's: slavery and death, rather than poverty and bereavement.[53] Either he will accompany her onto the ships of the Trojans, or he will be cast over the wall of the city by a Greek warrior who might have suffered from Hector's prowess in war.[54] The second option, along with Andromache's νήπια τέκνα (24.730, also at 22.484, 'infant children'), echoes Priam's description of infants (22.63, νήπια τέκνα) violently thrown over the city wall. Andromache's articulation of this fate for Astyanax shows that she has now linked her son's bodily fate to that of other children in second-phase warfare.[55]

The formula Andromache uses, ἀλόχους κεδνὰς καὶ νήπια τέκνα (24.730), 'usually occurs in the context of an absence of a husband/father and the dangers faced by his wife and children' (Brügger 2017: ad. 24.729b-30),[56] and is thus implicitly related to the horrors of second-phase warfare. The phrase νήπια τέκνα in particular is an associative unit indicating three things: a man is absent from the home and in mortal danger; his wife and children are left at home; and their fate hangs in the balance (Edmunds 1990: 30–3). The consequences of an absent man in a wartime context in the ancient world (and the entirety of the *Iliad* is a wartime context) is the murder, rape and enslavement of his womenfolk and children. In every passage in which Astyanax is referred to as νήπιος, his inevitable susceptibility to the murderous Achaeans looms, even when no narrator or character makes this explicit.[57]

As Andromache concludes her lament in Book 24, she conflates the future of her family with that of her enemy: both the 'lord' to whom she and Astyanax will be enslaved, and Hector, are 'not gentle' (24.734, ἀμειλίχου; 739, οὐ . . . μείλιχος); the 'mournful' (739, λυγρῇ) battle Hector fights echoes Astyanax's 'mournful death' (735, λυγρὸν ὄλεθρον);[58] and a Greek whose family member was killed by Hector will then kill his son. Andromache presents her and her child's fate as the direct consequence of Hector's combat (MacLeod 1982: ad. 24.739; Murnaghan 1999: 216), and she presents these experiences as common to both sides of the conflict. This consideration of the trauma experienced on both sides of war would affect a broad listening audience, and indeed, the issue of audience reaction is foregrounded by the presence and mournful response of Trojan bystanders, detailed directly after Andromache's speech (24.746).[59]

In this final lament, Andromache gives precisely the two alternatives for her son's future – slavery and death – that fit with second-phase warfare. But she still gives options for her son's fate,[60] and one is unrealistic. Greek martial forces in general captured young girls rather than boys (Gaca 2012; 2014; 2015: 286), and in Astyanax's case, it would be much more sensible for the Greeks to kill rather than enslave him, as he is a direct successor to Troy's best warrior and crown prince.[61] Indeed, murder during the city's sack is precisely what other sources specify as Astyanax's fate, beginning already in the Epic Cycle (see n. 30). Even in this final scene in the *Iliad*, however, Astyanax's death is not a certainty, and his remaining parent espouses two incompatible visions of his fate.

The *Iliad*'s inconsistent visions of Astyanax's destiny draw attention to each other and give the listener insight into the parents' turbulent emotions of terror, grief and final resignation as they contemplate their child's fate.[62] At the same time, these scenes and their verbal parallels foreground the strong possibility of corporeal violence inflicted on Astyanax's young body.[63] More broadly, they focus on the bodily suffering women and children face in second-phase warfare and the extremely high value placed on their physical bodies. The fear and violence of second-phase warfare comes through both in the *Iliad*'s explicit descriptions of it and, just as strikingly, in some characters' hesitation to describe it. Both the graphic descriptions and the visceral fear of this violence show how corporeally women and children experience war when violence moves from the fighting soldiers to the women and children they have died attempting to defend.

Further, the constellation of passages surrounding Astyanax's fate works with the other depictions of second-phase warfare throughout the poem to provide multiple images of the suffering and violence that the conquered experience in war, any of which might resonate with a given audience member. Astyanax's fate is represented by several options, just as there would have been for conquered children in reality. As Charles Segal writes of Andromache in particular, '[s]he is the bearer of the suffering of all the women in the war, and perhaps of all women in all war' (1971b: 55); similarly, Nicholas Richardson remarks, '[s]he and her son are archetypes of all widows and all orphans' (1993: ad. 22.487-99).[64] Although the Trojan War story is a myth, these bodies are paradigmatic of the suffering of all women and children in the atrocities of combat, including those who might have constituted the epic's audience.

This new reading of some of the *Iliad*'s best-known passages reveals how the epic refracts the realities of second-phase warfare. Conquered people who had suffered capture, rape, enslavement and loss would have seen themselves in Hector, Andromache and Astyanax. Mothers and fathers might have responded to Astyanax's parents' persistent avoidance of harsh reality, and the epic even grants space to the suffering of the conquerors, in Andromache's comparison of Hector with the man who kills Astyanax out of grief for his own male kin. The variations in these passages increase the poem's force, effectiveness and resonance for listeners: all survivors of war in the audience could have seen themselves and their families in the suffering of Hector and his family (Pratt 2007: 40). In the end, these characters are reduced to their bodies – killed, enslaved, violated, even unburied[65] – in a way that resonates broadly in the personal memories of the epic audience, for all of whom war was a part of life at varying removes and levels of bodily experience. Indeed, these physical sufferings are just as serious, prominent and prevalent as those endured by fighting soldiers. In contrast to Hector's words at 6.492, his family shows that war is not merely the concern of men;[66] nor is our consideration of the 'body of the combatant' limited to the bodies of fighting soldiers. Such an outlook qualifies an androcentric view of epic and of warfare in general, and takes into account that different members of a population, in the ancient as in the modern world, have varied experiences of and considerably different stakes in war.

Acknowledgements

This chapter has benefitted from feedback received at a 2019 conference held at King's College London, *Warfare in Antiquity: Perceptions, Realities and Reception in the 21st Century*, where I presented an earlier version of it. I would also like to acknowledge my debt to Patricia Rosenmeyer, Sharon James, Brian McPhee, Annemarie Ambühl, the anonymous reviewer and the editor of this volume, Hannah-Marie Chidwick, for their helpful comments, suggestions and support. Any remaining errors are mine alone.

Notes

1. Especially González (2013); see also Dalby (1995); Bakker and Kahane (1997); Scodel (1998a, 2004, 2009); Burgess (2004); Dowden (2004); Minchin (2014); Ready and Tsagalis (2018).

2. See n. 6 below.
3. Louise Pratt's discussion of Astyanax is a starting point for this study. She notes that Astyanax 'represents so effectively children's experience in war . . . . moreover, as his mother vividly imagines several different ends for the boy . . . she makes her son represent a range of childhood experiences in war' (2007: 27 n. 9). I argue that this range of experiences would resonate particularly with those receiving epic poetry who had experienced second-phase warfare.
4. Here I modify Tidy's phrase, used in the Introduction as a guiding force for this volume: 'Correcting the erasure of people in the theorization and study of war (previously understood and analysed purely as a domain of state institutions) involves putting bodies "back in"' (2019: 221).
5. For analysis and problematization of the term 'combat' and its conventional relationship to masculinity and the figure of the heroic soldier, see Millar and Tidy (2017).
6. See Gaca's work for a full introduction to the practice of martial ravaging and rape, from whom I take the term 'second-phase warfare'. She includes the *Iliad* and connects the practice through to the modern world (2008: especially 148–51; 2010; 2011; 2012; 2014; 2015; 2016a: especially 1041–8; 2016b; 2018a–b; 2021). See also Pritchett (1974: 238–42); Rihll (1993); de Sensi Sestito (1999); Rabinowitz (2014); Deacy and McHardy (2015); Solvang (2015); Fabre-Serris and Keith (2015). On second-phase warfare in Roman antiquity and literature, see Rollinger (2018), Pyy (2020).
7. I do not have adequate space to address global sexual and gender-based violence in modern conflicts and conflict studies here, but for critical considerations of rape as a weapon of war in modern conflicts, see, e.g., Enloe (2004: 99–121); Baaz Eriksson and Stern (2013); Holzner (2011). Enloe discusses women who suffer rape as a weapon of war as more than victims (118); for the complexities of the latter point regarding female combatants, see Holzner (2011).
8. For this scholarly trend and the innovations at which this volume aims, see the Introduction. For the body in Homer, see Snell (1953: 1–23); Vernant (1991); Segal (1971b); Renehan (1979); Pelliccia (1995); Clarke (1999); Bolens (2000: 19–99); Bain (2007); Holmes (2010: 41–83); Gavrylenko (2012).
9. The capture and enslavement of these bodies could be a primary motive of ancient warfare (Gaca 2008, 2016a). Perhaps Agamemnon's wish at 6.57-60 to kill even the unborn boys in their Trojan mothers' wombs indicates an excessive desire for revenge that supersedes this commercial interest in war.
10. The capture and possession of Briseis and Chryseis provide the impetus for the poem's narrative. See Gaca for Briseis' lament and second-phase warfare (2016a: *passim*, especially 1052–3); see also Murnaghan (1999: 206–9); Dué (2002); Feichtinger (2018). Gaca lists and discusses the many other captive women in Homeric epic (2015: 284–6); see also Wickert-Micknat (1983: 1–50).
11. For martial rape and ravaging as a regular order from superior officers to their soldiers in the ancient Mediterranean, see Gaca (2008: 149; 2015: 306; 2021: 47–8), citing also *Il.* 6.55-60; Deut. 20:10-20; Paus. 10.22.2-4; App. Pun. 553; Polyb. 10.15. For martial rape as an order in the *Iliad*, see 2.354-9, with Gaca (2021: 56); 3.301, with Gaca (2016a: 1043); Graziosi and Haubold (2010: 29). For consideration of this top-down order of sexual violence in modern warfare, along with unsanctioned and uncontrolled sexual violence in war, see Baaz Eriksson and Stern (2013: 64–87); Enloe also considers rape as sanctioned by military leadership (2000: 111, 120).
12. The two phases of warfare are explored in Book 9 (Gaca 2016a: 1043), when Meleager's wife Cleopatra begs him to save Calydon, or else men will be killed, the city destroyed and women

and children enslaved (9.593-4). For an analysis of Cleopatra's rhetoric, see Nappi (2015: 44–6). Other passages in Homer that refer to second-phase warfare, but are not discussed in this chapter, include *Il.* 8.164-6, 16.7-11 (argued for portraying second-phase warfare in Gaca 2008; see her references, as well as Porter 2010, for a refutation), 19.291-9; *Od.* 9.39-42 (Gaca 2008: 159 n. 41). See also Brügger (2017: ad. 24.731-5). For the sexual exploitation of captured women in Homeric epic, see *Il.* 8.287-91 and 9.663-8; for the labour of captive women, see *Il.* 6.450-61, 9.658-61, 11.624-41, 14.5-8, 23.262-3, 23.704-5, 24.643-8, and *Od.* 20.105-10 (Gaca 2008: 149 n. 11). On these topics, see also Ndoye (2010: 250–78); Wickert-Micknat (1983: 1–50, 150–209).

13. Per Scodel: 'Women are the enemies' most precious property' (1998b: 142). See also Ameis and Hentze (1908: ad. 6.426); Stoevesandt (2016: ad. 6.426).

14. In the *Iliad*, Achilles admits as much: he was 'fighting men for the sake of their wives' (9.327, ἀνδράσι μαρνάμενος ὀάρων ἕνεκα σφετεράων). Although the cause of the Trojan War is the rape of a single woman (Helen), Achilles' words, and his use of the plural 'wives' (ὀάρων), betray the purpose and meaning of war in the greater cultural context of Homeric poetry: the capture of fertile bodies to be enslaved by the conquering society. For more on this passage, see Gaca (2011: 81). Notably, the cause of the Trojan War, according to myth, is a woman's body – Helen's – following a contest judging female corporeal beauty. Thucydides' assertion that the Trojan War was motivated by acquisition (1.8) perhaps draws on hints throughout the *Iliad* that war in ancient Greece was motivated by capture; see particularly Rihll (1993). That war was about women is emphasized in Herodotus' prologue, in which their retaliatory capture leads to extensive conflict between Persia and Greece (1.1-6).

15. Although distinguishing women as noncombatants, Rabinowitz says, 'the violence and greed unleashed in war produces violence beyond the combat zone', and that women and children 'suffer their own form of combat trauma as a result of men's licensed warrior behavior . . . They are not simply collateral damage; although noncombatants they are part of the war effort' (2014: 201). For a consideration of women's roles in war in the *Iliad* specifically, see Nappi (2015).

16. Munteanu argues that Euripides' *Trojan Women* glorifies the suffering women and children rather than the epic hero (2010–11). For more on rape of captives in *Trojan Women*, see Scodel (1998b); Rabinowitz (2011; 2013; 2017).

17. Relatively little scholarship seriously engages with second-phase warfare in antiquity, but see n. 6. To focus only on fighting-age males is to eliminate discussion of warfare as it affects about 75 per cent of the population, including all women and girls, according to ancient demographic evidence (Gaca 2016a: 1046). As Rabinowitz puts it: 'War was about women. It was also a war on women' (2014: 185–6). See also Nappi (2015: 34–5).

18. For the depiction of enslaved people and violence against them, particularly enslaved women, in Homer, see Wilson (2021). She notes that Homeric epic was likely designed to be heard by those who were not free (35).

19. As Gaca notes, those living in the ancient Mediterranean were familiar with martial rape and required little explanation for the practice: it was a fact of life. See Gaca (2021: 47), citing Diod. Sic. 19.8.4-5; Polyb. 9.39.2, 16.32.4. For a consideration of how the depiction of women and war in tragedy would have affected an Athenian audience, see Rabinowitz (2014).

20. Easterling's notion of 'heroic vagueness' is useful: the heroic setting of Greek tragedy, vague on legal and political specifics, encourages varied audience members to enjoy and identify with the drama (1997). Gibert uses the concept to contend that members of widely different social classes in the audience of Euripides' *Ion* would have been compelled, regardless, to identify with Ion. As he notes, 'Greek poetry had always encouraged audiences to appropriate and

identify with mythical heroes and heroines in particular' (2019: 36). The multiple views and ambiguity of Astyanax's futures did the same for the epic audience.

21. Text of the *Iliad* is Monro and Allen's ([1902] 1963a; [1902] 1963b). Translations are my own. See also Hector to Diomedes: 'You will not lead away our women in your ships' (8.165-6, οὐδὲ γυναῖκας / ἄξεις ἐν νήεσσι), with Gaca (2008: 159, 159 n. 41).

22. Leaf attempts to locate the fountains ([1886] 1900: ad. 6.457); see also Ameis and Hentze (1908: ad. 6.456, 6.457); Graziosi and Haubold (2010: ad. 6.457).

23. Graziosi and Haubold note that πόλλ' ἀεκαζομένη connotes forced sex in addition to violent abduction (2010: ad. 6.454-83, 6.458, with bibliography). For the verb tenses in this passage, see Leaf ([1886] 1900: ad. 6.453); Ameis and Hentze (1908: ad. 6.456, 6.457); Graziosi and Haubold (2010: ad. 6.454-63, 6.462). On Hector's speech, see Murnaghan (1999: 208, 213); Leaf ([1886] 1900: ad. 6.446); Stoevesandt (2016: ad. 6.459-63). Stoevesandt notes, with bibliography, that Andromache's status here is so problematic because she will be subordinate to another woman (2016: ad. 6.546-8, see also Ameis and Hentze 1908: ad. 6.456; Graziosi and Haubold 2010: ad. 6.457); for Andromache's predicted subordination to a male enslaver, see 24.734.

24. Kirk says that Hector omits potential sexual violence against Andromache (1990: ad. 6.455, 20-1); see also Stoevesandt (2016: ad. 6.456-8), Graziosi and Haubold (2010: 46, ad. 6.454-63). But the word ἑλκηθμοῖο in particular can denote sexual violence (Leaf [1886] 1900: ad. 6.465, with comparanda; Graziosi and Haubold 2010: ad. 6.454-63, 6.465; Deacy and McHardy 2015: 264), and Gaca makes a strong case for the connotations of sexual violence in this passage, as the scholia confirm (T ad. *Il.* 6.454) (2015: 287–8). Later, Priam uses ἕλκω of second-phase warfare to describe the capture of his daughters and daughters-in-law (22.62-5, de Jong 2012: ad. 22.62; cf. Leaf [1888] 1902: ad. 22.62), recalling this passage in Book 6 (Graziosi and Haubold 2010: ad. 6.465; Stoevesandt 2016: ad. 6.465).

25. On these lines, especially their dissonance with Hector's later behaviour: Segal (1971a: 39); Kirk (1990: ad. 6.449); Graziosi and Haubold (2010: ad. 6.449); Stoevesandt (2016: ad. 6.449).

26. Achilles also uses the phrase ἐλεύθερον ἦμαρ ἀπούρας twice to describe the 'day of freedom' that the Greeks will take away from Trojan women (16.831, 20.193). On the significance of ἦμαρ and ἐλεύθερος, see Leaf ([1886] 1900: ad. 6.455); Kirk (1990: ad. 6.455); Graziosi and Haubold (2010: ad. 6.455, 6.463, with bibliography); de Jong (2012: ad. 22.212); Stoevesandt (2016: ad. 6.454-5, with bibliography).

27. Cf. Priam's mention of Achilles either killing his sons or selling them as slaves (22.45).

28. On Astyanax in this scene, see Herter (1973); Ingalls (1998: 22–3); Graziosi and Haubold (2010: 47, ad. 6.466-81, 6.466, 6.467, 6.492-3); Stoevesandt (2016: ad. 6.466-96, with bibliography). Stoevesandt also notes a proleptic hint at Astyanax's future here: Hector takes him from the bosom of his nurse, just as Neoptolemus does in order to kill him in the *Little Iliad* (fr. 21 Bernabé [frr. 18, 29, 30 West] 2016: ad. 6. 466-73).

29. Hector describes 'both power and physical force: Astyanax, like his father, is expected to be the strongest warrior and, eventually, the ruler of Troy' (Graziosi and Haubold 2010: ad. 6.478).

30. For Astyanax's death by being thrown over the walls of Troy by a Greek warrior, see *Little Iliad* fr. 21 Bernabé (frr. 18, 29, 30 West); *Sack of Ilion argumentum* 20-1, fr. 5 Bernabé (*argumentum* 4, fr. 3 West).

31. Notably, Deacy and McHardy link one visual depiction of his death to second-phase warfare (2015: 261). For Astyanax's depiction in ancient visual sources, especially vase-painting, see Laurens (1985); Gantz (1993: 652–7); Morris (1995); Anderson (1997: 192–9, 208–29, 231–4); Bouvier (2000–1).

32. For overviews and relevant visual and literary sources, see Stoll (1884–90: 660); Wagner (1896: 1866–7); Leaf ([1888] 1902: ad. 24.735); Frazer (1921: 239 n. 2); Robert ([1923] 1967: 1259–63); MacLeod (1982: ad. 24.734-8, 24.736-8); Touchefeu (1984: 929–37); Gantz (1993: 650–1); Anderson (1997: 53–9, 192–9, 208–29, 231–4); Burgess (2010); Brügger (2017: ad. 24.734b-8).

33. Although Astyanax's death does not occur in the *Iliad*, Burgess (2010) considers hypertextual links throughout the poem that hint at his eventual fate. On hints about his fate and other sources for it, see nn. 28, 30, 31, 32, 34, 37, 41, 49, 51, 54, 63.

34. The location of the meeting between Hector and Andromache on the wall (6.373, πύργῳ) hints at Astyanax's death, as he is thrown from a tower in the mythic tradition (Anderson 1997: 58; Burgess 2010: 216–19; Graziosi and Haubold 2010: ad. 6.481; Stoevesandt 2016: ad. 6.373). See *Little Iliad*, fr. 21 Bernabé [frr. 18, 29, 30 West], *Sack of Ilion, argumentum* 20-1, fr. 5 Bernabé [*argumentum* 4, fr. 3 West]. For more on Hector's prayer, see Ameis and Hentze 1908: ad. 6.478; Edmunds (1990: 32); Graziosi and Haubold (2010: ad. 6.466-81, 6.475-81, with bibliography); Stoevesandt (2016: ad. 6.475-81, 6.480-1, with bibliography). Ameis and Hentze note that after this prayer, Hector hands Astyanax not back to his nurse, but rather to his mother, in order to comfort her with the presence of the city's future protector (1908: ad. 6.482; see also Graziosi and Haubold 2010: ad. 6.482-93). This analysis does not take into account the unlikelihood of Astyanax surviving this ordeal.

35. Andromache uses the word χήρη, 'widow', extremely frequently compared to the rest of early epic (Graziosi and Haubold 2010: ad. 6.408; see also Brügger 2017: ad. 24.725); ὀρφανικόν is also notable for its rarity (Graziosi and Haubold 2010: ad. 6.432).

36. For Andromache's narration about her family, see Stoevesandt (2016: ad. 6.413-28), who notes the dire consequences for her after Hector's death: she will have no family to return to, as would be traditional (see also Wickert-Micknat 1983: 63; Graziosi and Haubold 2010: ad. 6.413, 6.425-8, 6.432; Brügger 2017: ad. 24.725). But of course, it is not realistic for her to be returned to her family: enslavement will be her fate if the city falls. As Graziosi and Haubold note, her true fate will be worse than her mother's (2010: ad. 6.414-28, 6.425-8), although in later mythic tradition her future brightens after Neoptolemus' death with marriage to Helenus (cf. Eur. *Andr.*; Virg. *Aen.* 3).

37. Le Meur-Weissman notes that Andromache's tears in this scene evoke the fatal future of the family and Astyanax in particular, but neither Hector nor Andromache mention his death outright (2009: 32). The scholia (bT ad. *Il.* 17.401-2) consider the overall theme of Andromache's ignorance of the truth as functioning to evoke sympathy (Richardson 1993: ad. 22.437-515).

38. That Andromache is accompanied by only one enslaved servant in this scene is unusual, and makes the encounter between her and her husband more intimate (Kurz 1966: 126; Stoevesandt 2016: ad. 6.372).

39. The nurse's presence is first noted at 6.372 and 6.389, for a total of four references in this scene.

40. The word νήπιος is difficult and has led to richly varied translation and scholarship. I translate with the widely used 'infant', 'child' (LSJ s.v. νήπιος), but see especially Edmunds (1990) for a complication of this translation. This debate lies beyond the scope of this chapter, but see Harris (1985); Edmunds (1990); Bakker (1997: 35–6); Ingalls (1998: 17–18, 32–4); Ready (2011: 181); Briand (2011); Allen-Hornblower (2016: 52 n. 143, with bibliography); Maravela (2019).

41. See Stoevesandt (2016: ad. 6.400) for further translations of and bibliography on this portrayal of Astyanax's vulnerability. See also Leaf ([1886] 1900: ad. 6.400); Ameis and Hentze

(1908: ad. 6.400; 1906: ad. 22.484, 24.726); Graziosi and Haubold (2010: ad. 6.400-3). The comparison of Astyanax to a star at 6.401 might contribute to a sense of foreboding about the child (Edmunds 1990: 31; Graziosi and Haubold 2010: ad. 6.400, with bibliography).

42. See Stoevesandt (2016: ad. 6.408); Graziosi and Haubold (2010: ad. 6.408). Andromache's vulnerability is accentuated in a similarly tragic context with a related word when she orders a bath to be prepared for Hector in Book 22, not knowing he is dead already (22.445, νηπίη).

43. Indeed, as the *homilia* closes, Andromache returns to lament the future death of her husband with many enslaved women, who had also likely experienced second-phase warfare (6.494-502). On these lines, see Ameis and Hentze (1908: ad. 6.500); Graziosi and Haubold (2010: ad. 6.494-502, ad. 6.499); Stoevesandt (2016: ad. 497-502, with bibliography).

44. *Il.* 22.487-99 was athetized by Aristarchus; Leaf thinks the athetization should proceed from 22.488-507 ([1888] 1902: ad. 22.487); against atheization, see de Jong (2012: ad. 22.487-506). Aristarchus' rationale, paraphrased by Leaf: 'the destitution predicted is absurd for a son of the royal race while Priam and Hector's brothers are still alive' and 'the whole passage is a commonplace on the sorrows of orphanage with no application to the case of Astyanax' ([1888] 1902). Cf. Richardson (1993: ad. 22.489). Aristarchus and I disagree on why this fate is absurd – in my opinion, the reality of second-phase warfare explains why Astyanax will never be an orphan in Troy – but his objections overall strengthen my argument. First, as discussed above, Andromache could be imagining here that Hector's death might not mean the fall of Troy, but this notion is unsupported. Further, Astyanax's predicted fate as an orphan might have resonated with audience members who had lost parents to war, and this strictly illogical passage is included in the epic for that very reason. Indeed, here Andromache describes 'a child' (22.490, παῖδα) and his experience of orphanhood in a general way.

45. On this phrase, see Leaf ([1888] 1902: ad. 22.489), de Jong (2012: ad. 22.489). De Jong (2012: ad. 22.482-507) and Murnaghan (1999: 215) remark on the differences and parallels between Andromache's two laments and Hector's earlier prayer for Astyanax. See Graziosi and Haubold, with bibliography, for how Andromache's lament for her family in Book 6 looks forward to her laments in 22 and 24 (2010: ad. 6.405-39).

46. I here follow Leaf in interpreting θαλέων as metaphorical, referring to Astyanax's 'good cheer' ([1888] 1902: ad. 22.504); de Jong takes θαλέων as referring to the food, 'sumptuous things' (2012: ad. 22.503-4), as do Ameis and Hentze (1906: ad. 22.504).

47. For the evocative effect of these lines, and parallels, see de Jong (2012: ad. 22.500-4).

48. Priam's strategy to keep Hector at home echoes Andromache's at 6.407-32. Here he modifies the general consequences of a fallen city to apply specifically to his own family (de Jong 2012: ad. 22.56-76).

49. The description of infants and women should make Hector think of his own wife and child (de Jong 2012: ad. 22.56-76); this outcome for infants will be Astyanax's fate in other epics and tragedy (Ameis and Hentze 1906: ad. 22.64). The visceral, somatic resonance of Priam's description of the fate of infants is heightened by the subsequent graphic description of his own death, in which his carcass will be ravaged and consumed by his own dogs (22.66-76).

50. As Segal says, 'Here in book 22 the agony of loss still has its rawest edges' (1971a: 37).

51. Rabinowitz also draws attention to the violence done to Astyanax's body in *Trojan Women*: 'The violent death of Astyanax makes it clear that innocent noncombatants pay with their lives for decisions they never made. Attention to his sweet flesh highlights both the cost to the defeated and the cruelty of the winning army' (2014: 200–1).

52. See Murnaghan (1999) for female lament and speech in the *Iliad*; see also Dué (2002), Nappi (2015). The audiences for lament in the *Iliad* include both free and captive women (19.301-2,

19.338-9, 24.509-12) (Alexiou 1974: 10). Female lament indicates the multilayered perspectives of the listeners of Homeric epic, and includes themes of social dislocation and captivity of women (Murnaghan 1999: 207–8). For more on Andromache's lament in particular, see Tsagalis ([2004] 2012: 75–108); Perkell (2008: 97–101). For the themes and traditions of lament, see Alexiou (1974). See also MacLeod (1982: ad. 24.721-2, with bibliography); de Jong (2012: ad. 22.477-514, with bibliography); Stoevesandt (2016: ad. 6.407-65, with bibliography); Brügger (2017: ad. 24.719-76, with bibliography).

53. For the greater accuracy of Andromache's final prediction, see MacLeod (1982: ad. 24.723-76); Richardson (1993: ad. 22.437-515, 24.725-45); Le Meur-Weissman (2009: 39); de Jong (2012: 12); Brügger (2017: ad. 24.725-45, with bibliography).

54. For the prediction as a reference to other mythic sources, see Anderson (1997: 53–9); Brügger (2017: ad. 24.734b-8, with bibliography).

55. For her address to Astyanax in this passage (24.732, τέκος), and whether the formulation might indicate the baby's presence, see Ameis and Hentze (1906: ad. 24.732); MacLeod (1982: ad. 24.732-40); de Jong (2012: ad. 22.485-6); Brügger (2017: ad. 24.732, with bibliography).

56. With this quote, Brügger summarizes the ideas developed by Edmunds in her discussion of νήπιος used of children in Homer (1990: 25–60).

57. Throughout the passages discussed in this chapter, Astyanax is denoted with variants of νήπιος (6.400, 6.408, 22.484, 22.502, 24.726).

58. Andromache's sufferings are referred to as ἄλγεα λυγρά (742, 'mournful pain') a few lines later. On these correspondences: MacLeod (1982: ad. 24.739); Richardson (1993: ad. 24.734-9); Murnaghan (1999: 216); Brügger (2017: ad. 24.725-45, 24.739).

59. For the audience of Andromache's lament see Brügger (2017: ad. 24.746).

60. MacLeod notes these two different scenarios (1982: ad. 24.732-4).

61. Indeed, this very dilemma is staged in tragedy (Euripides' and Seneca's *Trojan Women*). See also Philippo (2007).

62. On general correspondences and contrasts between the passages in Books 6, 22 and 24 that I discuss, see Lohmann (1988: 72–4); Segal (1971a: 57); MacLeod (1982: ad. 24.723-76); de Jong (2012: 11–13); Stoevesandt (2016: ad. 6.369-502, with bibliography); Brügger (2017: ad. 24.723-76, 24.725-45, 24.726-7).

63. Corporeal violence indeed inflicted on Astyanax in other epic, tragedy and visual sources (see nn. 30–3).

64. Similarly MacLeod (1982: ad. 24.723-76); de Jong (2012: ad. 22.483-5).

65. Lack of burial is not treated at length in this chapter, but it is Priam's fate after the fall of the city, as noted in n. 49. The fates of old men and women in second-phase warfare is indeed hinted at in the *Iliad* in such passages, but is beyond the scope of this chapter. See McHardy's and Lucia Nováková's chapters in this volume.

66. On this famous line of Hector, see in particular Rousseau (2015). As Graziosi and Haubold put it: 'To continue at home, as normal, while men see to the fighting, proves impossible because war destroys everybody; it is ultimately not just 'the concern of men' (2010: 492). It is above all through the perspectives of women that the poet exposes the brutality of war' (30).

CHAPTER 9
REPAIRING AND RECONSTRUCTING THE BODY OF THE COMBATANT: THE COMMISSION, DESIGN AND MANUFACTURE OF ASSISTIVE TECHNOLOGY FOR WOUNDED SOLDIERS AND VETERANS

Jane Draycott

Introduction

There is a considerable amount of ancient literary, documentary, archaeological and bioarchaeological evidence for the fact that an ancient combatant, whether soldier or sailor, could experience any number of different types of wound that could lead to the loss of a body part as a direct or indirect result of their military service (see also chapters by Fiona McHardy and Helen Snell in this volume). The act of combat itself could lead to the loss of a body part through traumatic injury, such as the loss of an eye due to a direct hit from a projectile, or the loss of a hand – usually the right hand – due to a strike from a sword (Salazar 2000); some examples of such injuries can be seen on the skeletons of the deceased from the Battle of Chaironeia and the Herulian sack of Athens (Liston 2020). However, there were additional ways that an ancient combatant might lose body parts: judicial punishment meted out for crimes such as desertion likewise involved the loss of body parts, such as the loss of a hand; extrajudicial punishment such as the treatment of prisoners of war often involved the loss of body parts, whether as a result of torture or as a means of attempting to prevent combatants from returning to the fray once their period of captivity was over, such as the loss of certain fingers or even an entire hand; self-mutilation was practiced by individuals hoping to avoid having to become combatants, such as the loss of thumbs; and one possible side effect of long periods of time spent on campaign in unforgiving climates and environments were conditions such as frostbite or trench foot that could lead to the loss of body parts, such as toes or even entire feet (for full discussion, see Draycott 2022). Consequently, focusing on the role of the body and its injury in antiquity can prompt us to change the way we think about ancient combatants' bodies in particular, and ancient war in general.

Yet, while there are many descriptions in works of ancient literature across a range of genres of combatants losing their body parts, there are relatively few descriptions of what happened to those combatants afterwards. It would appear combatants received medical treatment while on campaign, and that individuals with specific medical roles were involved in providing it, and that these individuals could possess extensive knowledge, skill, expertise and equipment (Scarborough 1968; Nutton 1969; Salazar 2000; Baker

2004). The corpus of ancient medical literature that survives from antiquity contains enough descriptions of amputation that it is possible to follow the development and refinement of ancient amputation techniques from the earliest accounts in the treatises of the Hippocratic Corpus through to the latest accounts in the compilations of late antiquity. According to the orthopaedic treatises of the Hippocratic Corpus, if one lost a finger or toe, one was in relatively little danger, and this could be treated like any other wound, while if one lost a hand or foot at the joint, one was still in relatively little danger but required rather more involved treatment and a lengthier recovery time (Hippoc. *Art.* 68). The real danger was considered to come from infection that could ultimately develop into gangrene (Hippoc. *Art.* 69). The implication of these writings is that the classical Greek surgeon was not particularly aggressive; he (or she) treated the loss of a body part as and when it occurred but did not necessarily seek to remove any proactively. Rather, they waited until it parted from the body of its own accord. Following research into the circulatory system and the discovery of how to ligate blood vessels in Alexandria during the Hellenistic period, later surgeons were much more proactive and undertook amputations as they considered necessary (Celsus, *Med.* 7.33; Majno 1975: 328). It is possible to gain an insight into the way that military surgery developed and advanced over the course of the Roman period by perusing the collection of medical instruments belonging to the retired military surgeon Eutyches, recovered from the House of the Surgeon at Rimini (De Carolis 2009; Jackson 2003; Bliquez 2014: 205, 244, 247). Clearly, in some respects the ancient combatant's body was considered to be a type of war technology that could, under certain circumstances, be repaired, and a considerable amount of time and material resources were dedicated to doing just that.

However, the corpus of ancient medical literature that survives from antiquity does not tend to contain accounts of the potentially extremely long process by which an individual would or could become rehabilitated after they had undergone medical treatment. Unlike modern medical practitioners, except for the author of Hippocratic treatise *Joints*, their ancient counterparts do not seem to have considered rehabilitation to be part of their remit or even their responsibility (Hippoc. *Art.* 58). Understandably, documentary and archaeological evidence is not exactly helpful either.[1]

Thus, we do not know exactly what happened to combatants after they were discharged from the infirmary.[2] We do not know about their rehabilitation, or the process by which they reintegrated into society, or even if they did, in fact, reintegrate into society.[3] We are not certain of the extent to which they were treated similarly or differently to civilians who had sustained similar injuries. We should consider the scope of the impact of such an injury – an individual would not only look physically different if they lost a hand, an arm, a foot, or a leg, but they would not be able to do some of the things that they could do before, and they would also hold themselves differently, and move differently, there would likely be psychological ramifications, and there were potentially negative consequences to all of these things, as ancient rhetorical instruction manuals and surviving speeches make clear.[4] Attitudes towards ancient military veterans were ambiguous, simultaneously positive and negative, laudatory yet critical.[5] Combatants were not immune from public mockery and humiliation as a direct result of their injuries (Draycott 2018). Not even extremely high-status individuals such as kings and generals were safe from what we might charitably and optimistically refer to as teasing.[6] Even those acknowledged as war heroes were not protected from attempts at social exclusion and marginalization (Pliny *HN* 7.29). Thus, a single moment in combat could change the course of someone's entire life, fundamentally and permanently altering their way of inhabiting their body and in doing so rendering them a combatant for life, albeit in various sorts of non-militaristic combative situations, mental and emotional, verbal and non-verbal. So, it should not come as a surprise to learn that some combatants sought to repair and reconstruct their damaged bodies using assistive technology such as extremity prostheses, or that they used this equipment to make specific statements about themselves and through doing so created new narratives that, we might imagine, they preferred to the ones that had previously been imposed upon them by others. And indeed, here we see the beginnings of the 'supercrip'[7] narrative that frequently appears in public media discourse (Garland [1995] 2010: 31–2). This combination of human body and technology and the 'technologically enhanced man' that results can be viewed as a type of cybernetic organism or cyborg for short (Clynes and Kline 1960).[8] As a result, expanding the parameters of when precisely we consider a body to be in combat, what we consider a combatant's body to be, and what we consider it to consist of, can facilitate an alternative approach to and perspective on the study of ancient war. And so, in this chapter, I shall use ancient literary, documentary, archaeological and bioarchaeological evidence to explore the ways in which the body of the combatant was repaired and reconstructed using assistive technology such as prosthetic limbs and other types of mobility aid, thus ensuring the creation of ancient cyborgs, and examine the relationship between this assistive technology and the armour and weapons utilized by combatants in their previous lives as soldiers and sailors. I shall argue that the influence of armour and weapons on the design, commission and manufacture of extremity prostheses was profound, not only due to the preferences of the individuals who were responsible for seeking to have them made in the first place, but also due to the knowledge, skills, experiences and, ultimately, occupations of those responsible for making them. Using comparative evidence from later historical periods, I shall also argue that these martial aspects of ancient extremity prostheses were crucial to the ways in which this specific category of ancient prosthesis-user sought to resituate himself in ancient society and empower himself as an impaired individual rather than a disabled one, and so going forward with his life as an ancient cyborg.

Prostheses in Antiquity

The earliest classical account of a combatant losing a body part and replacing it with an extremity prosthesis can be found in Herodotus' *Histories*, which dates from the middle of the fifth century BCE:

> This man had been put in prison and doomed to die by the Spartans for the much harm that he had done them. Being in this evil case, inasmuch as he was in peril of

> his life and like to be very grievously maltreated ere his death, he did a deed well nigh past believing: being made fast in iron-bound stocks, he got an iron weapon that was brought in some wise into his prison, and straightway conceived a plan of such hardihood as we have never known; reckoning how best the rest of it might get free, he cut off his own foot at the instep. This done, he burrowed through the wall out of the way of the guards that kept ward over him, and so escaped to Tegea; all night he journeyed and all day he hid and lay close in the woods, till on the third night he came to Tegea, while all the people of Lacedaemon sought him; and they were greatly amazed, seeing the half of his foot cut off and lying there, but not being able to find the man himself. Thus did he then escape from the Lacedaemonians and take refuge in Tegea, which at that time was unfriendly to Lacedaemon; and after he was healed and had made himself a foot of wood, he declared himself an open enemy of the Lacedaemonians. Yet the enmity that he bore them brought him no good at the last; for they caught him at his divinations in Zacynthus, and slew him. (9.37, trans. Godley)[9]

The diviner Hegesistratus of Elis lost part of his foot and replaced it with a wooden appliance, which perhaps resembled the Greville Chester Toe or the Cairo Toe, which comes from Egypt and dates to the first half of the first millennium BCE (Devecka 2019: 189–91). Wood seems to have been a popular choice for extremity prostheses in classical antiquity: while most ancient sources simply refer to 'wood' (ξύλινος in Greek, *ligneus* in Latin), one does mention olive wood specifically, and olive wood was commonly used for tools (*Anth. Pal.* 9.233). This is unsurprising since it would have been an easily accessible, relatively cheap material that, while being relatively easy to work, even by someone with only rudimentary carpentry skills, was also strong enough to bear the weight of an adult human without being too heavy for them to manipulate with ease. All these factors ensured that its popularity as a choice of material for extremity prostheses continued well into the twentieth century. Thus, we can see the possibility that prostheses were themselves likewise often seen as tools, much like the combatants themselves.[10] Herodotus directly connects Hegesistratus' acquisition of his prosthetic foot with the redoubling of his efforts against the Spartans, and certainly the prosthetic foot seems to have facilitated his militant activities, enabling him to travel around the Mediterranean and continue serving as a diviner. It is debatable whether the Greeks would have excluded Hegesistratus from religious rites because of his impairment, as seems to have been the case with certain Roman, Jewish and Egyptian religious rites, but in any case, his prosthetic foot seems to have ensured his re-entry into ancient society and facilitated his continuing efforts in the ongoing conflict against the Spartans. The second, and perhaps today the most famous, account of a combatant losing a body part and replacing it with an extremity prosthesis can be found in Pliny the Elder's *Natural History*, which dates from 77 CE:

> No person living, in my opinion at least, ever excelled M. Sergius, although his great-grandson, Catiline, tarnished the honours of his name. In his second campaign he lost his right hand; and in two campaigns he was wounded three and

> twenty times; so much so, that he could scarcely use either his hands or his feet; still, attended by a single slave, he afterwards served in many campaigns, though but an invalided soldier. He was twice taken prisoner by Hannibal (for it was with no ordinary enemy that he would engage), and twice did he escape from his captivity, after having been kept, without a single day's intermission, in chains and fetters for twenty months. On four occasions he fought with his left hand alone, two horses being slain under him. He had a right hand made of iron, and attached to the stump, after which he fought a battle, and raised the siege of Cremona, defended Placentia, and took twelve of the enemy's camps in Gaul. All this we learn from an oration of his, which he delivered when, in his praetorship, his colleagues attempted to exclude him from the sacred rites, on the ground of his infirmities. What heaps upon heaps of crowns would he have piled up, if he had only had other enemies! For, in matters of this nature, it is of the first importance to consider upon what times in especial the valour of each man has fallen. What civic crowns did Trebia, what did the Ticinus, what did Lake Thrasymenus afford? What crown was there to be gained at Cannæ, where it was deemed the greatest effort of valour to have escaped from the enemy? Other persons have been conquerors of men, no doubt, but Sergius conquered even Fortune herself. (7.29, trans. Rackham)[11]

The general Marcus Sergius Silus lost his right hand and replaced it with a metal appliance. Since the right hand was the sword hand and thus extended out in front of the combatant during hand-to-hand combat rather than remaining behind the protection of the shield with the rest of the body, it is not surprising that it seems to have been particularly vulnerable to injury, with numerous ancient literary accounts of battle detailing how individuals lost their right hands. Hands were also important in other types of martial activity, necessary for wielding a range of weapons apart from the sword and operating siege engines (Veg. *Mil.* 1.15-16, 20, 2.4-6, 3.14, 4.8, 21, 29).[12] While no prosthetic hands have so far been recovered during the excavation of archaeological sites dating to classical antiquity, there are literary references to prosthetic hands made from ivory and gold (Dio Chrys. *Or.* 8.28). Although Pliny describes Marcus Sergius Silus' hand as iron, the same material as his sword, it is questionable as to whether the entire prosthesis was made from solid iron, due to the weight of such an object (the same limitation would apply to a prosthesis made from solid gold, or any other metal). The extremity prostheses that have, to date, been recovered during the excavation of archaeological sites dating to classical and late antiquity demonstrate that other types of ancient extremity prostheses seem to have comprised a wooden core covered with metal sheeting or plating, so perhaps Marcus Sergius Silus' prosthesis was of this sort. Since there are so many ancient literary accounts of combatants losing their right hands, it seems unlikely that Marcus Sergius Silus was the only one in this position, particularly since his story seems to have been so well known during the Roman Republic and Empire, so potentially ancient prosthetic hands were as variable as ancient prosthetic feet and legs (Beagon 2002; van Lommel 2019).

While ancient literary accounts do not discuss the loss of legs in combat with the same frequency that they discuss the loss of hands, there is archaeological evidence for the use of prosthetic legs in classical and late antiquity, and this evidence implies that the individuals' legs may have been lost through combat, due to damage sustained to a part of the leg that was not protected by either greave or shield while the combatant was on foot or on horseback. A prosthetic right leg recovered from a tomb in Capua dating to approximately 300 BCE and subsequently known as the 'Capua Limb' was made from wood, iron and bronze, comprising a wooden core covered in bronze sheeting that it has been surmised was worn in conjunction with a leather and bronze belt to hold it in place and, assuming that the prosthesis could be securely fastened at the thigh and the waist, facilitate a limited amount of movement in conjunction with some sort of additional piece of assistive technology such as a crutch. Other finds recovered from the tomb were a bronze urn and some locally produced red-figure pottery, and these, in conjunction with the materials used in the limb's construction, indicate that it was likely worn by a high-status individual, or at the very least, whatever their status, one with considerable financial resources. Considering the social and cultural context of Capua during this period, this individual was perhaps a veteran of the Second Samnite War (327–304 BCE) or even a retired gladiator.[13] A prosthetic right leg recovered from a grave at Hemmaberg dating to approximately the sixth century CE comprised a wooden peg-leg set into a leather cup finished with an iron band. The skeleton to which it was attached demonstrated palaeopathological changes consistent with prolonged periods of time spent on horseback prior to receiving the injury, and the location and angle of the amputation on the lower right leg indicates that the individual was actually on horseback when the injury occurred, so here we have both archaeological and bioarchaeological evidence for a combatant subsequently using a prosthesis.

The Relationship Between Armour and Weapons and Prostheses

It is clear that there was a relationship between the armour and weapons, and the assistive technology utilized by combatants such as soldiers and sailors. In archaic, classical and Hellenistic Greece and Republican Rome, armour and weapons were highly personalized and individualized. Their creation involved extended collaboration between different types of artisan. An individual combatant was responsible for provisioning themselves. While those lower down the social scale might have had to make do with hand-me-downs from a relative or friend, members of the social elite could afford specially made, highly personalized items. Since a combatant who suffered the loss of a body part such as a hand, arm, foot, or leg was not necessarily automatically discharged from military service, if he was to continue to serve, he would need suitable armour and weapons, and his existing equipment would need to be either adapted or replaced. His new physical condition had to be considered by those responsible for providing these things.

There is archaeological evidence for the adaptation of an individual's armour to accommodate a physical impairment from one of the Macedonian tombs at Vergina.

One of the sets of human remains that has been recovered from the tomb is that of a woman aged around thirty years old who has been identified as Philip II's seventh wife, the daughter of the Scythian king Atheas.[14] Bioarchaeological examination of her remains has determined that she had Schmorl's nodes, indicative of horse-riding from an early age, and had suffered a Schatzker type IV fracture of the left tibia, leading to leg shortening, atrophy, knee disfiguration and mobility impairment (Antikas and Wynn-Antikas 2016: 689). Alongside her remains were a gold-plated quiver, seventy-four arrowheads, spearheads and a suit of ceremonial armour. The armour includes a pair of gilded bronze greaves that do not match because they were made for a pair of legs where one was shorter and thinner than the other, the result of a poorly healed fracture; the left greave is 39 centimetres high and its maximum diameter is 30 centimetres, while the right greave is 41.6 centimetres high and its maximum diameter is 33.5 centimetres. Thus the left greave is significantly smaller than the right greave (Antikas and Wynn-Antikas 2016: 687). Here is evidence of an artisan paying attention to the specific medical requirements of their patron, as well as, potentially, the individual in question continuing to go about their business despite their impairment. Since the armour has been designated as ceremonial by scholars, it is debatable whether it was worn for any sort of combat, although its wearer was posthumously styled as a warrior by her peers (it is, however, possible that she also possessed a suit of functional armour that she wore regularly but that was not considered grand enough to be included in her tomb). However, the point stands that the ceremonial armour was specially made for this impaired individual and those responsible for making it took her impairment into account during the process.

Did armour, with its focus on the physical body of the warrior and their specific requirements, inspire prostheses? And were armourers responsible for producing extremity prostheses, whether for combatants on active duty or veterans, in the first instance? Armour has been described as 'one of the earliest artificial human enhancements', and its form as a representation of an idealized version of the male (or, as we saw above, in some cases, the female) physique has been noted (Mayor 2018: 131–2). It is not beyond the realms of possibility that after seeking to idealize the complete physical form of a combatant, and providing them with a panoply, an armourer might also seek to idealize the incomplete physical form of a combatant, providing them with an extremity prosthesis. Or a combatant, having received a panoply that supplemented their own body in combat, and having appreciated this supplementation, might seek to acquire an extremity prosthesis that would supplement, perhaps even augment, their body out of combat. Thus, the study of ancient prostheses has a definite contribution to make to the study of ancient warfare.

It has been suggested that the 'Capua Limb' was inspired by the greave (Bliquez 1996: 2672). Since it was recovered from a tomb in Capua that has been dated to around 300 BCE, contemporaneous to the Second Samnite War (327–304 BCE), it has been suggested that it was made for some sort of combatant, perhaps either a soldier or a gladiator, who lost his leg in combat (ibid., 2671). Or it was potentially made for someone who wished to present themselves as having been, or having been like, a combatant. Since Capua was the chief centre for copper and bronze production in antiquity, it makes sense that an extremity prosthesis created there would utilize both the material and the creativity and skill of the artisans that made their livings working it (Pliny *HN* 34.95).

Slightly later in date, the iron hand of Marcus Sergius Silus seems to have been designed with military application in mind, because he used it as a means of holding his shield. Since he lost his right hand fighting in the Second Punic War (218–201 BCE) yet wished to continue as a soldier, it makes sense that his prosthesis would have been designed, commissioned and manufactured with that aim in mind (of course, that assumes that he only had one prosthesis, and that he only wore it in combat; he could, however, have had several different ones that he wore in different contexts). Since Roman soldiers of this period did not wear gauntlets, that piece of military equipment cannot have directly inspired Silus' hand the way that greaves could have directly inspired the 'Capua Limb'. If we consider the function of this particular prosthetic hand, it seems to have been intended primarily as a way of holding a shield. Shields actually have quite a lot in common with prostheses and are a good way in to thinking about them, since both need to be worn, carried and handled by their user; both need to be tough yet pliable; and both are made from multiple different materials working in conjunction with each other. Through repeated use, they mould themselves to the shape of their user's specific body and become literally infused with their blood, sweat, tears and other bodily fluids (Noel 2019). This leads us to question where the ancient combatant's body was considered to end, and it is worth noting that the ancient prostheses that have survived in the archaeological record have done so precisely because they were included in their users' graves, for the most part attached to the body in death as they had been in life, rather than simply placed next to them in the manner of other types of grave good. If Silus' prosthetic hand was intended to simply support his shield, we might imagine a basic cap placed over the stump to hold the shield in place. But Pliny does say 'iron hand' specifically, so potentially we need to envisage something more like a gauntlet or a boxing glove. While Greek and Roman soldiers may not have worn gauntlets, athletes who engaged in sports such as boxing and wrestling did, so it is possible that we should look to those for both inspiration for Silus' hand and an understanding of how it might have been possible to use it to bear a shield.

While it is from a slightly later historical period, an improvised prosthetic hand comprising of a knife strapped to the wrist has been recovered in a grave from the Veneto in northern Italy, alongside the skeleton of a male agreed between forty and fifty years, dating from the sixth to eighth centuries CE (Micarelli et al. 2018). This shows how an individual could weaponize their extremity prosthesis if they considered it necessary, and in turn be weaponized by them, thus being ever in a state of readiness for combat even outside military contexts such as battlefields (ibid.). The knife was attached to his right forearm with a leather strap and buckle, and the considerable wear on his right upper second incisor shows that he regularly used his teeth in conjunction with his left hand to attach and adjust it (ibid., 6, 8, 12). Thus, this prosthesis was literally imprinted by its wearer's teeth and infused with his saliva.

Ian Hodder's studies into the entanglement of humans and objects have led to considerations of men and weapons and what has been described as 'masculine body-weapon synergic entanglements' (Hodder 2014; Spiegel 2019: 267).[15] If weapons and wearable artefacts such as armour can be considered extensions of masculine human agency and having a part in male inner subjectivity, why not prostheses too (Spiegel 2019: 267)? If this perceived man/weapon entanglement arises from affixing weapons to a man, surely weaponized prostheses such as Marcus Sergius Silus' shield/hand, the Capuan's greave/leg and the Venetian's knife/hand can be considered in this light too? And we can in turn view these man/weapon entanglements as cyborgs (Devecka 2019: 189–91).

The surviving examples indicate that extremity prostheses were extremely personal: they were designed, commissioned and manufactured according to an individual's specific preferences or requirements. Due to the technical knowledge, skill and expertise required, not just in relation to the trade itself but also in relation to the capabilities of the human body, it is likely that the same artisans that produced personalized armour and weapons for ancient combatants produced personalized prostheses for wounded combatants and veterans. Considering the ancient association of the impaired and disabled with crafts such as metalwork, epitomized by the Greek god Hephaistos and his Roman counterpart Vulcan, it is even possible that ancient artisans drew on their own experiences of impairment to inspire their innovative creations (Draycott 2022). Thus, veterans such as Hegesistratus, the anonymous Capuan and Marcus Sergius Silus would have been able to defy their societies' expectations and continue to play noteworthy roles in it at moments of historical significance, in which case their prostheses may then have inspired those of others, and the process may then have repeated itself.

Comparative Discussion

Many of the themes discussed above are likewise present in more recent incidents of prostheses, and altogether looking at war through the frame of the body and its equipment allows us to see how war has been embodied throughout human history (see also Snell, this volume). The contemporary prosthesis in all its variability is the result of almost three centuries of medical and surgical advances such as the development of anaesthesia, antiseptic and amputation techniques, in conjunction with recognition of the necessity of dealing with the physically devastating after-effects of modern warfare.

Prostheses first came to prominence, albeit on a small scale, in the wake of the Napoleonic Wars (1803–15). Henry William Paget, the first Marquess of Anglesey, was shot in the right knee at the Battle of Waterloo, and his right leg was amputated. He was subsequently fitted with a prosthetic leg designed and patented by James Potts of London, and this prosthesis is thought to have been the world's first articulated prosthetic leg, comprising a wooden shank and socket, and a foot that could be controlled by catgut tendons running from the knee to the ankle. Flexion of the knee caused dorsiflexion of the foot, and extension of the knee caused plantar flexion of the foot. This type of prosthesis came to be known as the 'Anglesey Leg' after its most famous user. The amputation did not spell the end of his military career, and in fact, the adoption of an extremity prosthesis made the Marquess something of a celebrity in Victorian society (Warne 2008: 29; see also Anglesey 1961). It is worth noting, though, that he was in a privileged position as a member of the British aristocracy, being wealthy enough to afford to purchase a technologically advanced prosthetic leg, and having been operated on by a surgeon competent enough to leave him with a stump that enabled him to wear them. Many of his contemporaries had to make do with simple peg legs, or even just a pair of crutches. In some cases, the 'social condition' of a patient was considered by the surgeon performing the amputation, who was of the view that while a wealthy individual would be able to avail themselves of the highest possible quality prosthesis, which necessitated providing them with a suitable stump, a poor individual would not be able to, so it did not necessitate providing them with a suitable stump (Warne 2008: 31).

In the wake of the United States of America's Civil War (1861–5), during which an estimated sixty thousand limbs were amputated, a concerted effort was made by private companies to develop prosthetic technology to cater to the unprecedented demand for it, and many of these shamelessly drew upon the 'Anglesey Leg' for inspiration (Figg and Farrell-Beck 1993; Hasegawa 2012: xiii). This relationship between warfare and prosthesis manufacture on an industrial scale continued in the wake of the First World War (1914–18), the Second World War (1939–45) and the Vietnam War (1955–75) (Hermes 2002; Hasegawa 2012; Neumann 2010). It has been suggested that the soldiers who returned from these wars constituted the 'crippled soldier problem' (Brown 2002; Kinder 2015: 3).[16] To reintegrate a veteran into society fully and effectively, it was considered necessary to restore not only his economic productivity and thus prevent him from proving a drain on national resources, but also his masculinity, endangered by his new-found frailty and vulnerability. Anxiety about a former soldier's or sailor's ability to return to their previous occupation or at the very least earn a living was profound (Hasegawa 2012: 3–4). This anxiety was not only felt by the soldier or sailor in question, but also by society at large, which saw the primary problem with the veteran being not the impairment but the possibility that it would result in them becoming a burden on post-war society, crippling (and here I use that pejorative word deliberately) the nation's economic growth and well-being (Kinder 2015: 9). This was a problem that needed to be solved. Consequently, cooperation between surgeons who performed amputations and prosthetists who designed and manufactured prostheses increased exponentially.

Yet, we could consider these soldiers to be 'broken soldiers' rather than 'crippled', 'wounded', or 'injured', because what is 'broken' can be 'fixed', and it is clear that a concerted effort was made to 'fix' these 'broken soldiers' by crafting them prostheses of varying types so as to facilitate their reintegration into society (and the workforce) (Hasegawa 2012: ix). In the late nineteenth to twentieth centuries, the preferred prosthesis was one that facilitated easy and natural motion, appeared realistic, was durable, lightweight and affordable (ibid., 3–4). It promised its user the 'fiction of wholeness' and the ability to 'pass' (Kinder 2015: 35), although the extent to which either was true was entirely dependent upon the nature and extent of the individual's injury or injuries. While there was a certain amount of pride felt about a wound received in service to one's country, which could be viewed a sign of martial valour and manly courage, broadly speaking nineteenth- and twentieth-century society tended to stigmatize impaired bodies and characterize them as economically burdensome, psychologically unstable and socially objectionable (ibid., 30; Caddick et al. 2020). Yet ultimately, just as the Marquess of Anglesey popularized prostheses in Victorian society, the increasingly high profile of prostheses in Western Europe during the early twenty-first century is the result of the high profile – one might say celebrity, even – of veterans of the wars in Afghanistan (2001–21) and Iraq (2003–11), many of whom have added to their national and international reputations through participation in the Invictus Games and the Paralympic Games. This is despite the fact that prosthesis users account for less than 0.5 per cent of the wounded veteran population, at least in the United States of America, and there are far more veterans with invisible wounds than visible ones (Montgomery 2018: 30).

While some amputees prefer realistic prostheses, others prefer to use their prosthesis to make a statement about themselves. One such amputee is Ryan Seary, a former Royal Engineer Sapper, who specialized in explosive ordinance disposal, and who stepped on an LMC pressure plate while on active duty in Babaji, Helmand province, as part of Operation Herrick II. The detonation that resulted removed his left food and hand at the scene, and upon his return to Camp Bastion, surgeons amputated his left leg above the knee and his left arm above the elbow.[17] Sophie and Eduoardo de Oliveira Baratata of the Alternative Limb Project crafted the Anatomical Leg for Ryan, which comprises a realistic foot combined with an 'alternative' leg designed to resemble an anatomical drawing. According to Ryan, '[a]lthough my appearance has changed I don't notice it that much, occasionally someone tells me that I'm being stared at but it doesn't faze me. Often young children will be confused and want to ask about or touch my leg but that's only natural curiosity.'[18] Regarding his prosthesis, '[w]hen I received the leg it was better than I could have imagined, it can only be described as Awesome! The detail is amazing and I can't get over how realistic the foot looks, many people have taken a few seconds to get their mind round the fact that it's not real' (De Oliveira Barata 2021). Clearly, while technology may have advanced over the last two and a half millennia, humanity has remained consistent: the process of combatants personalizing their prostheses that began in classical antiquity with Hegasistratus of Elis, continued with the wearer of the Capua Limb and with Marcus Sergius Silus, is still ongoing today.

Conclusions

Ancient extremity prostheses were not standardized in any way, as is to be expected from items that were likely designed, commissioned and manufactured entirely at an individual's direction and discretion. They were practical and useful; they empowered their users and facilitated their return to some semblance of normality, in addition to making positive statements about the individuals wearing them, such as reminding others of their sacrifice, often referring directly to their military service through their resemblance to armour and weapons, and highlighting wealth, status and even technical sophistication. In the process, they can be considered to have made their users ancient cyborgs, ever after blurring the boundaries between their body and their military technology (whether armour or weapon), thereby rendering them not only permanently equipped for military combat, if the opportunity arose, but also for other types of potentially combative situations that they might subsequently find themselves in, such as an encounter in which they were expected to explain or even justify themselves and their impairment.

Notes

1. We could perhaps view the *pridiani* from Mons Claudianus and Vindolanda in this light, as they detail soldiers who were not fit for duty and give the reasons why, and presumably some of these men were convalescing after the acute phase of their illness or injury, see *O.Claud.* 212 (137–45 CE), *O.Claud.* 213 (127–45 CE), *O.Claud.* 217 (137–45 CE) and Tab. Vindol. 88. For discussion, see Cuvigny (1992) and (1997); Bowman and Thomas (1991).
2. Soldiers who were unfit for service due to a physical or mental impairment were entitled to a medical discharge, a *missio causaria*. For discussion, see van Lommel (2013).
3. See Horstmanshoff (2012) for acknowledgement of rehabilitation, although it is not discussed in detail. It was not a certainty that a wounded soldier would be automatically discharged from the military; according to Sen. *Tranq.* 4.5: *praecisis quoque manibus ille in proelio invenit, quod partibus conferat, qui stat tamen et clamore iuvat* (even with his hands cut off, a man finds he can contribute something to his side in battle if he stands and helps with the shouting). Lucan takes this idea to its logical ultimate extension by depicting a soldier who has been blinded by a lead sling bullet requesting that his comrades use him as a human shield: *ingentem militis usum hoc habet ex magna defunctum parte cadaver: viventis feriere loco* (3.713-21, This body, defunct for the most part, might have remarkable use as a soldier: you will be struck in place of a living man). See also Rance (2020: 181–2) for discussion of veterans who had been medically discharged in the late empire.
4. For discussion of the importance of the way an individual held and moved his body while speaking in public, see Corbeill (1996: 14–56); for discussion of the importance of the way an individual walked, see Corbeill (2004: 107–39); O'Sullivan (2011).
5. For discussion, see van Lommel (2015). See also Melchior (2011) for discussion of PTSD specifically. For a study of the varied ways that contemporary British newspapers report on military personnel, which show significant overlap with ancient attitudes, see Caddick et al. (2020).
6. See, for example, the case of Philip II of Macedon, Demetrius, *Eloc.* 293; and the case of Antigonus I, Ael. *VH* 12.43; Plut. *The Education of Children* 14b; for discussion, see Draycott (2018). The mistreatment of the disabled could also often go beyond the verbal, see Draycott (2015).
7. For explanation and discussion of the concept or stereotype of the 'supercrip', see Harnett (2000: 21): an impaired person who 'triumph[s] over the tragedy of their condition . . . who, through astounding personal endeavour overcomes their disability'. As far as these ancient examples are concerned, they often follow a pattern in which an individual triumphs over several different types of adversity: first, the actual loss of their extremities; second, their imprisonment by their enemies; third, attempts made by their peers to marginalize them because of their impairment. Thus, they could be said to be setting positive examples and potential sources of inspiration for amputees regarding the possibilities of prostheses; the

ancient equivalent of the contemporary 'inspiration porn'. For explanation and discussion of the concept of 'inspiration porn', see Grue (2016). See also Caddick et al. (2020: 13–14) for discussion of this in relation to contemporary military veterans.

8. For the way that this definition became increasing nuanced and expansive over the course of the twentieth century, and as a result could potentially be used to refer to 'anyone with an artificial organ, limb or supplement (like a pacemaker), anyone reprogrammed to resist disease (immunized) or drugged to think/behave/feel better (psychopharmacology)', see Gray, Figueroa-Sarriera and Mentor (1995: 22). For discussion of ancient cyborgs, see Liveley (2008); Devecka (2019).

9. τε καὶ τῶν Τελλιαδέων ἐόντα λογιμώτατον, τὸν δὴ πρότερον τούτων Σπαρτιῆται λαβόντες ἔδησαν ἐπὶ θανάτῳ ὡς πεπονθότες πολλά τε καὶ ἀνάρσια ὑπ᾽ αὐτοῦ. ὁ δὲ ἐν τούτῳ τῷ κακῷ ἐχόμενος, ὥστε τρέχων περὶ τῆς ψυχῆς πρό τε τοῦ θανάτου πεισόμενος πολλά τε καὶ λυγρά, ἔργον ἐργάσατο μέζον λόγου. ὡς γὰρ δὴ ἐδέδετο ἐν ξύλῳ σιδηροδέτῳ, ἐσενειχθέντος κως σιδηρίου ἐκράτησε, αὐτίκα δὲ ἐμηχανᾶτο ἀνδρηιότατον ἔργον πάντων τῶν ἡμεῖς ἴδμεν· σταθμησάμενος γὰρ ὅκως ἐξελεύσεταί οἱ τὸ λοιπὸν τοῦ ποδός, ἀπέταμε τὸν ταρσὸν ἑωυτοῦ. ταῦτα δὲ ποιήσας, ὡς φυλασσόμενος ὑπὸ φυλάκων, διορύξας τὸν τοῖχον ἀπέδρη ἐς Τεγέην, τὰς μὲν νύκτας πορευόμενος, τὰς δὲ ἡμέρας καταδύνων ἐς ὕλην καὶ αὐλιζόμενος, οὕτω ὡς Λακεδαιμονίων πανδημεὶ διζημένων τρίτῃ εὐφρόνῃ γενέσθαι ἐν Τεγέῃ, τοὺς δὲ ἐν θώματι μεγάλῳ ἐνέχεσθαι τῆς τε τόλμης, ὁρῶντας τὸ ἡμίτομον τοῦ ποδὸς κείμενον, κἀκεῖνον οὐ δυναμένους εὑρεῖν. τότε μὲν οὕτω διαφυγὼν Λακεδαιμονίους καταφεύγει ἐς Τεγέην ἐοῦσαν οὐκ ἀρθμίην Λακεδαιμονίοισι τοῦτον τὸν χρόνον· ὑγιὴς δὲ γενόμενος καὶ προσποιησάμενος ξύλινον πόδα κατεστήκεε ἐκ τῆς ἰθέης Λακεδαιμονίοισι πολέμιος. οὐ μέντοι ἔς γε τέλος οἱ συνήνεικε τὸ ἔχθος τὸ ἐς Λακεδαιμονίους συγκεκυρημένον· ἥλω γὰρ μαντευόμενος ἐν Ζακύνθῳ ὑπ᾽ αὐτῶν καὶ ἀπέθανε. See also Plut. *Mor. De frat. amor* 3.1.

10. It is worth noting that there are debates in the Jewish Talmud about whether an extremity prosthesis is a shoe, in which case it is permissible to wear it on the Shabbat, or a tool, in which case it is not permissible to use it on the Shabbat; for discussion, see Lehmhaus (2019).

11. *Verum in his sunt quidem virtutis opera magna, sed maiora fortunae: M. Sergio, ut equidem arbitror, nemo quemquam hominum iure praetulerit, licet pronepos Catilina gratiam nomini deroget. secundo stipendio dextram manum perdidit, stipendiis duobus ter et vicies vulneratus est, ob id neutra manu, neutro pede satis utilis, animo tantum salvo, plurimis postea stipendiis debilis miles. bis ab Hannibale captus—neque enim cum quolibet hoste res fuit—bis vinculorum eius profugus, in viginti mensibus nullo non die in catenis aut compedibus custoditus. sinistra manu sola quater pugnavit, duobus equis insidente eo suffossis. dextram sibi ferream fecit, eaque religata proeliatus Cremonam obsidione exemit, Placentiam tutatus est, duodena castra hostium in Gallia cepit, quae omnia ex oratione eius apparent habita cum in praetura sacris arceretur a collegis ut debilis, quos hic coronarum acervos constructurus hoste mutato! etenim plurimum refert in quae cuiusque virtus tempora inciderit. quas Trebia Ticinusve aut Trasimenus civicas dedere? quae Cannis corona merita, unde fugisse virtutis summum opus fuit? ceteri profecto victores hominum fuere, Sergius vicit etiam fortunam.*

12. See also the correlation between *dextra* as 'hand' and 'sword', as noted in the Introduction to this volume.

13. The gladiator graveyard discovered in Ephesus included the skeleton of a gladiator who had undergone the amputation of his right lower leg, so clearly such injuries could be sustained in combat, see Kanz and Grossschmidt (2009: 216). See Gal. *Comp.* 203-8 for his account of working as a physician to a team of gladiators.

14. See Hdt. 4.71.2 for the Scythian custom of sacrificing one of the king's concubines upon his death and burying her beside him; Antikas and Wynn-Antikas (2016: 682) identifies this woman as the first case of a royal woman with a documented disability in ancient Greece.

15. We should not forget, however, that women used weapons and armour too, not just the Scythian princess mentioned above; see Mayor (2014).

16. We do see some instances of this in antiquity, one example is when Alexander III of Macedon encounters a sizeable group of Greeks who had been kidnapped and held captive by the Persians and, during their captivity, tortured and mutilated beyond recognition, with some having lost their ears, others their noses, others their hands, others their feet. They are given the choice of either returning home, where it is uncertain how their families and friends (and society as a whole) will greet them, or staying in Persia and starting their own colony, and ultimately, they decide on the latter course of action; see Diod. Sic. 17.69.2–7.

17. De Oliveira Barata (2021). For more information about Ryan Seary's life since his amputation, see the short film *Ryan's Leg* (2014). He subsequently competed in the 2014 Invictus Games, winning gold for the Men's 50 m Backstroke and bronze for the Men's 100 m Freestyle.

18. De Oliveira Barata (2021). For discussion of the experiences of military veterans with visible injuries when out in public, see Montgomery (2018).

CONCLUSION
BEYOND THE BODY OF THE COMBATANT

Annemarie Ambühl

The present volume focuses on the soldierly body, applying a holistic approach to studying ancient warfare that goes beyond just the strategic. But is it even possible for us to reconstruct the bodily and mental experiences of ancient combatants, when still today war veterans struggle with the difficulty of sharing their (often traumatic) experiences? If we cannot approach directly the minds of past societies, can we at least get some grip on their bodies? In the end, bodily experiences, too, need to be constructed and mediated in order to communicate them to audiences, just as an iconic photograph can capture a moment, a fragment of a body or an object, and transport it through time. As a fitting opening for the volume (Chapter 1), Helen Snell's artworks heighten this transtemporal quality of photographs by explicitly grafting the present onto the past. Through her digital manipulation of archive images from WW1 and WW2 and recent military propaganda, historical and contemporary imagery is merged, highlighting the inherent close links between war technology and the body. The images feature equipment designed either for saving or for destroying soldierly bodies (Figures 1.1–1.2, 1.3–1.7, 1.10 and 1.11), and the carcasses of monstrous warships evoke eerie analogies to the fragile human body (Figures 1.8 and 1.9). To me, her monumental installation *Fighting Fit Triptych* with its filigree elements that extend towards the ground suggests a system of vessels transporting bodily fluids such as blood or tears – anticipating the potentially tragic future of the body that is being declared fit for combat – or a network of roots symbolizing connections that reach deep into the past of humankind. In the accompanying text by Helen Snell and Hannah-Marie Chidwick, this process of transtemporal layering is taken one step further, as the artworks are juxtaposed with quotes from Vegetius' fourth-century CE handbook of military science, evoking striking continuities in the view of the soldierly body as currency to be exchanged in war politics and war-making. In this way, both the visual and the textual archives of the past are opened towards the future of understanding war through an intermedial approach.

Similar issues regarding war and the soldierly body apply to all sorts of media and items of material culture, be they from modern or from ancient contexts. They act as interfaces that challenge us to enter into a dialogue. In the case of our ancient sources, however, the methodological problems sketched above are further complicated by the specific nature of the material. On the one hand, considering the well-known biases of ancient literature in general and especially war literature, which is mostly written from a male, elite, top-down perspective, how can we recover the perspective of common soldiers and other persons affected by warfare and bodily violence?[1] Here the comparatively rare personal documents such as letters or inscriptions are of great value,[2] even though they may to some degree be shaped by the same literary conventions and social expectations that underly the depiction of fictional characters in epics or tragedies, whose experiences can nevertheless be as gripping as the ones of 'real' individuals.[3] On the other hand, what can the scarce (bio)archaeological evidence from ancient battlefields and the remains of personal items tell us about the bodily experiences of actual soldiers?[4]

All these divergent threads have to be brought together in order to answer the question that lies at the heart of the present volume (cf. Chidwick's Introduction): how was war embodied in the ancient world? The contributors have found various ways to deal with such methodological obstacles and to turn them into fruitful challenges, often by applying contemporary theories to their respective source materials and thus opening up interdisciplinary and cross-temporal perspectives. As a result, the prevailing images of the Greek hoplite or the Roman legionary as the perfect prototypes of the ancient soldierly body – male, strong, unyielding – have made room for diversity. Stereotypical dichotomies of gender, of ability and disability, and of the body as an active agent or a passive victim of violence have been overcome in favour of a more flexible model, where the boundaries of the body become as fluid as the boundaries of war itself. Warfare manifests itself in many different forms, on the battlefield and beyond, in and outside the fighter's body.

This blurring of boundaries extends to the definition of the body itself. In constructions of past identities, what role is played by the body and how is the body circumscribed? For the conceptualization of the soldierly body, extensions of the body such as armour, weapons, dress and emblems are crucial elements. In the ancient Mediterranean, the construction of identities often follows an ethnographical model, whereby definitions of the self and the other rely not so much on innate bodily characteristics but rather on cultural markers like peculiar equipment or fighting habits. The present volume brings into focus regions situated at the margins of the ancient Mediterranean beyond Greece and Rome, such as ancient Parthia or border provinces of the Roman Empire. Through the case studies of Redfern and Pearce's Roman Britain (Chapter 3), or Llewellyn-Jones' 'Shami Prince' (Chapter 4), each contribution, respectively, emphasizes the crucial role of a culturally embedded reading of bodies and their paraphernalia. Moreover, Džino and Coopey (Chapter 5) foreground how the same symbolic bodies can assume multiple meanings and functions in creating group identities, when viewed from the opposing perspectives of conqueror or conquered. Similar negotiations can even take place within the body of a single (fictive) warrior, as Chidwick (Chapter 6) shows using the example of Virgil's Camilla, who paradoxically combines feminine and masculine, Italian and Roman, traits and thus calls into question norms of gender and nationality. A similarly complex character is the Massagetan queen Tomyris from Herodotus' *Histories*; as McHardy (Chapter 2) convincingly argues, her seemingly appalling treatment of the dead body of the Persian king Cyrus cannot be dismissed as a typically 'barbarian' or 'female' practice but actually fits right into a Greek cultural discourse on pain, anger and revenge, whereby battlefield decapitation and mutilation function as a 'bodily code' intended to humiliate the defeated opponent. Such case studies demonstrate the need for further work on the synergies of gender and ethnicity regarding the identities of bodies in the context of ancient warfare.[5]

As suggested by the last example, another imaginary border to be crossed is the distinction between body and mind. The idea that cognition is always embodied and that combat experience may produce lasting mental effects was certainly recognized in one way or another long before modern cognitive studies and trauma theory, approaches that have gained much attention in classical studies in the past years.[6] For the present volume, trauma is especially relevant in its etymological, corporeal sense that is closely bound up with the (wounded) body. Several contributions focus on sensory experiences of and emotional responses to battle and other forms of violence as they are reflected in ancient visual and textual sources. War narratives and war monuments offer various strategies of coping with the bodily sensations and emotions processed by the combatants themselves, the victims and their respective communities.[7] Through social practices or mediated reflections, through shocking confrontations or comforting rituals, different audiences can relate to the traumatic experiences of historical or fictional individuals across time and space, up to the present day. Yet in another sense, such spatial and temporal boundaries are already transcended by the combatant himself, who embodies experiences on and off the battlefield. While the narrative structuring of battle accounts and war manuals often gives the impression of a neat series of subsequent stages that can be anticipated through military training (preparation for combat, battle proper and post-combat), ancient warfare comes in many more shapes (marches through difficult or dangerous terrain, skirmishes, night raids, sieges, etc.) that all challenge the body of the combatant in their own different and often unforeseen ways. In second-phase warfare (if not already before) the lines between combatants and non-combatants become blurred when acts of violence are perpetrated against all the parties involved, a practice that continues to this day. As Nováková (Chapter 7) and Sorscher (Chapter 8) show in their discussions of the ancient literary and (much fewer) visual representations, during and after the sack of a city, the bodies of women and children are reduced to booty and thus raped or enslaved.[8]

However, it is not only the bodies of the targeted victims but crucially also the soldierly body itself that is transformed by the battlefield, mentally as well as physically. When the bodily integrity of the combatant is at stake, other versions such as the fragmented, dismembered body or the fluid, dissolving body take over.[9] Apart from the well-known literary resonances and metaphorical implications of such images,[10] the veteran's maimed body has a very real place in ancient life, too.[11] A soldier's wounded body can be 'fixed' by various means in order to incorporate it back into society, as Draycott (Chapter 9) demonstrates. The prosthetic body points to concepts of the ancient body as the sum of its parts that can be disassembled and reassembled.[12] A special case of disassembling is treated by McHardy (Chapter 2), who in her study of representations of the mutilation of the face and head in Greek war narratives interprets these destructive acts as the literal defacement of the enemy, based on the central role of these parts in a 'cultural coding' of the body. Conversely, if the prosthetic body can be seen as a mechanical device, it also forms an analogue to and prolongation of the former 'combat machine'. Such a close association between the human body and war technology might seem a typically modern conceit, but in fact it appears in Greek and Roman literature, too,[13] as exemplified by the quote from Vegetius 'the heavy infantry stood, so to speak, like a wall of iron' (*Mil.* 2.17, *tamquam murus, ut ita dicam, ferreus stabat*), which so surprisingly correlates with the photograph of the iron hull of a war ship resembling a ruptured front line or injured flesh (Chapter 1, *ILLUSTRIOUS*). But from the Homeric epics on, it is not only the front line that can be described as a wall or tower (cf. Hom. *Il.* 4.299, ἕρκος ἔμεν πολέμοιο; cf. Alcaeus fr. 112.10 L-P, ἄνδρες γὰρ πόλιος πύργος ἀρεύιος), but also the single warrior, who can, through conceptual metaphors, be stylized as a wall (Telamonian Ajax in Hom. *Il.* 6.5, ἕρκος Ἀχαιῶν; cf. the super-soldier Scaeva in Luc. 6.201, *stat non fragilis pro Caesare murus*) or as a tower (the fallen Trojan Echepolus in Hom. *Il.* 4.462, ἤριπε δ' ὡς ὅτε πύργος ἐνὶ κρατερῇ ὑσμίνῃ), an embodied offensive or defensive weapon. In a further twist, in the context of the description of the naval battle near Massilia in Lucan's *Civil War*, a dying soldier who is reduced to an amputated torso turns himself into a weapon, first acting passively as a human shield to protect his fellow soldiers, then actively as a dead weight in order to sink an enemy ship (3.609-26), while another blinded combatant is repurposed by his comrades as an engine for hurling missiles (3.709-22).[14]

Even though such examples from Greek and Roman epic might be interpreted as hyperbolic fictions, they reflect the notion that even dying or dead bodies are not defunct but enact their own agency. All the contributions to this volume prompt us to more rigorously examine the notion of agency, itself a contested issue in the context of war: who has a choice about becoming a combatant and throwing their body into the fight? But clearly the end of the battle does not mean the end for the soldierly body, which lives on in one form or another. In a positive sense, the bodies of the fallen act as an anchor of remembrance, to be mourned by their relatives and idealized as heroic fighters on behalf of their communities. But there is also a more ambivalent, corporeal side to the bodies of the fallen.[15] Nováková (Chapter 7) interestingly argues for a reciprocal relationship between the depiction of dead bodies and the destruction of the natural environment, in the sense that the body can stand for the landscape and the landscape for the body. In her view, the wounded body on Greek war memorials reflects a landscape scarred by war, while in the literary imagination Nature herself can actively take part in the fighting as a combatant. The havoc wrought against human bodies thus spills over into the natural world, crossing yet another boundary. In aftermath narratives dead bodies indeed persist and literally transform the landscape, not only as a vector of the pollution caused by war, but also as a fertilizer of the killing fields that will be used as agricultural spaces again in the future (e.g. Luc. 7.847-68).[16]

Moreover, acts of excessive violence imagined or actually perpetrated in the aftermath of battle also threaten to break down the critical distinction between humans and animals. It is on the battlefield that human bodies intermingle with animal bodies, and not only in the palpable sense that the fallen on the battlefield are reduced to carrion to be devoured by scavengers (e.g. Luc. 7.825-46). Even more worryingly, the victor himself can be compared to a wild beast, when in his desire for revenge he transgresses the boundaries of the humane and turns to 'beastly' behaviour such as cannibalism; here, mental suffering leads to a quasi-bodily metamorphosis, a literal dehumanizing of the warrior (cf. McHardy in Chapter 2).[17] This mixing up of human bodies and bodies of the natural world is an intriguing topic that deserves to be further explored in line with the ecocritical approaches recently adopted in Classical Studies.

After reading all the contributions that individually and in dialogue across the volume address so many crucial aspects of the body of the combatant, for me an overall question remains: how can we bridge the gap between representations of the often extreme, shocking treatments of the body in written sources, especially in 'high' genres such as epic, tragedy or historiography aimed at producing dramatic effects, or the idealized bodies in visual monuments, and the everyday bodily experiences of countless combatants and other groups involved in warfare that can only partly be reconstructed through archaeological research or the study of documentary texts? From the perspective of a literary scholar, it strikes me that such questions might always have been present in ancient texts and that some answers can be teased out by reading them against the grain. Ancient audiences, too, may have responded in multiple ways to the same narratives and images, depending on their respective biographical backgrounds, and picked up nuances that lie just below the surface. For instance, the role of weather conditions and natural features of landscapes and their impact on the outcome of military campaigns has been recognized as an ideologically charged motif in literary battle accounts that can serve as a justification of defeat.[18] But at the same time such literary motifs reflect real-life factors that immediately affect the body of the combatant, its accessories and its relation to its natural environment.[19] From the Homeric epics to Latin historiography, soldierly bodies have to deal with all kinds of hardships that shape their experience of war, perhaps in an even deeper way than combat itself. Here I am not so much thinking of spectacular scenes such as Livy's accounts of Hannibal's army crossing the icy Alps or the battle fought in the thick mist at Lake Trasimene, although in these narratives, too, occasionally the body of the common soldier takes centre stage.[20] Rather it is the small, seemingly casual details that convey a tangible sense of the routine of the body at war, the importance of training, equipment, suitable garments, nourishment and rest and the mental states that go with it, ranging from confidence and comradeship to panic and despair.[21]

So, in Book 14 of the *Odyssey* (14.457-522), the disguised Odysseus tells an allegedly autobiographical tale from the Trojan war to test the loyalty of his swineherd Eumaeus: During a nightly ambush in a swamp he had forgotten his warm cloak and would have frozen to death in his bare tunic under his ice-covered shield, were it not for his comrade Odysseus' presence of mind. It is precisely the sensory details that render the cunningly fabricated tale credible in the eyes of the audience. In Roman war narratives, too, marshes are testing grounds for soldierly bodies. In Livy, shortly before the above-mentioned battle of Lake Trasimene, the Gauls in Hannibal's army suffer most from the exhausting march through the flooded territory and the concomitant lack of sleep (22.2); the devastating effects on body and mind of such prolonged deprivations are expressed in the memorable phrase 'they could not uphold their bodies with their spirits nor their spirits with hope' (22.2.6, *neque aut corpora animis aut animos spe sustinebant*),[22] and the dehumanizing demoralization is conveyed through the graphic images of men of who just let themselves fall to die next to the scattered bodies of pack animals or who use the heaps of cadavers to get some sleep at last. In Tacitus' account of Germanicus' visit to the battleground of the Teutoburg Forest (*Ann.* 1.61-8), now it is the Romans' turn to be hindered by swamps, their bodies weighed down by their heavy cuirasses (*corpora gravia loricis*) and their weapon technique being unsuited to the environment, whereas the tireless Germanic Cherusci are at an advantage thanks to their tall bodies (*procera membra*) and their familiarity with fighting in marshes (1.64).[23] After a bloody battle during which the horses are caught in the slippery mud, causing havoc among their riders, the deprivations again affect the soldiers' minds as well, leading to sleeplessness, despair and a mass panic; there are no tents, no bandages for the wounded and all the food is tainted with mud and blood (1.65.7: *non tentoria maniplis, non fomenta sauciis: infectos caeno aut cruore cibos* . . .). On the next day, after a successful battle, although the same dire conditions remain, the Romans miraculously regain their strength and sanity (*sanitas* surely referring to mental as well as bodily soundness) through victory (1.68.5). Passages like these, despite their obvious literariness, convey a strong sense of the combatants' bodies and minds as crucial factors in warfare.

To conclude, therefore strictly speaking the body does not need to be *written into* the history of war, as Chidwick postulates in her Introduction, for it is already inscribed right at its centre, waiting to be *read out* of it. Taken together, the contributions gathered in this volume demonstrate the crucial role of the body as a starting point for a study of ancient warfare. Through innovative synergies of thematic and methodological approaches, the ancient Mediterranean body at war turns out to be a multi-faceted, dynamic research topic that transcends any static divisions between subject and object, aggressor and victim, between the body as a strategic agent or a mere instrument of state power, between the fighting body and the wounded or dead body, between different ethnic or gendered identities, and between the body, its culturally marked accessories, and its natural environment. The countless bodies we have come across on every page of this volume open up just as many avenues of exploration. The body of the combatant with its many interfaces and entanglements thus offers a rich potential for future research, beyond the boundaries of this volume.

Notes

1. A recent example of such a reading is Richlin (2018), focusing on soldiers and war captives in Roman comedy.
2. An interesting case study has been presented by Reinard and Rollinger (2020: abstract, 163), who interpret a late antique Greek letter from Egypt with a first-person account of mental disorder resulting from a military attack as 'the only reliable documentary evidence for psychological war trauma from the ancient world known so far'.
3. For the application of the concept of immersion to the experiences of ancient audiences, see most recently Grethlein, Huitink and Tagliabue (2020).
4. Various combinations of the above mentioned sources and approaches are found in recent handbooks on ancient warfare that can be used as a starting point for further research, among

them: Sabin, van Wees and Whitby (2007); Campbell and Tritle (2013); Brice (2019); A. D. Lee (2020); Heckel et al. (2021). From a reception perspective, cf. also Caston and Weineck (2016) and Berti, Castello and Scilabra (2020).

5. On the roles of gender and ethnicity for the construction of ancient identities (not specifically in the context of war), see Fabre-Serris, Keith and Klein (2021).
6. Embodied cognition in antiquity is discussed in Lauwers, Schwall and Opsomer (2018) and Anderson, Cairns and Sprevak (2019). For the identification of war trauma in ancient Greece and Rome as well as the appeal of ancient war literature to contemporary veterans, see, e.g., Cosmopoulos (2007), Meineck and Konstan (2014) and Karanika and Panoussi (2020), for a critical review of such approaches, see Reinard and Rollinger (2020) (cf. above n. 2). For reflections of transgenerational trauma in Roman civil war poetry, cf. Walde (2011), for ancient communities' strategies of coping with crisis and trauma, see Klooster and Kuin (2020).
7. For such sensory experiences, cf. also the examples in Ambühl (2016a, 2019). On the role of emotions (especially fear) on the battlefield as reflected in Roman war manuals, see recently Różycki (2021) and generally on emotions conveyed by ancient (war) narratives, de Bakker, van den Berg and Klooster (2022); cf. above n. 3 on the concept of immersion.
8. On ancient war's impact on the 'home front', see Cecchet, Degelmann and Patzelt (2019), on gendered practices in ancient war, Fabre-Serris and Keith (2015) and Walde and Wöhrle (2018). On the fate of children, see the comparative volume by Denzler, Grüner and Raasch (2016).
9. Among recent studies of dismemberment, see McClellan (2019), Estèves (2020: 209–303) and Gherchanoc and Wyler (2020). For conceptualizations of various bodily fluids in the ancient Mediterranean, see Bradley, Leonard and Totelin (2021), especially the sections 'Dissolving and Liquefying Bodies' and 'Wounded and Putrefying Bodies'; specifically on blood, see also Bodiou and Mehl (2017). There are also some entries concerning the body of the combatant in the *Dictionnaire du corps dans l'Antiquité* (Bodiou and Mehl 2019), which derives from the French school of historical anthropology.
10. A well-studied example is the metaphor of the healthy vs the diseased or mutilated body politic; for Roman culture, cf., e.g., Gardner (2019) and Walters (2020).
11. On this topic, see the research by van Lommel (2015) and Laes (2018) with further bibliography.
12. See Bolens (2000) and Müller and Wagner (2014) for comparative 'synthetic' body concepts.
13. Chidwick (2021) traces poststructuralist concepts of the 'war machine' in Latin epic, especially in Lucan. On the body as machine, cf. also Gerolemou and Kazantzidis (2023).
14. On this passage, cf. also Draycott (Chapter 9, n. 3). In Ambühl (forthcoming), I discuss more examples of Lucan's de(con)struction of body concepts in the context of civil war.
15. On the treatment of war dead, see, e.g., Hope (2015), and from a cross-cultural perspective Clauss, Reiß and Rüther (2019).
16. For various views on landscapes of war, see the contributions in Reitz-Joosse, Makins and Mackie (2021); more specifically, cf. Ambühl (2016b) and Zientek (2021) for Thessaly as a literary landscape of civil war and the nexus between battlefields and agriculture (on the last, see also Plut. *Vit. Mar.* 21.3; cf. Hughes 2013: 130, and 2014: 152).
17. On cannibalism and 'beastialization', see Andò (2013); cf. also Ambühl (2019: 127–30).
18. See Östenberg (2018), Van Broeck (2018) and Fabrizi (2021); on forests and mountains, cf. also Estèves (2020: 143–87).

19. The Vindolanda tablets provide insight into such experiences, too (e.g. 234.6-8 on dealing with winter storms). On living conditions affecting the soldiers' health, see also Redfern and Pearce in Chapter 3.
20. During the difficult descent to Italy, bodies of men and animals topple over each other and the Punic soldiers cling to the slippery rocks with their hands and knees to get some hold (Liv. 21.35-6). At Lake Trasimene, the Roman soldiers get disoriented in the fog and are attacked before they are able to properly prepare themselves for battle, so that they are encumbered rather than protected by their own armour and only hear the clattering of weapons and the sounds of bodies being hit all around them (Liv. 22.4-5); on the 'battle of perceptions' in this narrative, see recently Feldherr (2021).
21. See, e.g., Chrissanthos (2007: 227–9) and J. W. I. Lee (2019) on the everyday experiences of ancient soldiers and the hardships of camp life. Heidenreich and Roth (2019) approach battlefield panic from the perspective of neuroscience.
22. Cf. a similar phrase in the opposite situation, during the march of Alexander's soldiers through the desert: *Itaque primum animi, deinde corpora deficere coeperunt* (Curt. 7.5.6, first their minds, then their bodies began to give up).
23. A similar description is found in the context of the Batavian revolt in Tac. *Hist.* 5.14-18.

REFERENCES

Abbasi, V. R., R. Mehrafarin and M. Saadatiyan (2013), 'Clothing of People in Sistan during Parthian Period with Reference to the Frescos of Koh-e Khajeh', *International Journal of Business and Social Science*, 4 (8): 308–19.

Abramić, M. (1937), 'O predstavama Ilira na antiknim spomenicima', *Časopis za zgodovino in narodnopisje*, 32: 7–19.

Adkins, A. W. H. (1960), *Merit and Responsibility: A Study in Greek Values*, Oxford: Oxford University Press.

Adkins, A. W. H. (1977), 'Callinus 1 and Tyrtaeus 10 as Poetry', *HSCP*, 81: 59–97.

Agamben, G. (1998), *Homo Sacer: Sovereign Power and Bare Life*, trans. D. Heller-Roazen, Stanford, CA: Stanford University Press.

Alexiou, M. (1974), *The Ritual Lament in Greek Tradition*, Cambridge: Cambridge University Press.

Allen-Hornblower, E. (2016), *From Agent to Spectator: Witnessing the Aftermath in Ancient Greek Epic and Tragedy*, Berlin: De Gruyter.

Allison, P. M. (2006), 'Mapping for Gender: Interpreting Artefact Distribution Inside 1st- and 2nd-century A.D. Forts in Roman Germany', *Archaeological Dialogues*, 13: 1–20.

Allison, P. M. (2013), *People and Spaces in Roman Military Bases*, Cambridge: Cambridge University Press.

Alofs, E. (2015), 'Studies on Mounted Warfare in Asia III: The Iranian Tradition – Cavalry Equipment, Infantry, and Servants, *c.* CE 550–1350', *War in History*, 22 (2): 132–54.

Alston, R. (2010), 'Arms and the Man: Soldiers, Masculinity and Power in Republican and Imperial Rome', in L. Foxhall and J. Salmon (eds), *When Men Were Men*, 205–23, London: Routledge.

Amaral, A. L. (1994), 'Duas rainhas em Heródoto: Tómiris e Artemísia', *Humanitas*, 46: 17–41.

Ambühl, A. (2016a), 'Preface', in A. Ambühl (ed.), 'Krieg der Sinne – Die Sinne im Krieg: Kriegsdarstellungen im Spannungsfeld zwischen antiker und moderner Kultur / War of the Senses – The Senses in War: Interactions and Tensions between Representations of War in Classical and Modern Culture', *Thersites*, 4: i–xxiv.

Ambühl, A. (2016b), 'Thessaly as an Intertextual Landscape of Civil War in Latin Poetry', in McInerney, J. and I. Sluiter (eds), *Valuing Landscape in Classical Antiquity: Natural Environment and Cultural Imagination*, 297–322, Leiden and Boston, MA: Brill.

Ambühl, A. (2018), 'Frauen als Akteurinnen in Kriegssituationen in der antiken Historiographie und weiteren Gattungen: das Fallbeispiel Tomyris', in C. Walde and G. Wöhrle (eds), *Gender Studies in den Altertumswissenschaften: Gender und Krieg*, 19–49, Trier: Wissenschaftlicher Verlag.

Ambühl, A. (2019), 'The Touch and Taste of War in Latin Battle Narrative', *Trends in Classics*, 11 (1): 119–38.

Ambühl, A. (forthcoming), 'Kämpfende und verwundete, sterbende und gefallene Körper: Der Krieg als Prüfstein für Körperkonzepte in der griechischen und römischen Antike', in J. Althoff and T. Pommerening (eds), *Lebendig oder tot, gesund oder krank: Der menschliche Körper in vormodernen Kulturen*, Darmstadt: Philipp von Zabern.

Ameis, K. F. and C. Hentze (1906), *Homers* Ilias *für den Schulgebrauch erklärt, vol. 2.4, Gesang XII–XIV*, 4th edn, Leipzig: Teubner.

References

Ameis, K. F. and C. Hentze (1908), *Homers* Ilias *für den Schulgebrauch erklärt, vol. 1.2, Gesang IV–VI*, 6th edn, Leipzig: Teubner.

Andersen, L. M. (2009), 'The Roman Military Community as Expressed in Its Burial Customs during the First to Third Centuries CE', PhD thesis, Brown University, Providence, RI.

Anderson, M. (1997), *The Fall of Troy in Early Greek Poetry and Art*, Oxford: Clarendon.

Anderson, M., D. Cairns and M. Sprevak (eds) (2019), *Distributed Cognition in Classical Antiquity*, Edinburgh: Edinburgh University Press.

Andò, V. (2013), *Violenza bestiale: Modelli dell'umano nella poesia greca epica e drammatica*, Caltanissetta and Roma: Salvatore Sciascia Editore.

Andreasen, J. O., L. K. Bakland, M. T. Flores, F. M. Andreasen and L. Andersson (2011), *Traumatic Dental Injuries – A Manual*, Oxford: Wiley-Blackwell.

Andrews, M. C. (1992), 'Virgil's Camilla and the Death of Hector', *Shakespeare Quarterly*, 43 (2): 219–21.

Anglesey, G. C. H. V. P. (1961), *One-leg: The Life and Letters of Henry William Paget, 1st Marquess of Anglesey*, London: Cape.

Antikas, T. G. and L. K. Wynn-Antikas (2016), 'New Finds from the Cremains in Tomb II at Vergina Point to Philip II and a Scythian Princess', *International Journal of Osteoarchaeology*, 26 (4): 682–92.

Arrington, N. T. (2015), *Ashes, Images and Memories: The Presence of the War Dead in Fifth-Century Athens*, Oxford: Oxford University Press.

Atherton, S. (2016), 'Researching Military Men', in A. J. Williams, K. N. Jenkings, M. F. Rech and R. Woodward (eds), *The Routledge Companion to Military Research Methods*, 243–55, Oxford: Routledge.

Avery, H. C. (1972), 'Herodotus' Picture of Cyrus', *AJP*, 93: 529–46.

Baaz Eriksson, M. and M. Stern (2013), *Sexual Violence as a Weapon of War?: Perceptions, Prescriptions, Problems in the Congo and Beyond*, London: Zed Books.

Bain, D. (2007), 'Low Words in High Places: Sex, Bodily Functions, and Body Parts in Homeric Epic and Other Higher Genres', in P. J. Finglass, C. Collard and N. J. Richardson (eds), *Hesperos: Studies in Ancient Greek Poetry Presented to M. L. West on His Seventieth Birthday*, 40–57, Oxford: Oxford University Press.

Baker, P. (2004), *Medical Care for the Roman Army on the Rhine, Danube and British Frontiers from the First through Third Centuries AD*, Oxford: Archaeopress.

Bakker, E. (1997), 'Storytelling in the Future: Truth, Time, and Tense in Homeric Epic', in E. Bakker and A. Kahane (eds), *Tradition, Performance, and the Epic Text*, 11–36, Cambridge, MA: Harvard University Press.

Bakker, E. and A. Kahane (eds) (1997), *Tradition, Performance, and the Epic Text*, Cambridge, MA: Harvard University Press.

Barkawi, T. and S. Brighton (2011), 'Powers of War: Fighting, Knowledge and Critique', *International Political Sociology*, 5: 126–43.

Barnett, C. (2014), 'Alcoholic Beverages and Resistance to Roman Imperialism in Dalmatia', *Croatian Studies Review*, 10: 13–34.

Barrett, F. J. (2002), 'Gender Strategies of Women Professionals: The Case of the US Navy', in M. Dent and S. Whitehead (eds), *Managing Professional Identities*, 157–73, London: Routledge.

Barthes, R. (2006), *The Language of Fashion*, trans. A. Stafford, Oxford: Berg.

Bartsch, S. (1997), *Ideology in Cold Blood*, Cambridge, MA: Harvard University Press.

Basham, V. M. (2013), *War, Identity and the Liberal State*, London and New York: Routledge.

Basham, V. M. (2018), 'Liberal Militarism as Insecurity, Desire and Ambivalence: Gender, Race and the Everyday Geopolitics of War', *Security Dialogue*, 49 (1–2): 32–43.

Bassett, S. E. (1933), 'Achilles' Treatment of Hector's Body', *TAPA*, 64: 41–65.

Bassett, S. R. (1999), 'The Death of Cyrus the Younger', *CQ*, 49: 473–83.

Basson, W. P. (1986), 'Vergil's Camilla: A Paradoxical Character', *Acta Classica*, 29: 57–68.

BBC (2016), 'Transgender soldier becomes first woman on Army front line', 17 September. Available online: https://www.bbc.co.uk/news/uk-37391807 (accessed 14 February 2022).
Beagon, M. (2002), 'Beyond Comparison: M. Sergius Silus, Fortunae Victor', in G. Clark and T. Rajak (eds), *Philosophy and Power in the Graeco-Roman World: Essays in Honour of Miriam Griffin*, 111–32, Oxford: Oxford University Press.
Beck, J. (2016), 'A Military Definition of Reality: Researching Literature and Militarization' in A. J. Williams, N. Jenkings, R. Woodward and M. F. Rech (eds), *The Routledge Companion to Military Research Methods*, 84–94, Oxford: Routledge.
Beck, U., M. Wagner, X. Li, D. Durkin-Meisterernst and P. E. Tarasov (2014), 'The Invention of Trousers and Its Likely Affiliation with Horseback Riding and Mobility: A Case Study of Late 2nd Millennium BC Finds from Turfan in Eastern Central Asia', *Quaternary International*, 12: 224–35.
Behr, F. D'A. (2007), *Feeling History: Lucan's Stoicism and the Poetics of Passion*, Columbus, OH: Ohio State University Press.
Bekić, L. (2011), 'Andetrij, rimsko vojno uporište. Topografske odrednice', in A. Librenjak and D. Tončinić (eds), *Arheološka istraživanja u Cetinskoj krajini*, 315–26, Zagreb and Sinj: Hrvatsko arheološko društvo, Muzej Cetinske krajine and Filozofski Fakultet Sveučilišta u Zagrebu.
Bell, S. (2021), 'Images and Interpretation of Africans in Roman Art and Social Practice', in L. Cline and N. T. Elkins (eds), *The Oxford Handbook of Roman Imagery and Iconography*, 425–63, New York: Oxford University Press.
Benn Torres, J. (2020), 'Anthropological Perspectives on Genomic Data, Genetic Ancestry, and Race', *Yearbook of Physical Anthropology*, 171: 74–86.
Bennett, D., S. S. Frere and S. Stow (eds) (1982), *Excavations at Canterbury Castle: Volume 1*, Canterbury: Canterbury Archaeological Trust.
Berger, R. W. (1979), 'Rubens's "Queen Tomyris with the Head of Cyrus"', *Museum of Fine Arts Bulletin*, 77: 4–35.
Bernabé, A. ([1987] 1996), *PEG*, 1, Stuttgart: Teubner.
Bernand, A. (1999), *Guerre et violence dans la Grèce antique*, Paris: Hachette.
Bernard, M. (ed.) (2020), *Fashion Theory: A Reader*, London: Routledge.
Berti, I., M. G. Castello and C. Scilabra (eds) (2020), *Ancient Violence in the Modern Imagination: The Fear and the Fury*, London: Bloomsbury Academic.
Bethard, J. D. and E. A. DiGangi (2020), 'Letter to the Editor—Moving beyond a Lost Cause: Forensic Anthropology and Ancestry Estimates in the United States', *Journal of Forensic Sciences*, 65: 1791–2.
Birke, L. and K. Thompson (2018), *(Un)stable Relations: Horses, Humans, and Social Agency*, London and New York: Routledge.
Birley, A. R. (1979), *The People of Roman Britain*, London: Batsford.
Birley, A. R. (1998), 'A New Tombstone from Vindolanda', *Britannia*, 29: 299–306.
Birley, E., J. Charlton and W. P. Hedley (1933), 'Excavations at Housesteads in 1932', *Archaeologia Aeliana*, 4 (10): 82–96.
Bishop, M. C. (1992), 'The Early Imperial "Apron"', *Journal of Roman Military Equipment Studies*, 3: 81–104.
Bishop, M. C. and J. C. N. Coulston (2006), *Roman Military Equipment from the Punic Wars to the Fall of Rome*, 2nd edn, Oxford: Oxbow Books.
Blakey, M. L. (2020), 'Archaeology under the Blinding Light of Race', *Current Anthropology*, 61: S183–S197.
Bliquez, L. J. (1996), 'Prosthetics in Classical Antiquity: Greek, Etruscan, and Roman Prosthetics', *ANRW*, II 37(3): 2640–76.
Bliquez, L. J. (2014), *The Tools of Asclepius: Surgical Instruments in Greek and Roman Times*, Leiden: Brill.
Blümel, C. (1923), *Der Fries des Tempels der Athena Nike*, Berlin: Altmann.

Bodiou, L. and V. Mehl (eds) (2017), *L'Antiquité écarlate: Le sang des Anciens*, Rennes: Presses Universitaires de Rennes.

Bodiou, L. and V. Mehl (eds) (2019), *Dictionnaire du corps dans l'Antiquité*, Rennes: Presses Universitaires de Rennes.

Bolens, G. (2000), *La logique du corps articulaire: Les articulations du corps humain dans la littérature occidentale*, Rennes: Presses universitaires de Rennes.

Booth, P. (2014), 'A Late Roman Military Burial from the Dyke Hills, Dorchester on Thames, Oxfordshire', *Britannia*, 45: 243–73.

Booth, P., A. Boyle, S. Clough, H. E. M. Cool and D. Poore (eds) (2010), *The Late Roman Cemetery at Lankhills, Winchester: Excavations 2000–2005*, Oxford Monograph No. 10, Oxford: Oxford Archaeology.

Borchhardt, J. (1976), *Die Bauskulptur des Heroon von Limyra: Das Grabmal des lykischen Königs Perikles*, Berlin: Gebr. Mann.

Bosworth, B. (2009), 'Thucydides and the Unheroic Dead', in O. Palagia (ed.), *Art in Athens during the Peloponnesian War*, 168–87, Cambridge: Cambridge University Press.

Bousquet, A., J. Grove and N. Shah (2020), 'Becoming War: Towards a Martial Empiricism', *Security Dialogue*, 5 (1): 1–20.

Bouvier, D. (2000–1), 'Morts de Priam et d'Astyanax: deux scènes interdites dans les poèmes homériques', *Classica (Brasil)*, 13 (13–14): 37–57.

Bowie, A. (2003), 'Fate May Harm Me, I have Dined Today: Near-Eastern Royal Banquets and Greek Symposia in Herodotus', *Pallas*, 61: 99–109.

Bowman, A. (1994), *Life and Letters from the Roman Frontier*, New York: Routledge.

Bowman, A. K. and J. D. Thomas (1991), 'A Military Strength Report from Vindolanda', *JRS*, 81: 62–73.

Bowman, A. K. and J. D. Thomas (1994), *The Vindolanda Writing Tablets II*, London: British Museum.

Boyd, B. W. (1992), 'Virgil's Camilla and the Traditions of Catalogue and Ecphrasis (*Aeneid* 7.803-17)', *AJP*, 113 (2): 213–34.

Bradley, M., V. Leonard and L. Totelin (eds) (2021), *Bodily Fluids in Antiquity*, Oxford: Routledge.

Breeze, D., J. Close-Brooks and G. Ritchie (1976), 'Soldiers' Burials at Camelon, Stirlingshire 1922 and 1975', *Britannia*, 7: 73–95.

Brewster, E. H. (1927), 'A Weaver of Oxyrhynchus: Sketch of a Humble Life in Roman Egypt', *TAPA*, 58: 132–54.

Briand, M. (2011), 'À propos de νήπιος dans l'*Iliade* et l'*Odyssée*: ambiguïtés et variations auctoriales, entre récit et performativité', in E. Raymond (ed.), *Vox poetae: Manifestations auctoriales dans l'épopée gréco-latine*, 195–213, Paris: De Boccard.

Brice, L. L. (ed.) (2019), *New Approaches to Greek and Roman Warfare*, Hoboken, NJ: Wiley-Blackwell.

Briggs, D. (2003), 'Babies in Homer's Epic Poems *c.*750–725 B.C.', *Infant Observation*, 6 (3): 24–9.

Brommer, F. (1967), *Die Metopen des Parthenon*, Mainz: von Zabern.

Brown, E. (2002), 'The Prosthetics of Management: Motion Study, Photography, and the Industrialized Body in World War 1 America', in K. Ott, D. Serlin and S. Mihm (eds), *Artificial Parts, Practical Lives: Modern Histories of Prosthetics*, 241–81, New York: New York University Press.

Brügger, C. (2017), *Homer's* Iliad: *The Basel Commentary, Book XXII*, ed. A. Bierl and J. Latacz, trans. B. W. Millis, S. Strack; ed. and trans. S. D. Olson, Berlin: De Gruyter.

Brunet, S. (2014) 'Women with Swords', in S. Christesen and D. G. Kyle (eds), *A Companion to Sport and Spectacle in Greek and Roman Antiquity*, 544–58, Chichester: John Wiley & Sons.

Buck, T., E. M. Greene, A. Meyer, V. Barlow and E. Graham (2019), 'The Body in the Ditch: Alternative Funerary Practices on the Northern Frontier of the Roman Empire?', *Britannia*, 50: 203–24.

Bunn, S. (2010), *Nomadic Felts*, London: British Museum Press.

Buora, M. (2008), 'Diffusione delle fibule Aucissa nell'area altoadriatica', in M. Buora and S. Seidel (eds), *Fibule antiche del Friuli, Cataloghi e monografie archeologiche dei Civici musei di Udine 9*, 30–32, Udine/Roma: L'Erma di Bretschneide.

Burgess, J. S. (2004), 'Performance and the Epic Cycle', *CJ*, 100 (1): 1–23.

Burgess, J. S. (2010), 'The Hypertext of Astyanax', *Trends in Classics*, 2: 211–24.

Burn, A. R. (1953), '*Hic breve vivitur*: A Study of the Expectation of Life in the Roman Empire', *Past & Present*, 4: 2–31.

Burnham, D. (1973), *Cut My Cote*, Toronto: Royal Ontario Museum.

Butler, J. (1990), *Gender Trouble: Feminism and the Subversion of Identity*, London: Routledge.

Butler, J. (2009), *Frames of War*, London: Verso.

Butler, S. and A. Purves (eds) (2013), *Synaesthesia and the Senses*, Durham: Acumen.

Byers, S. N. (2010), *Introduction to Forensic Anthropology*, New York: Allyn and Bacon.

Caddick, N., L. Cooper, L. Godier-McBard and M. Fossey (2020), 'Hierarchies of Wounding: Media Framings of "Combat" and "Non-Combat" Injury', *Media, War & Conflict*, 14 (4): 503–21.

Caffell, A. and M. Holst (2012), *Osteological analysis of 3 and 6 Driffield Terrace, York, North Yorkshire*, York: York Osteoarchaeology, No. 0212.

Caffell, A. and M. Holst (2016), *Osteological Analysis of 40–46 Western Road, Leicester*, York: York Osteoarchaeology, No. 0716.

Calament, F. and M. Durand (eds) (2014), *Antinoé à la vie, à la mode*, Lyons: Fage.

Callava, T. (2019), *Silk Through the Ages*, London: LID Publishing.

Cambi, N. (1984), 'Gardunski tropej', in Ž. Rapanić (ed.), *Cetinska krajina od prethistorije do dolaska Hrvata*, 77–92, Split: Hrvatsko arheološko društvo.

Cambi, N. (1994), 'Stele iz kasnoantičke grobnice u Dugopolju', *VAHD*, 86: 147–81.

Cambi, N. (2005), *Kiparstvo Rimske Dalmacije*, Split: Književni Krug.

Cambi, N. (2013), 'Roman Military Tropaea from Dalmatia', in M. Sanader, A. Rendić-Miočević, D. Tončinić and I. Radman-Livaja (eds), *Proceedings of the XVIIth Roman Military Equipment Conference: Weapons and Military Equipment in a Funerary Context*, 9–22, Zagreb: FF Press.

Cambi, N. (2020), 'Salona and Tilurium, Two Interconnected Stone Carving Centres', *Arheološki radovi i rasprave*, 19: 7–30.

Campanile, D., F. Carlà-Uhink and M. Facella (eds) (2017), *TransAntiquity: Cross-dressing and Transgender Dynamics in the Ancient World*, London: Routledge.

Campbell, B. and L. A. Tritle (eds) (2013), *The Oxford Handbook of Warfare in the Classical World*, Oxford: Oxford University Press.

Carroll, M. (2006), *Spirits of the Dead*, Oxford: Oxford University Press.

Carroll, M. (2009), 'Dead Soldiers on the Move: Transporting Bodies and Commemorating Men at Home and Abroad', *Gladius*, 13: 823–32.

Caston, V. and S.-M. Weineck (eds) (2016), *Our Ancient Wars: Rethinking War through the Classics*, Ann Arbor, MI: University of Michigan Press.

Cecchet, L., C. Degelmann and M. Patzelt (eds) (2019), *The Ancient War's Impact on the Home Front*, Newcastle-upon-Tyne: Cambridge Scholars Publishing.

Chamberlain, A. (2006), *Demography in Archaeology*, Cambridge: Cambridge University Press.

Chaniotis, A. (2005), *War in the Hellenistic World: A Social and Cultural History*, Oxford: Blackwell Publishing.

Chenery, C., H. Eckardt and G. Müldner (2011), 'Cosmopolitan Catterick? Isotopic Evidence for Population Mobility on Rome's Northern Frontier', *Journal of Archaeological Science*, 38: 1525–36.

Chesi, C. M. and F. Spiegel (eds) (2020), *Classical Literature and Posthumanism*, London: Bloomsbury Academic.

Chidwick, H.-M. (2017), '*Quidquid homo est*: Military Manliness in Lucan's *Civil War*', *HARTS & Minds*, 3 (2): 1–18.

Chidwick, H.-M. (2018), 'The Military Step: Theorising the Mobilisation of the Roman Army', *Critical Military Studies*, 7 (2): 107–31.
Chidwick, H.-M. (2021), 'One or Many *Milites*? Military Multiplicity in Latin Epic', *Ramus*, 49 (1–2): 111–32.
Chrissanthos, S. G. (2007), 'Aeneas in Iraq', in M. B. Cosmopoulos (ed.), *Experiencing War: Trauma and Society in Ancient Greece and Today*, 225–57, Chicago, IL: Ares.
Clarke, M. (1999), *Flesh and Spirit in the Songs of Homer: A Study of Words and Myth*, Oxford: Clarendon.
Clauss, M., A. Reiß and S. Rüther (eds) (2019), *Vom Umgang mit den Toten: Sterben im Krieg von der Antike bis zur Gegenwart*, Paderborn: Ferdinand Schöningh.
Clynes, M. E. and N. S. Kline (1960), 'Cyborgs and Space', *Astronautics*, 14 (9): 26–7 and 74–6.
Cohen, B. (2012), 'The Non-Greek in Greek Art', in T. J. Smith and D. Plantzos (eds), *A Companion to Greek Art*, 456–79, Chichester: Wiley Blackwell.
Coker, C. (2013), *Warrior Geeks: How 21st Century Technology is Changing the Way We Fight and Think about War*, London: C. Hurst & Co.
Colledge, M. A. R. (1977), *Parthian Art*, London: Paul Elek.
Collingwood, R. G. and R. P. Wright (1995), *The Roman Inscriptions of Britain, Volume I: Inscriptions on Stone* (rev. edn), Stroud: Sutton.
Collins, R. (2006), 'Late Roman Frontier Communities in Northern Britain: A Theoretical Context for the "End" of Hadrian's Wall', *TRAC*, 2005: 1–11.
Connolly, J. (2016), 'A Theory of Violence in Lucan's *Bellum Civile*', in P. Mitsis and I. Ziogas (eds), *Wordplay and Powerplay in Latin Poetry*, 273–97, Berlin and New York: De Gruyter.
Constandse-Westermann, T. S. (1982), 'A Skeleton Found in a Roman Well at Velsen (Province North Holland. The Netherlands)', *Helinium*, 22: 135–67.
Cook, B. F., B. Ashmole and D. E. Strong (2005), *Relief Sculpture of the Mausoleum at Halicarnassus*, Oxford: Oxford University Press.
Coopey, E. S. (2020), '*Fratres Romani* in Dalmatia: The Social Dynamics of *Legio* VII and the Construction of Community and Identity through Roman Funerary Monuments', MRes diss., Macquarie University, Sydney.
Corbeill, A. (1996), *Controlling Laughter: Political Humour in the Late Roman Republic*, Princeton, NJ: Princeton University Press.
Corbeill, A. (2004), *Nature Embodied: Gesture in Ancient Rome*, Princeton, NJ: Princeton University Press.
Cosmopoulos, M. B. (ed.) (2007), *Experiencing War: Trauma and Society in Ancient Greece and Today*, Chicago, IL: Ares.
Coulston, J. C. N. (1998), 'How to Arm a Roman Soldier', *Bulletin of the Institute of Classical Studies*, 42 (71): 167–90.
Coulston, J. C. N. (2004), 'Military Identity and Personal Self-Identity in the Roman Army', in L. de Ligt, E. A. Hemelrijk and H. W. Singor (eds), *Roman Rule and Civic Life: Local and Regional Perspectives*, 133–52, Amsterdam: J.C. Gieben.
Coulston, J. (2007), 'Art, Culture and Service: The Depiction of Roman Soldiers on Funerary Monuments of the Third Century AD', in L. de Blois and E. Lo Cascio (eds), *Impact of Empire VI, The Impact of the Roman Army (200 BC–AD 476): Economic, Social, Political, Religious and Cultural Aspects*, 529–61, Leiden: Brill.
Coulston, J. and H. Dodge (2000), '"Armed and Belted Men": The Soldiery in Ancient Rome', in J. Coulston and H. Dodge (eds), *Ancient Rome: The Archaeology of the Eternal City*, 112–75, Oxford: Oxford School of Archaeology.
Čović, B. (1987), 'Srednjodalmatinska grupa', in A. Benac (ed.), *Praistorija Jugoslavenskih zemalja*, Vol. 5, 442–80, Sarajevo: Akademija nauka i umjetnosti BiH.
Crerar, B. (2015), 'Deviancy in Late Romano-British Burial', in M. Millett, L. Revell and A. Moore (eds), *The Oxford Handbook of Roman Britain*, 381–405, Oxford: Oxford University Press.

Crossland, Z. (2012), 'Materiality and Embodiment', in D. Hicks and M. C. Beaudry (eds), *The Oxford Handbook of Material Culture Studies*, 386–404, Oxford: Oxford University Press.

Crummy, N., P. Crummy and C. Crossan (eds) (1995), *Excavations of Roman and Later Cemeteries, Churches and Monastic Sites in Colchester, 1971–88*, Colchester Archaeological Report 9, Colchester: Colchester Archaeological Trust.

Curtis, V. S. (1993), 'A Parthian Statuette from Susa and the Bronze Statue from Shami', *Iran*, 31: 63–9.

Curtis, V. S. (2017), 'The Parthian Haute-couture at Palmyra', in T. Long and A. Højen Sørensen (eds), *Positions and Professions in Palmyra*, 52–67, Copenhagen: Specialtrykkeriet Arco.

Cuvigny, H. (1992), 'Morts et maladies', in L. Rubinstein, J. Bingen, A. Bülow-Jacobsen, W. Cockle, H. Cuvigny and W. Van Rengen (eds), *Mons Claudianus Ostraca Graeca et Latina I (O.Claud.1-190)*, 75–110, Cairo: Documents de fouilles de l'Institut français d'archéologie orientale 29.

Cuvigny, H. (1997), 'Morts et maladies', in J. Bingen and A. Bülow-Jacobsen (eds), *Mons Claudianus Ostraca Graeca et Latina II (O.Claud. 191-414)*, 191–223, Cairo: Documents de fouilles de l'Institut français d'archéologie orientale 32.

Dalby, A. (1995), 'The *Iliad*, the *Odyssey*, and their Audiences', *CQ*, 45 (2): 269–79.

Davies, M. (1994), 'Odyssey 22.474-7: Murder or Mutilation?', *CQ*, 44: 534–6.

Davies, S. M., P. S. Bellamy, M. J. Heaton and P. J. Woodward (eds) (2002), *Excavations at Alington Avenue, Fordington, Dorchester, Dorset, 1984–87*, Dorset Natural History and Archaeological Society Monograph Series 15, Dorchester: Dorset Natural History and Archaeological Society.

de Bakker, M., B. van den Berg and J. Klooster (eds) (2022), *Emotions and Narrative in Ancient Literature and Beyond: Studies in Honour of Irene de Jong*, Leiden and Boston, MA: Brill.

De Carolis, S. (2009), *Ars Medica: I ferri del mestiere*, Guaraldi: Rimini.

de Jong, I. J. F. (2012), *Homer,* Iliad: *Book XXII*, Cambridge: Cambridge University Press.

de Oliveira Barata, S. (2021), *The Alternative Limb Project*. Available online: https://thealternativelimbproject.com/limbs/anatomical-leg/ (accessed 11 February 2022).

de Sensi Sestito, G. (1999), 'Schiave di Guerra tra δίκαιον ed ὠμότης', in F. Reduzzi Merola and A. Storchi Marino (eds), *Femmes-Esclaves: Modèles d'Interprétation Anthropologique, Économique, Juridique,* 111–28, Naples: Jovene Editore.

de Spinoza, B. (1996), *Ethics*, trans. E. Curley, London: Penguin.

De Vivo, J. S. (2014), 'The Memory of Greek Battle: Material Culture and/as Narrative of Combat', in P. Meineck and D. Konstan (eds), *Combat Trauma and the Ancient Greeks*, 163–84, Berlin: Springer.

Deacy, S. and F. McHardy (2015), 'Ajax, Cassandra, and Athena: Retaliatory Warfare and Gender Violence at the Sack of Troy' in G. Lee, H. Whittaker and G. Wrightson (eds), *Ancient Warfare: Introducing Current Research*, vol. 1, 252–72, Newcastle-upon-Tyne: Cambridge Scholars Press.

DeBoer, K. R. (2019), 'Arms and the Woman: Discourses of Militancy and Motherhood in Vergil's *Aeneid*', *Arethusa*, 52 (2): 129–64.

Deleuze, G. (2004), *Desert Islands and Other Texts 1953–1974*, ed. D. Lapoujade, trans. M. Taormina, South Pasadena, CA: Semiotext(e).

DeMello, M. (2014), *Body Studies: An Introduction*, London: Routledge.

Deminion, M. (2020), 'Manly and Monstrous Women: (De-)Constructing Gender in Roman Oratory', in A. Surtees and J. Dyer (eds), *Exploring Gender Diversity in the Ancient World*, 197–208, Edinburgh: Edinburgh University Press.

Denzler, A., S. Grüner and M. Raasch (eds) (2016), *Kinder und Krieg: Von der Antike bis in die Gegenwart*, Berlin and Boston, MA: Walter de Gruyter.

Department of Defense (DoD) (2020), 'DTM 19-004, Military Service by Transgender Persons and Persons with Gender Dysphoria', *Military Health System*. Available online: https://health.mil/Reference-Center/Policies/2020/03/17/Military-Service-by-Transgender-Persons-and-Persons-with-Gender-Dysphoria (accessed 5 January 2022).

Derrick, M. (2009), 'The Excavation of a Roman Cemetery at 21-33 Newarke Street, Leicester', *Transactions of the Leicestershire Archaeology and History Society*, 83: 63–102.
Desmond, W. (2004), 'Punishments and the Conclusion of Herodotus' *Histories*', *Greek, Roman and Byzantine Studies*, 44: 19–40.
Devecka, M. (2019), 'The Seer's Two Bodies: Some Early Greek Histories of Technology', in G. M. Chesi and F. Spiegel (eds), *Classical Literature and Posthumanism*, 185–92, London: Bloomsbury.
Dietert, M. and D. Dentice (2023), 'Transgender Military Experiences: From Obama to Trump', *Journal of Homosexuality*, 70 (6): 993–1010.
Dimopoulou, A. (2019), 'Η μέθη ως αδίκημα στην αρχαία Ελλάδα', in M. Giouni (ed.), *Δίκαιο και Ιστορία 4*, 1–15, Athens: Εκδόσεις Σάκκουλα.
Dinsmoor, W. B. (1956), 'The Sculptured Frieze from Bassae', *AJA*, 60 (4): 401–52.
Dinter, M. (2012), *Anatomizing Civil War: Studies in Lucan's Epic Technique*, Ann Arbor, MI: University of Michigan Press.
Dodig, R. (2007), 'Rimski Vojni Pečati Na Crijepu Iz Ljubuškoga', *VAHD*, 31 (1): 143–64.
Dornan-Fish, J. (2012), 'Motive Matters: Intentionality, Embodiment, and the Individual in Archaeology', *Time & Mind* 5 (3): 279–98.
Dowden, K. (2004), 'The Epic Tradition in Greece', in R. Fowler (ed.), *The Cambridge Companion to Homer*, 188–205, Cambridge: Cambridge University Press.
Down, C. and M. Rule (1978), *Chichester Excavations I*, West Sussex: Phillimore.
Down, C. and J. Magilton (eds) (1993), *Chichester Excavations VIII*, Chichester: Chichester District Council.
Draycott, J. (2015), 'Reconstructing the Lived Experience of Disability in Antiquity: A Case Study from Roman Egypt', *Greece & Rome*, 62 (2): 189–205.
Draycott, J. (2018), 'Life as a Cyclops: Mythology and the Mockery of the Visually Impaired', in G. Kazantzidis and N. Tsoumpra (eds), *Morbid Laughter: Exploring the Comic Dimensions of Disease in Classical Antiquity*, 404–9, Chicago, IL: Illinois Classical Studies.
Draycott, J. (2022), *Prosthetics and Assistive Technology in Ancient Greece and Rome*, Cambridge: Cambridge University Press.
Draycott, J. (2023), 'Automata, Cyborgs, and Hybrids: Bodies and Machines in Antiquity', in M. Gerolemou and G. Kanzantzidis (eds), *Body and Machine in Classical Antiquity*, 48–69, Cambridge: Cambridge University Press.
Drews, R. (2004), *Early Riders: The Beginnings of Mounted Warfare in Asia and Europe*, London: Routledge.
Dronke, U. (1969), *The Poetic Edda, Volume 1: Heroic Poems*, Oxford: Oxford University Press.
duBois, P. (1991), *Centaurs and Amazons: Women and the Pre-history of the Great Chain of Being*, Ann Arbor, MI: University of Michigan Press.
Dué, C. (2002), *Homeric Variations on a Lament by Briseis*, Lanham, MD: Rowman and Littlefield.
Duffy, X. (2018), *Commemorating Conflict: Greek Monuments of the Persian Wars*, Oxford: Archaeopress.
Džino, D. (2006), 'Delmati, vino i formiranje etničkog identiteta u predrimskom Iliriku', *VAPD*, 99: 71–80.
Dzino, D. (2010), *Illyricum in Roman Politics, 229 BC–AD 68*, Cambridge: Cambridge University Press.
Džino, D. (2021), *From Justinian to Branimir: The Making of the Middle Ages in Dalmatia*, London and New York: Routledge.
Džino, D. and A. Domić Kunić (2018), 'A View from the Frontier Zone: Roman Conquest of Illyricum', in M. Milićević Bradač and D. Demicheli (eds), *The Century of the Brave: Roman Conquest and Indigenous Resistance in Illyricum during the Time of Augustus and His Heirs*, 77–87, Zagreb: FF Press.
Easterling, P. E. (1997), 'Constructing the Heroic', in C. Pelling (ed.), *Greek Tragedy and the Historian*, 21–37, Oxford: Oxford University Press.

Eckardt, H. (2005), 'The Social Distribution of Roman Artefacts: The Case of Nail-cleaners and Brooches in Britain', *Journal of Roman Archaeology*, 18: 139–60.
Eckardt, H. (ed.) (2010), *Roman Diasporas: Archaeological Approaches to Mobility and Diversity in the Roman Empire*, *JRA* Suppl., 78, Portsmouth: JRA.
Eckardt, H., G. Müldner and G. Speed (2015), 'The Late Roman Field Army in Northern Britain? Mobility, Material Culture and Multi-isotope Analysis at Scorton (N. Yorks)', *Britannia*, 46: 191–223.
Edgar, H. J. and K. L. Hunley (2009), 'Race Reconciled?: How Biological Anthropologists View Human Variation', *AJPA*, 139: 1–4.
Edmondson, J. (2015a), 'Inscribing Roman Texts: Officinae, Layout, Carving Techniques', in J. Edmondson and C. Bruun (eds), *Oxford Handbook of Roman Epigraphy*, 111–30, Oxford: Oxford University Press.
Edmondson, J. (2015b), 'Roman Family History', in J. Edmondson and C. Bruun (eds), *Oxford Handbook of Roman Epigraphy*, 559–81, Oxford: Oxford University Press.
Edmunds, S. (1990), *Homeric* nêpios, New York: Garland.
Edwards, C. (2007), *Death in Ancient Rome*, New Haven, CT: Yale University Press.
Egan, R. B. (1983), 'Arms and Etymology in *Aeneid* 11', *Vergilius*, 29: 19–26.
Egging Dinwiddy, K. and P. Bradley (2011), *Prehistoric Activity and a Romano-British Settlement at Poundbury Farm, Dorchester, Dorset*, Salisbury: Wessex Archaeology.
Elders, M. J., G. R. Brown, E. Coleman, T. A. Kolditz and A. M. Steinman (2015), 'Medical Aspects of Transgender Military Service', *Armed Forces & Society*, 41 (2): 199–220.
Ellerbrock, U. (2021), *The Parthians: The Forgotten Empire*, London: Routledge.
Emery, M. V., A. T. Duggan, T. J. Murchie, R. J. Stark, J. Klunk, J. Hider, K. Eaton, E. Karpinski, H. P. Schwarcz, H. N. Poinar and T. L. Prowse (2018), 'Ancient Roman Mitochondrial Genomes and Isotopes Reveal Relationships and Geographic Origins at the Local and Pan-Mediterranean Scales', *Journal of Archaeological Science: Reports*, 20: 200–9.
Enloe, C. (2000), *Maneuvers: The International Politics of Militarizing Women's Lives*, Berkeley, CA: University of California Press.
Enloe, C. (2004), *The Curious Feminist: Searching for Women in a New Age of Empire*, Oakland, CA: University of California Press.
Enloe, C. (2013), 'Combat and "Combat": A Feminist Reflection', *Critical Studies on Security*, 1 (2): 260–3.
Entwistle, J. (2000), *The Fashioned Body*, Cambridge: Polity Press.
Estèves, A. (2020), *Poétique de l'horreur dans l'épopée et l'historiographie latines*, Bordeaux: Ausonius.
Euripides (2008), *Fragments*, ed. and trans. C. Collard and M. Cropp, Cambridge, MA, and London: Harvard University Press.
Evans, D., N. Holbrook and E. R. McSloy (2006), 'A Later Iron Age Cemetery and Roman Settlement at Henbury School, Bristol: Excavations in 2004', *Bristol and Gloucestershire Archaeological Report No. 4*, 1–50, Gloucestershire: Cotswold Archaeology.
Fabre-Serris, J. and A. Keith (eds) (2015), *Women and War in Antiquity*, Baltimore, MD: Johns Hopkins University Press.
Fabre-Serris, J., A. Keith and F. Klein (eds) (2021), *Identities, Ethnicities and Gender in Antiquity*, Berlin and Boston, MA: Walter de Gruyter.
Fabrizi, V. (2021), 'War, Weather and Landscape in Livy's *Ab Urbe Condita*', in B. Reitz-Joosse, M. W. Makins and C. J. Mackie (eds), *Landscapes of War in Greek and Roman Literature*, 38–61, London: Bloomsbury Academic.
Fagan, G. G. (2011), *The Lure of the Arena: Social Psychology and the Crowd at the Roman Games*, Cambridge: Cambridge University Press.
Fantuzzi, M. (2020), *Rhesus*, Cambridge: Cambridge University Press.
Farrokh, K. (2005), *Sasanian Elite Cavalry*, Oxford: Oxbow.

Farwell, D. E. and T. L. Molleson (1993), *Excavations at Poundbury 1966–80. II: The Cemeteries*, Dorset Natural History and Archaeological Society Monograph 11, Dorchester: Dorset Natural History and Archaeological Society.
Feder, E. K. (2007), *Family Bonds: Genealogies of Race and Gender*, Oxford: Oxford University Press.
Feichtinger, B. (2018), '*Rapta* – ein weibliches (Nach-)Kriegsschicksal', in C. Walde and G. Wöhrle (eds), *Gender Studies in den Altertumswissenschaften: Gender und Krieg*, 69–90, Trier: Wissenschaftlicher Verlag.
Feldherr, A. (2021), 'The Challenge of Historiographic *Enargeia* and the Battle of Lake Trasimene', in B. Reitz-Joosse, M. W. Makins and C. J. Mackie (eds), *Landscapes of War in Greek and Roman Literature*, 62–88, London: Bloomsbury Academic.
Felton, D. (2014), 'The Motif of the Mutilated Hero in Herodotus', *Phoenix*, 68: 47–61.
Ferris, I. M. (2000), *Enemies of Rome: Barbarians through Roman Eyes*, Stroud: Sutton Publishing.
Fields, N. (2005), 'Headhunters of the Roman Army', in A. Hopkins and M. Wyke (eds), *Roman Bodies: Antiquity to the Eighteenth Century*, 55–66, London: Oxbow.
Figg, L. and J. Farrell-Beck (1993), 'Amputation in the Civil War: Physical and Social Dimensions', *Journal of the History of Medicine and the Allied Sciences*, 48: 454–75.
Fine, C. (2018), *Testosterone rex: Unmaking the Myths of Our Gendered Minds*, London: Icon.
Fisher, G. and D. D. P. Loren, (2003), 'Embodying Identity in Archaeology: Introduction', *Cambridge Archaeological Journal*, 13 (2): 225–30.
Fletcher, J. (2007), 'Horkos in the *Oresteia*,' in A. Sommerstein and J. Fletcher (eds), *Horkos: The Oath in Greek Society*, 102–12, Exeter: Bristol Phoenix Press.
Fluck, C. and G. Vogelsang-Eastwood (eds) (2004), *Riding Costume in Egypt: Origin and Appearance*, Leiden: Brill.
Fontein, J. (2010), 'Between Tortured Bodies and Resurfacing Bones: The Politics of the Dead in Zimbabwe', *Journal of Material Culture*, 15 (4): 423–48.
Ford, S. and C. Schnitzlein (2017), 'Gender Dysphoria in the Military', *Current Psychiatry Reports*, 19, 102: 1–8.
Forrest, S. (2016), *The Age of the Horse: An Equine Journey through Human History*, London: Atlantic Books.
Foucault, M. (1991), *Discipline and Punish: The Birth of the Prison*, trans. A. Sheridan, London: Penguin.
Foucault, M. (2004), *Society Must Be Defended*, trans. D. Macey, London: Penguin.
Fowler, B. H. (1991), 'The Creatures and the Blood', *Illinois Classical Studies*, 16: 85–100.
Foxhall, L. (2013), *Studying Gender in Classical Antiquity*, Cambridge: Cambridge University Press.
Foxhall, L. and G. Neher (2011), 'Gender and the City before Modernity: An Introduction', *Gender & History*, 23 (2): 491–509.
Fratantuono, L. (2007), *Madness Unchained: A Reading of Virgil's Aeneid*, Plymouth: Lexington Books.
Frazer, J. (1921), *Apollodorus: The Library*, vol. 1, Cambridge, MA: Harvard University Press.
Frembgen, J. W. (2006), 'Honour, Shame, and Bodily Mutilation: Cutting off the Nose among Tribal Societies in Pakistan', *Journal of the Royal Asiatic Society*, 16: 243–60.
Gaardbo Kuhn, K., D. H. Campbell-Lendrum, B. Armstrong and C. R. Davies (2003), 'Malaria in Britain: Past, Present, and Future', *Proceedings of the National Academy of Sciences USA*, 100 (17): 9997–10001.
Gabričević, B. (1984), 'Iz antičkog perioda Cetinjske krajine', in Ž. Rapanić (ed.), *Cetinska krajina od prethistorije do dolaska Hrvata*, 93–106, Split: Hrvatsko arheološko društvo.
Gaca, K. (2008), 'Reinterpreting the Homeric Simile of *Iliad* 16.7-11: The Girl and Her Mother in Ancient Greek Warfare', *AJP*, 129 (2): 145–71.
Gaca, K. (2010), 'The Andrapodizing of War Captives in Greek Historical Memory', *TAPA*, 140: 116–71.

Gaca, K. (2011), 'Girls, Women, and the Significance of Sexual Violence in Ancient Warfare', in E. Heineman (ed.), *Sexual Violence in Conflict Zones from the Ancient World to the Era of Human Rights*, 73–88, Philadelphia, PA: University of Pennsylvania Press.

Gaca, K. (2012), 'Telling the Girls from the Boys and the Children: Interpreting *paides* in the Sexual Violence of Populace-Ravaging Warfare', *Illinois Classical Studies*, 35–6: 85–109.

Gaca, K. (2014), 'Martial Rape, Pulsating Fear, and the Sexual Maltreatment of Girls (*paides*), Virgins (*parthenoi*), and Women (*gunaikes*) in Antiquity', *AJP*, 135: 303–57.

Gaca, K. (2015), 'Ancient Warfare and the Ravaging Martial Rape of Girls and Women: Evidence from Homeric Epic and Greek Drama', in M. Masterson, N. Sorkin Rabinowitz and J. Robson (eds), *Sex in Antiquity*, 278–97, London: Routledge.

Gaca, K. (2016a), 'Continuities in Rape and Tyranny in Martial Societies from Antiquity Onward', in S. L. Budin and J. Macintosh Turfa (eds), *Women in Antiquity: Real Women across the Ancient World*, 1041–56, London: Routledge.

Gaca, K. (2016b), 'Martial Religion, Ravaging Warfare, and Rape: Polytheistic Greek and Monotheistic Israelite Views', in M. Paludi and J. H. Ellens (eds), *Feminism in Religion*, 173–204, London: Palgrave Macmillan.

Gaca, K. (2018a), 'Ancient Warfare Beyond the Battle: Populace Ravaging and Heterosexual Rape', in N. Williams and N. Foote (eds), *Civilians and Warfare*, 42–66, London: Routledge.

Gaca, K. (2018b), 'The Martial Rape of Girls and Women in Antiquity and Modernity', in F. Ní Aoláin, N. Cahn, D. Haynes and N. Valji (eds), *The Oxford Handbook of Gender and Conflict*, 305–15, Oxford: Oxford University Press.

Gaca, K. (2021), 'Controlling Female Slave Sexuality and Men's War-Driven Sexual Desires', in D. Kamen and C. W. Marshall (eds), *Slavery and Sexuality in Classical Antiquity*, 40–65, Madison, WI: University of Wisconsin Press.

Gannett, L. (2014), 'Biogeographical Ancestry and Race', *Studies in History and Philosophy of Biological and Biomedical Sciences*, 47 (A): 173–84.

Gantz, T. (1993), *Early Greek Myth: A Guide to Literary and Artistic Sources*, Baltimore, MD: Johns Hopkins University Press.

Gardner, H. H. (2019), *Pestilence and the Body Politic in Latin Literature*, Oxford: Oxford University Press.

Garland, R. ([1995] 2010), *The Eye of the Beholder: Deformity and Disability in the Graeco-Roman World*, London: Bristol Classical Press.

Gavrylenko, V. (2012), 'The "Body without Skin" in the Homeric Poems', in M. Horstmanshoff, H. King and C. Zittel (eds), *Blood, Sweat and Tears: The Changing Concepts of Physiology from Antiquity into Early Modern Europe*, 481–502, Leiden: Brill.

Geber, J. (2017), 'Human Remains', in N. Holbrook, J. Wright, E. R. McSloy and J. Geber (eds), *The Western Cemetery of Roman Cirencester: Excavations at the Former Bridges Garage Tetbury Road, Cirencester, 2011–2015*, 98–118, Cirencester Excavations VII, Dorchester: Cotswold Archaeology.

Gerolemou, M. and G. Kanzantzidis (eds) (2023), *Body and Machine in Classical Antiquity*, Cambridge: Cambridge University Press.

Gervers, M. and V. Gervers (1974), 'Felt-making Craftsmen of Anatolian and Iranian Plateaux', *Textile Museum Journal*, 4 (1): 14–29.

Gheibi, M. (2009), *A History of Persian Costume*, Tehran: Hirmand Publishing [in Persian].

Gherchanoc, F. and S. Wyler (eds) (2020), *Corps en morceaux: Démembrer et recomposer les corps dans l'Antiquité classique*, Rennes: Presses Universitaires de Rennes.

Ghirshman, R. (1962), *Persian Art: The Parthian and Sasanian Dynasties*, New York: Golden Press.

Gibert, J. C. (2019), *Euripides: Ion*, Cambridge: Cambridge University Press.

Gildenhard, I. and J. Henderson (2018), *Virgil,* Aeneid *11 (Pallas & Camilla) 1–224, 498–521, 532–96, 648–89, 725–835*, Cambridge: Open Book Publishers.

Gilliver, K. (2007), 'Battle', in P. Sabin, H. van Wees, M. Whitby (eds), *The Cambridge History of Greek and Roman Warfare, Vol. II: Rome from the Late Republic to the Late Empire*, 122–57, Cambridge: Cambridge University Press.
Gleba, M. (2008), 'You are What You Wear: Scythian Costume as Identity', in M. Gleba, C. Munkholt and M.-L. Nosch (eds), *Dressing the Past*, 13–28, Oxford: Oxbow Books.
Godard, A. (1937), 'Les statues parthes de Shami', *Athar-e Iran*, II: 285–305.
Godard, A. (1965), *The Art of Iran*, London: Allen and Unwin.
Goff, B. (ed.) (2005), *Classics and Colonialism*, London: Duckworth.
Goldstein, J. S. (2006), *War and Gender*, Cambridge: Cambridge University Press.
Goldsworthy, A. (2003), *The Complete Roman Army*, London: Thames and Hudson.
Goldsworthy, A. and I. Haynes (eds) (1999), *The Roman Army as a Community*, JRA Supplementary Series 34, Portsmouth, RI: Journal of Roman Archaeology.
Golubović, S., N. Mrdjić and C. S. Speal (2009), 'Killed by the Arrow: Grave No. 152 from Viminacium', in A. W. Busch and H.-J. Schalles (eds), *Waffen in Aktion: Akten der 16 Internationalen Roman Military Equipment Conference (ROMEC). Xanten, 13–16 Juni 2007*, 55–64, Mainz: Philipp von Zabern.
González, J. M. (2013), *The Epic Rhapsode and His Craft*, Washington, DC: Center for Hellenic Studies.
Gransden, K. W. (ed.) (1998), *Virgil Aeneid Book XI*, Cambridge: Cambridge University Press.
Graver, M. (1995), 'Dog-Helen and Homeric Insult', *Classical Antiquity*, 14: 41–61.
Gray, C. H., H. J. Figueroa-Sarriera and S. Mentor (eds) (1995), *The Cyborg Handbook*, London: Routledge.
Graziosi, B. and J. Haubold (2010), *Homer,* Iliad: *Book VI*, Cambridge: Cambridge University Press.
Greene, E. (2013), 'Female Networks in the military communities of the Roman West: a view from the Vindolanda tablets', in E. A. Hemelrijk and G. Woolf (eds), *Women and the Roman City in the Latin West*, 369–90, Amsterdam: Brill.
Greene, E. M. (2016), 'Identities and Social Roles of Women in Military Settlements in the Roman West', in S. L. Budin and J. M. Turfa (eds), *Women in Antiquity*, 942–53, London and New York: Routledge.
Greene, E. M. (2020), 'Roman Military Communities and the Families of Auxiliary Soldiers', in L. L. Brice (ed.), *New Approaches to Greek and Roman Warfare*, 149–59, New York: Wiley.
Greener, B. K. (2023), '"Man(ne)'s World": Explaining the Paradoxical Nature of Attitudes towards Women in Combat', *Critical Military Studies*: 9 (3): 306–23.
Grethlein, J., L. Huitink and A. Tagliabue (eds) (2020), *Experience, Narrative, and Criticism in Ancient Greece: Under the Spell of Stories*, Oxford: Oxford University Press.
Grömer, K. and A. Aali (2016), 'Durchdacht gemacht: Die Kleidung eines Salzmannes aus Chehrabad, Iran', *Universum Magazin*, 3: 108–9.
Grömer, K. and N. Bagherpour (2018), 'The Salt Mummies of Zanjan: Textile Research in Iran in a Conservation and Exhibition Project', *Archaeological Textiles Review*, 60: 110–13.
Grue, J. (2016), 'The Problem with Inspiration Porn: A Tentative Definition and a Provisional Critique', *Disability & Society*, 31: 838–49.
Gruen, E. S. (2011), *Cultural Identity in the Ancient Mediterranean*, Los Angeles: Getty.
Gui, M. and D. Petruţ (2018), 'Soldiers and Their Monuments for Posterity: Manifestations of Martial Identity in the Funerary Iconography of Roman Dacia', in G. Cupcea and R. Varga (eds), *Social Interactions and Status Markers in the Roman World*, 115–38, Oxford: Archaeopress.
Gunderson, E. (2003a), *Staging Masculinity*, Ann Arbor: University of Michigan Press.
Gunderson, E. (2003b), *Declamation, Paternity, and Roman Identity*, Cambridge: Cambridge University Press.
Guyomarc'h, P., M. Campagna-Vaillancourt, C. Kremer and A. Sauvageau (2010), 'Discrimination of Falls and Blows in Blunt Head Trauma: A Multi-criteria Approach', *Journal of Forensic Sciences*, 55: 423–7.

Hall, E. (2018), 'Why are the Erinyes Female? or What is so Feminine about Revenge?', in L. Dawson and F. McHardy (eds), *Revenge and Gender in Classical, Medieval and Renaissance Literature*, 33–52, Edinburgh: Edinburgh University Press.
Hallett, J. (2015), 'Fulvia: The Representation of an Elite Roman Woman Warrior', in J. Fabre-Serris and A. Keith (eds), *Women and War in Antiquity*, 247–65, Baltimore, MD: Johns Hopkins University Press.
Hanson, V. D. (1993), *Hoplites: The Classical Greek Experience*, London: Routledge.
Haralambieva, A. (2013), 'Aucissafibeln mit Inschriften aus den Provinzen Moesia inferiorund Thracia', in G. Grabherr, B. Kainrath and T. Schierl (eds), *Relations Abroad*, 102–8, Innsbruck: Innsbruck University Press.
Hardwick, L. and C. Gillespie (eds) (2007), *Classics in Post-colonial Worlds*, Oxford: Oxford University Press.
Harnett, A. (2000), 'Escaping the "Evil Avenger" and the "Supercrip": Images of Disability in Popular Television', *Irish Communications Review*, 8: 21–9.
Harris, D. (1985), 'A Challenge to the Definition of *nêpios* in the *Iliad*', *Tropos*, 11: 23–6.
Harris, W. V. (2003), 'The Rage of Women', *Yale Classical Studies*, 32: 121–43.
Harrison, E. B. (1997), 'The Glories of the Athenians: Observations on the Program of the Frieze of the Temple of Athena Nike', *Studies in the History of Art*, 49: 108–25.
Harrison, E. H. (1966), 'The Composition of the Amazonomachy on the Shield of Athena Parthenos', *Hesperia*, 35 (2): 107–33.
Harrison, T. (2000), *Divinity and History: The Religion of Herodotus*, Oxford: Oxford Classical Monographs.
Harvey, J. (1996), *Traditional Textiles of Central Asia*, London: Thames and Hudson.
Hasegawa, G. R. (2012), *Mending Broken Soldiers: The Union and Confederate Programs to Supply Artificial Limbs*, Carbondale and Edwardsville, IL: Southern Illinois University Press.
Haynes, I. (2013), *Blood of the Provinces: The Roman Auxilia and the Making of Provincial Society from Augustus to the Severans*, Oxford: Oxford University Press.
Haynes, I. (2014), 'Identity and the Military Community in Roman Britain', in M. Millett, L. Revell and A. Moore (eds), *The Oxford Handbook of Roman Britain*, 448–63, Oxford: Oxford University Press.
Head, H. and G. Holmes (1911), 'Sensory Disturbances in Cerebral Lesions', *Brain*, 34: 102–254.
Hearne, V. (2007), *Adam's Task: Calling Animals by Name*, New York: Skyhorse Publishing.
Heckel, W., F. S. Naiden, E. E. Garvin and J. Vanderspoel (eds) (2021), *A Companion to Greek Warfare*, Hoboken, NJ: Wiley-Blackwell.
Hefner, J. T. (2016), 'Biological Distance Analysis, Cranial Morphoscopic Traits, and Ancestry Assessments in Forensic Anthropology', in M. A. Pilloud and J. T. Hefner (eds), *Biological Distance Analysis: Forensic and Bioarchaeological Perspectives*, 297–310, Cambridge: Academic Press.
Heidenreich, S. M. and J. P. Roth (2019), 'The Neurophysiology of Panic on the Battlefield', in L. L. Brice (ed.), *New Approaches to Greek and Roman Warfare*, 127–38, Hoboken, NJ: Wiley-Blackwell.
Hermes, L. (2002), 'Military Lower Extremity Amputee Rehabilitation', *Physical Medicine and Rehabilitation Clinics of North America*, 13: 45–66.
Herodotus (1925), *The Persian Wars*, trans. A. D. Godley, Cambridge, MA: Harvard University Press.
Herrmann, G. (1977), *The Iranian Revival*, Oxford: Elsevier-Phaidon.
Hershkowitz, D. (1998), *The Madness of Epic*, New York: Oxford University Press.
Herter, H. (1973), 'Der weinende Astyanax', *Grazer Beiträge*, 1: 157–64.
Hesiod (1914), *Homeric Hymns and Homerica*, ed. and trans. H. G. Evelyn-White, London: W. Heinemann.
Higate, P. R. (ed.) (2003), *Military Masculinities*, Westport, CT: Praeger.

Higate, P., A. Dawes, T. Edmunds, K. N. Jenkings and R. Woodward (2019), 'Militarization, Stigma, and Resistance: Negotiating Military Reservist Identity in the Civilian Workplace', *Critical Military Studies*, 7 (2): 173–91.
Higate, P. and M. Henry (2010), 'Space, Performance and Everyday Security in the Peacekeeping Context', *International Peacekeeping*, 17 (1): 32–49.
Hildebrandt, B. and C. Gillis (eds) (2017), *Silk: Trade and Exchangea along the Silk Road between China and Rome in Antiquity*, Oxford: Oxbow.
Hingley, R. (2014), 'Postcolonial and Global Rome', in M. Pitts and M. J. Versulys (eds), *Globalisation and the Roman World*, 32–46, Cambridge: Cambridge University Press.
Hinz, W. (1969), *Altiranische Funde und Forschungen*, Berlin: De Gruyter.
Hodder, I. (2014), 'The Entanglements of Humans and Things: A Long-Term View', *New Literary History*, 45: 19–36.
Hodgson, N. and D. Breeze (2020), 'Plague on Hadrian's Wall? Interpreting Evidence for Pandemics in Roman Britain', *Current Archaeology*, 365: 28–35.
Holbrook, N. (2015), 'The Towns of South-west England', in M. Fulford and N. Holbrook (eds), *The Towns of Roman Britain: The Contribution of Commercial Archaeology since 1990*, 90–116, London: Roman Society.
Holbrook, N., J. Wright, E. R. McSloy and J. Geber (2017), *The Western Cemetery of Roman Cirencester: Excavations at the former Bridges Garage Tetbury Road, Cirencester, 2011–2015*, Cirencester Excavations Vol. VII, Dorchester: Cotswold Archaeology.
Holder, P. (2017), 'Auxiliary Recruitment as Reflected in Military Diplomas Issued 71–168', in D. Dana and B. Rossignol (eds), *Entrer dans l'armée romaine: bassins de recrutement desunités auxiliaires (Ier-IIe s. ap. J.-C.) HiMA*, 13–33, Paris: Société d'Édition de Revues.
Holmes, B. (2010), *The Symptom and the Subject: The Emergence of the Physical Body in Ancient Greece*, Princeton, NJ: Princeton University Press.
Holmes, B. (2015), 'Situating Scamander: "Natureculture" in the *Iliad*', *Ramus*, 44: 29–51.
Holmqvist, C. (2013), 'Undoing War: War Ontologies and the Materiality of Drone Warfare', *Millennium*, 41 (3): 535–52.
Hölscher, T. (2003), 'Images of War in Greece and Rome: Between Military Practice, Public Memory, and Cultural Symbolism', *JRS*, 93: 1–17.
Hölscher, T. (2006), 'The Transformation of Victory into Power', in S. Dillon and K. E. Welch (eds), *Representations of Warfare in Ancient Rome*, 27–48, Cambridge: Cambridge University Press.
Holzner, B. (2011), 'Wars, Bodies and Development', in C. Sylvester (ed.), *Experiencing War*, 42–63, London: Routledge.
Homer ([1902] 1963a), *Homeri Opera, Volume I: Iliad Books I–XII*, 3rd edn, ed. D. B. Monro and T. W. Allen, Oxford: Clarendon Press.
Homer ([1902] 1963b), *Homeri Opera, Volume II: Iliad Books XIII–XXIV*, 3rd edn, ed. D. B. Monro and T. W. Allen, Oxford: Clarendon Press.
Hope, V. M. (1997), 'Words and Pictures: The Interpretation of Romano-British Tombstones', *Britannia*, 28: 245–58.
Hope, V. M. (2003), 'Trophies and Tombstones: Commemorating the Roman Soldier', *World Archaeology*, 35 (1): 79–97.
Hope, V. M. (2015), 'Bodies on the Battlefield: The Spectacle of Rome's Fallen Soldiers', in A. Bakogianni and V. M. Hope (eds), *War as Spectacle: Ancient and Modern Perspectives on the Display of Armed Conflict*, 157–77, London: Bloomsbury Academic.
Hope, V. M. (2016), 'Inscriptions and Identity', in M. Millett, A. Moore and L. Revell (eds), *The Oxford Handbook of Roman Britain*, 285–302, Oxford: Oxford University Press.
Hornblower, S. (2007), 'Warfare in Ancient Literature: The Paradox of War', in P. Sabin, H. van Wees and M. Whitby (eds), *The Cambridge History of Greek and Roman Warfare*, 22–53, Cambridge: Cambridge University Press.
Hornblower, S. (2013), *Herodotus: Histories Book V*, Cambridge: Cambridge University Press.

Horsfall, N. (2016), *The Epic Distilled: Studies in the Composition of the Aeneid*, Oxford: Oxford University Press.
Horsfall, N. (ed.) (2000), *Virgil,* Aeneid *7: A Commentary*, Leiden: Brill.
Horstmanshoff, M. (2012), 'Disability and Rehabilitation in the Graeco-Roman World', in R. Breitwieser (ed.), *Behinderungen und Beeinträchtigungen/Disability and Impairment in Antiquity*, 1–9, Oxford: Archaeopress.
Hoss, S. (2012), 'The Roman Military Belt', in M.-L. Nosch (ed.), *Wearing the Cloak: Dressing the Soldier in Roman Times*, 29–44, Oxford: Oxbow Books.
Hoss, S. (2013), 'A Theoretical Approach to Roman Military Belts', in M. Sanader, A. Rendić-Miočević, D. Tončinić and I. Radman-Livaja (eds), *Proceedings of the XVIIth Roman Military Equipment Conference: Weapons and Military Equipment in a Funerary Context*, 317–26, Zagreb: FF Press.
Hoss, S. (2017), 'Sharp Dressed Men: The Roman Military Belt as a Fashion Item', *JRMES*, 18: 85–99.
Houghtaling, L. and T. L. Osypuk (2023), 'Health Disparities among Millennial Veterans by Sexual Orientation', *Military Psychology*, 35 (3): 204–14.
House of Commons Defence Committee (HoC) (2021), 'Protecting Those Who Protect Us: Women in the Armed Forces from Recruitment to Civilian Life: Second Report of Session 2021–22', 25 July. Available online: https://committees.parliament.uk/publications/6961/documents/72773/default/ (accessed 14 February 2022).
Hovey, J. E. (2018), 'Gallantry and its Discontents: Joan of Arc and Virtuous Transmasculinity in Radclyffe Hall and Vita Sackville-West', *Feminist Modern Studies*, 1 (1–2): 113–37.
Hughes, J. D. (2013), 'Warfare and Environment in the Ancient World', in B. Campbell and L. A. Tritle (eds), *The Oxford Handbook of Warfare in the Classical World*, 128–39, Oxford: Oxford University Press.
Hughes, J. D. (2014), *Environmental Problems of the Greeks and Romans: Ecology in the Ancient Mediterranean*, 2nd edn, Baltimore, MD: Johns Hopkins University Press.
Hurst, H., P. Garrod, H. Robinson, J. Toynbee and D. Brown (eds) (1975), 'Excavations at Gloucester: third interim report: Kingsholm 1966–75', *The Antiquaries Journal* 55 (2): 267–94.
i Lacomba, A. R. (2014), 'La destrucción de Valentia (75 a.c.) y la cultura material de la época de sertorio (82-75 a.c.)', in J. M. Jávega and F. S. Sellés (eds), *Las guerras civiles romanas en Hispania: una revisión histórica desde la Contestania*, 65–78, Alicante: Diputación de Alicante.
i Lacomba, A. R. and M. C. Galvez (1995), 'La primera evidencia arqueológica de la destrucción de Valentia por Pompeyo', *Journal of Roman Archaeology*, 8: 19–40.
Ibarra, Á. (2009), 'Legions and Locals: Roman Provincial Communities and Their Trophy Monuments', PhD thesis, University of Texas, Austin.
Immerwahr, H. R. (1966), *Form and Thought in Herodotus*, Cleveland, OH: Press of Western Reserve University.
Ingalls, W. B. (1998), 'Attitudes towards Children in the *Iliad*', *Echos du monde classique: Classical Views*, 42 (1): 13–34.
Inskip, S. A. (2020), pers. comm.
Isaac, B. (2017), *Empire and Ideology in the Graeco-Roman World. Selected Papers*, Cambridge: Cambridge University Press.
Ivčević, S. (2002), 'Fibule', in E. Marin (ed.), *Longae Salonae*, vol. 1, 229–76, vol. 2, 124–47, Split: AMS.
Ivčević, S. (2004), 'Dijelovi opreme rimskog vojnika iz Garduna / Components of Roman Military Equipment from Gardun', *Opuscula archaeologica*, 28: 159–76.
Ivčević, S. (2010), 'Dijelovi vojničke opreme iz Garduna u Arheološkom muzeju u Splitu', *Opuscula archaeologica*, 34: 127–44.
Ivčević, S. (2011), 'Antičke fibule iz vojnog logora Tilurij (Gardun) / Roman fibulae from the Military Camp Tilurium (Gardun)', in A. Librenjak and D. Tončinić (eds), *Arheološka*

Istraživanja u cetinskoj krajini, Izdanja HAD 27, 161–86, Zagreb: HAD, MCK and Filozofski fakultet Sveučilišta u Zagrebu.
Ivčević, S. (2013a), 'Project Tilurium – Roman Military Equipment', in M. Sanader, A. Rendić-Miočević, D. Tončinić and I. Radman-Livaja (eds), *Proceedings of the XVIIth Roman Military Equipment Conference. Weapons and Military Equipment in a Funerary Context*, 435–53, Zagreb: FF Press.
Ivčević, S. (2013b), 'First-century Military Gear from Salona', in M. Sanader, A. Rendić-Miočević, D. Tončinić and I. Radman-Livaja (eds), *Proceedings of the XVIIth Roman Military Equipment Conference: Weapons and Military Equipment in a Funerary Context*, 299–316, Zagreb: FF Press.
Ivčević, S. (2014), 'Metalni nalazi', in M. Sanader, D. Tončinić, S. Ivčević, Z. Buljević and T. Šeparović, *Tilurium III: Istraživanja 2002–2006. godine*, 147–223, Zagreb: FF Press.
Ivčević, S. (2016), 'Tipologija i kronologija rimske vojne opreme u doba principata na području između rijeka Krke i Cetine', PhD thesis, University of Zagreb.
Ivčević, S. (2017), 'Metalni Nalazi', M. Sanader, D. Tončinić, S. Ivčević, Z. Buljević, T. Šeparović and I. Miloglav, *Tilurium IV: Arheološka istraživanja 2007–2010. godine*. 239–317, Zagreb: FF Press.
Ives, R. (2015), *Osteological Analysis of the Human Remains from America Street*, London: AOC Archaeology Group.
Ivleva, T. (2016), 'The Role of Brooches in Roman Military Dress of the Late 1st–3rd Centuries AD in the Roman West', *JRMES*, 17: 121–7.
Ivleva, T. and R. Collins (2020), *Un-Roman Sex: Gender, Sexuality, and Lovemaking in the Roman Provinces and Frontiers*, London: Routledge.
Jackson, R. (2005), 'Holding on to Health? Bone Surgery and Instrumentation in the Roman Empire', in H. King (ed.), *Health in Antiquity*, 97–119, London: Routledge.
Jäger, U. (2019), 'A Unique Alxon-Hunnic Horse-and-Rider Statuette (Late Fifth Century CE) from Ancient Bactria / Modern Afghanistan in the Pritzker Family Collection, Chicago', *Sino-Platonic Papers*, 290: 2–72.
James, S. (1999), 'The Community of Soldiers: A Major Identity and Centre of Power in the Roman Empire', in P. Baker, C. Forcey, S. Jundi and R. Witcher (eds), *TRAC*, 1998: 14–25, Oxford: Oxbow Books.
James, S. (2001), 'Soldiers and Civilians: Identity and Interaction in Roman Britain', in S. James and M. Millett (eds), *Britons and Romans: Advancing an Archaeological Agenda: CBA Research Report*, 125: 187–209, York: Council for British Archaeology.
James, S. (2002), 'Writing the Legions: The Development and Future of Roman Military Studies in Britain', *Archaeological Journal*, 159: 1–58.
James, S. (2006), 'Engendering Change in Our Understanding of the Structure of Roman Military Communities', *Archaeological Dialogues*, 13: 31–6.
James, S. (2009), 'The Point of the Sword: What Roman-era Weapons Could Do to Bodies – and Why They Often Didn't', in A. W. Busch and H. J. Schalles (eds), *Waffen in Aktion: Akten der 16. Internationalen Roman Military Equipment Conference (ROMEC)*, 41–54, Mainz: Verlag Philipp von Zabern.
James, S. (2011), *Rome and the Sword: How Warriors and Weapons Shaped Roman History*, London: Thames and Hudson.
James, S. (2013), 'The Archaeology of War', in B. Campbell and L. Tritle (eds), *The Oxford Handbook of Warfare in the Classical World*, 91–127, Oxford: Oxford University Press.
Janssen, A. J. (1957), *Het antieke tropaion*, Gent: Erasmus.
Jarrett, M. (1994), 'Non-legionary Troops in Roman Britain: Part One, The Units', *Britannia*, 25: 3–77.
Jenkins, I. (2006), *Greek Architecture and Its Sculpture*, Cambridge, MA: Harvard University Press.
Jones, C. P. (1987), 'Stigma: Tattooing and Branding in Graeco-Roman Antiquity', *JRS*, 77: 139–55.

Jones, C. P. (2005), 'Ten Dedications to the Gods and Goddesses and the Antonine Plague', *Journal of Roman Archaeology*, 18: 293–301.

Jones, M., D. Stocker and A. Vince (2003), *The City by the Pool: Assessing the Archaeology of Lincoln*, Oxford: Oxbow.

Jones, R. (1984), 'Death and Distinction in Military and Civilian Roman Britain', in T. F. C. Blagg and A. C. King (eds), *Military and Civilian in Roman Britain: Cultural Relationships in a Frontier Province*, 219–26, Oxford: BAR British Series.

Jovanova, L. (2017), 'Mass Grave', in T. Mikulcic (ed.), *A Glance into the Dark Side of the Roman History: Scupi – A Mass Grave*, 13–20, Skopje: Museum of the City of Skopj.

Joyce, R. A. (1998), 'Performing the Body in Prehispanic Central America', *Res*, 33: 14765.

Judd, M. A. (2002), 'Ancient Injury Recidivism: An Example from the Kerma Period of Ancient Nubia', *International Journal of Osteoarchaeology*, 12: 89–106.

Judd, M. A. (2004), 'Trauma in the City of Kerma: Ancient versus Modern Injury Patterns', *International Journal of Osteoarchaeology*, 14: 34–51.

Judd, M. A. (2008), 'The Parry Problem', *Journal of Archaeological Science*, 35: 1658–66.

Judd, M. A. (2017), 'Injury Recidivism Revisited: Clinical Research, Limitations, and Implications for Bioarchaeology', in C. E. Tegtmeyer and D. L. Martin (eds), *Broken Bones, Broken Bodies: Bioarchaeological and Forensic Approaches for Accumulative Trauma and Violence*, 1–24, London: Lexington Books.

Justinus (1953), *Epitome of the Philippic History of Pompeius Trogus*, trans. J. S. Watson, London: Henry G. Bohn.

Kakoschke, A. (2011), *Die Personennamen im römischen Britannien*, Hildesheim: Olms-Weidmann.

Kamash, Z. (2021), 'Rebalancing Roman Archaeology: From Disciplinary Inertia to Decolonial and Inclusive Action', *TRAJ*, 4 (1.4): 1–41.

Kanz, F. and K. Grossschmidt (2009), 'Dying in the Arena: The Osseous Evidence from the Ephesian Gladiators', in T. Wilmot (ed.), *Roman Amphitheatres and Spectacula, a 21st-century Perspective: Papers from an International Conference Held at Chester, 16th–18th February, 2007*, 211–20, Oxford: British Archaeological Reports.

Karanika, A. and V. Panoussi (eds) (2020), *Emotional Trauma in Greece and Rome: Representations and Reactions*, London and New York: Routledge.

Karasulas, A. (2004), *Mounted Archers of the Steppe, 600 BC–AD 1300*, Oxford: Oxbow.

Kawami, T. (1987), *Monumental Art of the Parthian Period*, Acta Iranica 26, Leiden: Brill.

Kawami, T. S. (1992), 'Archaeological Evidence for Textiles in Pre-Islamic Iran', *Iranian Studies*, 25 (1–2): 7–18.

Keegan, J. (1976), *The Face of Battle*, London: Jonathan Cape.

Keeling, M., H. Williamson, V. Williams, J. Kiff and D. Harcourt (2021), 'Body Image Concerns and Psychological Wellbeing among Injured Combat Veterans with Scars and Limb Loss: A Review of the Literature', *Military Behavioral Health*, 9 (1): 1–10.

Kéi, N. (2019), 'Du corps mutilé aux membres magnifiés', *Métis*, 17: 77–97.

Keith, A. (2012), 'Women in Augustan Literature', in S. L. James and S. Dillon (eds), *A Companion to Women in the Ancient World*, 385–99, Chichester: Wiley-Blackwell.

Keith, A. M. (2000), *Engendering Rome: Women in Latin Epic*, Cambridge: Cambridge University Press.

Keppie, L. (2003), '"Having Been a Soldier": The Commemoration of Military Service on Funerary Monuments of the Early Roman Empire', in J. J. Wilkes (ed.), *Documenting the Roman Army: Essays in Honour of Margaret Roxan*, 31–53, London: Institute of Classical Studies.

Kidd, F. (2003), 'Costume of the Samarkand Region of Sogdiana between the 2nd/1st Century B.C.E. and the 4th Century C.E.', *Bulletin of the Asia Institute*, 17: 35–69.

Kinder, J. M. (2015), *Paying with Their Bodies: American War and the Problem of the Disabled Veteran*, Chicago, IL, and London: University of Chicago Press.

Kinnee, L. (2018), *The Greek and Roman Trophy: From Battlefield Marker to Icon of Power*, London and New York: Routledge.
Kirchner, J. (1935), *Inscriptiones Graecae II et III: Inscriptiones Atticae Euclidis anno posteriors*, Berlin: De Gruyter.
Kirk, G. S. (1990), *The* Iliad*: A Commentary, Vol. II: Books 5–8*, Cambridge: Cambridge University Press.
Klooster, J. and I. N. I. Kuin (eds) (2020), *After the Crisis: Remembrance, Re-anchoring and Recovery in Ancient Greece and Rome*, London: Bloomsbury Academic.
König, A. (2022), *Visualising War and Peace*, podcast, University of St Andrews. Available online: https://arts.st-andrews.ac.uk/visualising-war/podcasts/ (accessed 26 May 2022).
Krebs, A. (2021), 'Open Wounds, Liquid Bodies and Melting Selves in Early Imperial Literature', in M. Bradley, V. Leonard and L. Totelin (eds), *Bodily Fluids in Antiquity*, 338–51, Oxford: Routledge.
Kremer, C., S. Racette, C. A. Dionne and A. Sauvageau (2008), 'Discrimination of Falls and Blows in Blunt Head Trauma: Systematic Study of the Hat Brim Line Rule in Relation to Skull Fractures', *Journal of Forensic Science*, 53: 716–19.
Krentz, P. M. (2000), 'Deception in Archaic and Classical Greek Warfare', in H. Van Wees (ed.), *War and Violence in Ancient Greece*, 167–200, Swansea: Classical Press of Wales.
Kucewicz, C. (2016), 'Mutilation of the Dead and the Homeric Gods', *CQ*, 66: 425–36.
Kucewicz, C. (2020), *The Treatment of the War Dead in Archaic Athens*, London: Bloomsbury.
Kümmel, G. (2003), 'A Soldier is a Soldier is a Soldier!? The Military and its Soldiers in an Era of Globalization', in G. Caforio (ed.), *Handbook of the Sociology of the Military*, 417–33, New York: Kluwer Academic/Plenum.
Kurilić, A. and Z. Serventi (2015), 'Buried Far Away: Easterners in Roman Liburnia', *Annales Universitatis Apulensis. Series Historica*, 19 (2): 13–36.
Kurz, G. (1966), *Darstellungsformen menschlicher Bewegung in der Ilias*, Heidelberg: Carl Winter.
Kwaspen, A. and A. De Moor (2013), 'The Pattern-cutting of Linen Trousers in Late Antiquity', in A. De Moor, C. Fluck and P. Linscheid (eds), *Drawing the Threads Together: Textiles and Footwear of the 1st Millennium* AD *from Egypt*, 252–63, Amsterdam: TerraLannoo.
Laes, C. (2018), *Disabilities and the Disabled in the Roman World: A Social and Cultural History*, Cambridge: Cambridge University Press.
Laforse, B. (2013), 'Fighting the Other: Part 1: Greeks and Achaemenid Persians', in B. Campbell and L. A. Tritle (eds), *The Oxford Handbook of Warfare in the Classical World*, 569–87, Oxford: Oxford University Press.
Langlands, R. (2018), *Exemplary Ethics in Ancient Rome*, Cambridge: Cambridge University Press.
Lateiner, D. (1990), 'Deceptions and Delusions in Herodotus', *Classical Antiquity*, 9: 230–46.
Lattimore, S. (2006), 'From Classical to Hellenistic Art', in K. H. Kinzl (ed.), *A Companion to the Classical World*, 456–79, Oxford: Blackwell Publishing.
Laurens, A.-F. (1985), 'Modalités iconographiques des meurtres d'enfants dans la céramique grecque: Variations iconiques sur le corps mutilé. Astyanax et Priam à Troie d'après la céramique attique', *Cahiers du GITA*, 1: 57–63.
Lauwers, J., H. Schwall and J. Opsomer (eds) (2018), *Psychology and the Classics: A Dialogue of Disciplines*, Berlin and Boston, MA: Walter de Gruyter.
Le Bohec, Y. (1994), *The Imperial Roman Army*, London: Batsford.
Le Meur-Weissman, N. (2009), 'Astyanax: Les enfants et la guerre dans l'*Iliade*', *Gaia*, 12: 29–43.
Leach, P. (1982), *Ilchester. Volume 1: Excavations 1974–1975*, Western Archaeological Trust Excavation Monograph 3, Bristol: Western Archaeological Trust.
Leach, S., H. Eckardt, C. Chenery, G. Müldner and M. Lewis (2010), 'A "Lady" of York: Migration, Ethnicity and Identity in Roman York', *Antiquity*, 84: 131–45.

Leach, S., M. Lewis, C. Chenery, G. Müldner and H. Eckardt (2009), 'Migration and Diversity in Roman Britain: A Multidisciplinary Approach to the Identification of Immigrants in Roman York, England', *AJPA*, 140: 546–61.
Leaf, W. ([1886] 1900), *The* Iliad, *Vol. I: Books I–XII*, London: Macmillan.
Leaf, W. ([1888] 1902), *The* Iliad, *Vol. II: Books XIII–XXIV*, London: Macmillan.
Lee, A. D. (2020), *Warfare in the Roman World*, Cambridge: Cambridge University Press.
Lee, G., H. Whittaker and G. Wrightson (eds) (2015), *Ancient Warfare: Introducing Current Research, Volume I*, Newcastle-upon-Tyne: Cambridge Scholars Publishing.
Lee, J. W. I. (2006), 'Warfare in the Classical Age', in K. H. Kinzl (ed.), *A Companion to the Classical World*, 480–508, Oxford: Blackwell Publishing.
Lee, J. W. I. (2013), 'The Classical Greek Experience', in B. Campbell and L. A. Tritle (eds), *The Oxford Handbook of Warfare in the Classical World*, 143–61, Oxford: Oxford University Press.
Lee, J. W. I. (2019), 'Daily Life in Classical Greek Armies, *c.* 500–330 BCE', in L. L. Brice (ed.), *New Approaches to Greek and Roman Warfare*, 39–51, New Jersey: Wiley-Blackwell.
Lee, M. (2015), *Body, Dress, and Identity in Ancient Greece*, Cambridge: Cambridge University Press.
Lehmhaus, L. (2019), '"An Amputee May Go Out with His Wooden Aid on Shabbat": Dynamics of Prosthetic Discourse in Talmudic Traditions', in J. Draycott (ed.), *Prostheses in Antiquity*, 97–124, London: Routledge.
Leigh, M. (1995), 'Wounding and Popular Rhetoric at Rome', *Bulletin of the Institute of Classical Studies*, 40 (1): 195–221.
Leightley, D., D. Pernet, S. Velupillai, R. J. Stewart, K. M. Mark, E. Opie, D. Murphy, N. T. Fear and S. A. M. Stevelink (2020), 'The Development of the Military Service Identification Tool: Identifying Military Veterans in a Clinical Research Database Using Natural Language Processing and Machine Learning', *Journal of Medical Internet Research Medical Informatics*, 8 (5): e15852.
Lendon, J. E. (2000), 'Homeric Vengeance and the Outbreak of Greek Wars', in H. van Wees (ed.), *War and Violence in Ancient Greece*, 1–30, Swansea: Classical Press of Wales.
Lendon, J. E. (2017), 'Battle Description in the Ancient World, Part 1: Structure, Array and Fighting', *Greece & Rome*, 61 (1): 39–64.
Lethaby, W. R. (1915), 'The Nereid Monument Re-examined', *JHS*, 35: 208–24.
Levithan, J. (2019), 'Roman Siege Warfare: Moral and Morale', in L. L. Brice (ed.), *New Approaches to Greek and Roman Warfare*, 139–48, Hobboken, NJ: Wiley-Blackwell.
Lippman, M., D. Scahill and P. Schultz (2006), 'Knights 843–59, the Nike Temple Bastion, and Cleon's Shields from Pylos', *AJA*, 110 (4): 551–63.
Liston, M. A. (2020), 'Skeletal Evidence for the Impact of Battle on Soldiers and Non-combtants', in L. L. Brice (ed.), *New Approaches to Greek and Roman Warfare*, 81–94, Malden, MA: Wiley-Blackwell.
Liveley, G. (2008), 'Science Fictions and Cyber Myths: Or, Do Cyborgs Dream of Dolly the Sheep?', in V. Zajko and M. Leonard (eds), *Laughing with Medusa: Classical Myth and Feminist Thought*, 275–94. Oxford: Oxford University Press.
Llewellyn-Jones, L. (2020), 'The Royal "Gaunaka": Dress, Identity, Status and Ceremony in Achaemenid Iran', in S.-J. Simpson and S. Pankova (eds), *Scythians and Other Early Eurasian Nomads*, 248–57, Oxford: Archeopress.
Lloyd, A. B. (ed.) (1996), *Battle in Antiquity*, London: Duckworth.
Loades, M. (2016), *The Composite Bow*, Oxford: Oxbow.
Loe, L. (2001), 'Appendix 1: Specialist Report on the Human Skull (8658) from Vindolanda, Northumberland', Vindolanda Charitable Trust, 2001, A. Birley pers. comm.
Lohmann, D. (1988), *Die Andromache-Szenen der* Ilias, Hildesheim: G. Olms.
Lomas, D. W. B. (2021), '#ForgetJamesBond: Diversity, Inclusion and the UK's Intelligence Agencies', *Intelligence and National Security*, 36 (7): 995–1017.

Lovano, M. (2013), 'Writers on War Part 2 Rome: A Story of Conflict', in B. Campbell and L. A. Tritle (eds), *The Oxford Handbook of Warfare in the Classical World*, 74–90, Oxford: Oxford University Press.
Luff, R. (1984), 'The Human Remains from the Legionary Ditch', in P. Crummy (ed.), *Colchester Archaeological Report 3: Excavations at Lion Walk, Balkerne Lane, and Middleborough, Colchester, Essex*, 97–8, Colchester: Colchester Archaeological Trust Ltd and Department of the Environment.
Ma, J. (2013), *Statues and Cities: Honorific Portraits and Civic Identity in the Hellenistic World, Oxford Studies in Ancient Culture and Representation*, Oxford: Oxford University Press.
Machado, D. (2021), 'Deconstructing Disciplina: Disentangling Ancient and Modern Ideologies of Military Discipline in the Middle Republic', *AJP*, 142 (3): 387–424.
Mackay, F. (2021), *Female Masculinities and the Gender Wars*, London: Bloomsbury.
MacLeish, K. T. (2012), 'Armor and Anaesthesia: Exposure, Feeling and the Soldier's Body', *Medical Anthropology Quarterly*, 26 (1): 49–68.
MacLeish, K. (2016), 'Ethnography and the Embodied Life of War-making', in A. J. Williams, N. Jenkings, R. Woodward and M. F. Rech (eds), *The Routledge Companion to Military Research Methods*, 219–30, Oxford: Routledge.
MacLeod, C. W. (1982), *Homer,* Iliad: *Book XXIV*, Cambridge: Cambridge University Press.
MacMullen, R. (1984), 'The Legion as a Society', *Historia*, 33 (4): 440–56.
Majno, G. (1975), *The Healing Hand: Man and Wound in the Ancient World*, Cambridge, MA: Harvard University Press.
Mann, J. C. (1983), *Legionary Recruitment and Veteran Settlement during the Principate*, London: Institute of Archaeology.
Mann, J. C. (1993), 'Cavalry Promotion at Kirkby Thore', *Britannia*, 24: 261–4.
Mann, J. C. (2002), 'The Settlement of Veterans Discharged from Auxiliary Units Stationed in Britain', *Britannia*, 33: 183–8.
Maravela, A. (2019), 'The Judgemental Narrator: Narratorial *Nepios*-Comments from Homer to Nonnus, *Symbolae Osloenses*, 93 (1): 58–105.
Marcus, J. (1989), 'Corpus/Corps/Corpse: Writing the Body in/at War', in H. M. Cooper, A. Auslander Munich and S. Merrill Squier (eds), *Arms and the Woman: War Gender and Literary Representation*, 124–67, Chapel Hill, NC: University of North Carolina Press.
Marić, A. (2019), 'Portretne stele na prostoru današnje Hercegovine – Društveni kontekst', *Hercegovina: Časopis za kulturno i historijsko naslijeđe*, 18: 7–33.
Marović, I. (1961), 'Fibeln mit Inschrift vom Typus AVCISSA in den Archäologischen Museen von Zagreb, Zadar und Split', *JRGZM*, 8: 106–20.
Martin, B. (2020), *Harmful Interaction between the Living and the Dead in Greek Tragedy*, Liverpool: Liverpool University Press.
Mathiesen, H. E. (1992), *Sculpture in the Parthian Empire*, Aarhus: Aarhus University Press.
Matijević, I. (2017), 'Neobjavljena stela veterana legije Sedma Claudia pia fidelis iz Sikula', *Miscellanea Hadriatica et Mediterranea*, 3 (1): 125–38.
Matijević, I. (2019), 'Životni Suputnici Rimskih Vojnika Na Salonitanskim Nadgrobnim Natpisima Iz Doba Principata', *Radovi Zavoda Za Hrvatsku Povijest*, 51: 61–116.
Mattingly, D. (1999), 'War and Peace in Roman North Africa', in B. Ferguson and N. L. Whitehead (eds), *War in the Tribal Zone*, 31–60, Oxford: James Currey.
Mayor, A. (2014), *The Amazons*, Princeton, NJ: Princeton University Press.
Mayor, A. (2018), *Gods and Robots: Myths, Machines and Ancient Dreams of Technology*, Princeton, NJ: Princeton University Press.
McClellan, A. M. (2016), 'The Death and Mutilation of Imbrius in Iliad 13', *Yearbook of Ancient Greek Epic Online*, 1: 159–74.
McClellan, A. M. (2019), *Abused Bodies in Roman Epic*, Cambridge: Cambridge University Press.
McClure, L. K. (1996–7), 'Clytemnestra's Binding Spell (*Ag*. 958–974)', *CJ*, 92: 123–40.

McCoy, M. B. (2013), *Wounded Heroes: Vulnerability as a Virtue in Ancient Greek Literature and Philosophy*, Oxford: Oxford University Press.

McDonnell, M. (2006), *Roman Manliness: Virtus and the Roman Republic*, New York: Cambridge University Press.

McKenzie, M. and C. Thomas (2020), *In the Northern Cemetery of Roman London. Excavations at Spitalfields Market, London E1, 1991–2007*, MoLA Monograph 58, London: Museum of London Archaeology.

McKinley, J. I. and K. Egging Dinwiddy (2009), '"Deviant" Burials from a Late Romano-British Cemetery at Little Keep, Dorchester', *Proceedings of the Dorset Natural History and Archaeological Society*, 130: 43–61.

McLoughlin, K. (2011), *Authoring War: The Literary Representation of War from the Iliad to Iraq*, Cambridge: Cambridge University Press.

McSorley, K. (2014), 'Towards an Embodied Sociology of War', *Sociological Review*, 62 (2): 107–28.

McSorley, K. (ed.) (2013), *War and the Body*, London and New York: Routledge.

McWhirr, A., L. Viner and C. Wells (1982), *Romano-British Cemeteries at Cirencester*, Cirencester: Cirencester Excavation Committee.

Meineck, P. (2016), 'Combat Trauma and the Tragic Stage: Ancient Culture and Modern Catharsis', in M. Caston and S-M. Weineck (eds), *Our Ancient Wars: Rethinking War Through the Classics*, 184–205, Ann Arbor, MI: University of Michigan Press.

Meineck, P. and D. Konstan (eds) (2014), *Combat Trauma and the Ancient Greeks*, New York: Palgrave Macmillan.

Melchior, A. (2011), 'Caesar in Vietnam: Did Roman Soldiers Suffer from Post-Traumatic Stress Disorder', *Greece & Rome*, 58 (2): 209–23.

Meskell, L. (1996), 'The Somatization of Archaeology: Institutions, Discourses, Corporeality', *Norwegian Archaeological Review*, 29 (1): 1–16.

Micarelli, I., R. Paine, C. Giostra, M. A. Tafuri, A. Profico, M. Boggioni, F. Di Vincenzo, D. Massani, A. Papini and G. Manzi (2018), 'Survival to Amputation in Pre-antibiotic Era: A Case Study from a Longobard Necropolis (6th–8th Centuries AD)', *Journal of Anthropological Sciences*, 96: 1–16.

Migotti, B. (2013), 'Can Soldiers on 3rd-Century Stelae in Pannonia be Recognized Only by the Sagum?', in M. Sanader, A. Rendić-Miočević, D. Tončinić and I. Radman-Livaja (eds), *Proceedings of the XVIIth Roman Military Equipment Conference: Weapons and Military Equipment in a Funerary Context*, 259–71, Zagreb: FF Press.

Mihajlović, V. D. (2020), 'Roman Epigraphic Funerary Markers, Ontological Transition, and Relational Work-nets', in I. Selsvold and L. Webb (eds), *Beyond the Romans: Posthuman Perspectives in Roman Archaeology*, 25–40, Oxford: Oxbow Books.

Miletić, Ž. (1990), 'Sjeverna Salonitanska Nekropola / Northern Necropolis of Salona', *RFFZd*, 29 (16): 163–94.

Miletić, Ž. (1991), 'Istočna i jugoistočna nekropola Salone / Eastern and South-Eastern Necopolis in Salona', *RFFZd*, 30 (17): 21–50.

Miletić, Ž. (2015), 'Identification of Oriental Figures Depicted on Military Stelae from a Tilurium Workshop', in A. Nikoloska and S. Müskens (eds), *Romanising Oriental Gods? Religious Transformations in the Balkan Provinces in the Roman Period: New Finds and Novel Perspectives*, 357–70, Skopje: Macedonian Academy of Sciences and Arts and University of Leiden.

Millar, K. and J. Tidy (2017), 'Combat as a Moving Target: Masculinities, the Heroic Soldier Myth, and Normative Martial Violence', *Critical Military Studies*, 3 (2): 142–60.

Millett, P. C. (2013), 'Writers on War Part 1 Greece: Winning Ways in Warfare', in B. Campbell and L. A. Tritle (eds), *The Oxford Handbook of Warfare in the Classical World*, 46–73, Oxford: Oxford University Press.

Milošević, A. (2009), '*Pons Tiluri*', *Histria Antiqua*, 17: 169–82.
Milošević, A. (2017), *Arheologija Sinjskoga polja*, Split: Muzej hrvatskih arheoloških spomenika.
Minchin, E. (2014), 'Poet, Audience, Time, and Text: Reflections on Medium and Mode in Homer and Virgil', in R. Scodel (ed.), *Between Orality and Literacy: Communication and Adaptation in Antiquity*, 267–88, Brill: Leiden.
Ministry of Defence (2023), 'UK Armed Forces Biannual Diversity Statistics: October 2023', National Statistics. Available online: https://www.gov.uk/government/statistics/uk-armed-forces-biannual-diversity-statistics-october-2023/uk-armed-forces-biannual-diversity-statistics-october-2023 (accessed 23 February 2024).
Ministry of Defence (2022), 'LGBT+', *Army*. Available online: https://apply.army.mod.uk/what-we-offer/what-we-stand-for/lgbt (accessed 5 January 2022).
Mitchell, L. (2023), *Cyrus the Great: A Biography of Kingship*, London: Routledge.
Maloney, C. and J. Schofield (1998), *Archaeology in the City of London, 1907–1991: A Guide to Records of Excavations by the Museum of London and its Predecessors*, London: Museum of London.
Montgomery, J., C. J. Knüsel and K. Tucker (2011), 'Identifying the Origins of Decapitated Male Skeletons from 3 Driffield Terrace, York, through Isotope Analysis: Reflections of the Cosmopolitan Nature of Roman York in the Time of Caracalla', in M. Bonogofsky (ed.), *The Bioarchaeology of the Human Head: Decapitation, Decoration, and Deformation*, 141–78, Gainesville, FL: University of Florida Press.
Montgomery, S. (2018), 'Invisible Inequality among "Wounded Warriors"', *Contexts*, 17 (3): 28–33.
Moore, M. B. (1972), 'The Gigantomachy of the Siphnian Treasury', *BCH Suppl.* IV: 305–35.
Morel, J.-M. A. W. and A. V. A. J. Bosman (1989), 'An Early Roman Burial in Velsen I', in C. van Driel-Murray (ed.), *Roman Military Equipment: The Sources of Evidence, Proceedings of the Fifth Roman Military Equipment, Conference*, vol. 476, 167–91, Oxford: British Archaeological Reports, International Series.
Morello, R. (2008), '*Segregem eam efficit:* Vergil's Camilla and the Scholiasts' in S. Casali and F. Stok (eds), *Servio: Stratificazioni esegetiche e modelli culturali*, 38–57, Brussels: Latomus.
Morgan, C. H. (1962), 'The Sculptures of the Hephaisteion: II. The Friezes', *Hesperia*, 31 (3): 221–35.
Morris, S. P. (1995), 'The Sacrifice of Astyanax: Near Eastern Contributions to the Siege of Troy', in J. B. Carter and S. P. Morris (eds), *The Ages of Homer: A Tribute to Emily Townsend Vermeule*, 221–46, Austin, TX: University of Texas Press.
Morris, M. (2010), 'An Archaeological Evaluation of Land at 40–46 Western Road, Leicester', *University of Leicester Archaeological Services* (*ULAS*), unpublished report series, pers. comm.
Morrison, J. V. (1992), *Homeric Misdirection: False Predictions in the* Iliad, Ann Arbor, MI: University of Michigan Press.
Mouton, A. (2020), *Flesh and Bones: The Individual and His Body in the Ancient Mediterranean Basin. Semitica et Classica: Supplementa (SUPSEC) 2*, Turnhout: Brepols.
Müldner, G., C. Chenery and H. Eckardt (2011), 'The "Headless Romans": Multi-isotope Investigations of an Usual Burial Ground from Roman Britain', *Journal of Archaeological Science*, 38: 280–90.
Müller, K. and A. Wagner (eds) (2014), *Synthetische Körperauffassung im Hebräischen und den Sprachen der Nachbarkulturen*, Münster: Ugarit-Verlag.
Muller, Y. (2014), 'La mutilation de l'ennemi en Grèce classique: pratique barbare ou préjugé grec?' in A. Allély (ed.), *Corps au supplice et violences de guerre dans l'Antiquité, Ausonius Scripta Antiqua 67*, 41–72, Bordeaux: Ausonius.
Muller, Y. (2015), 'Couper, trancher, amputer en Grèce ancienne', in P. Guisard and C. Laizé (eds), *Le Corps*, 381–402, Paris: Ellipses.
Muller, Y. (2017), 'Le sang dans les récits de mutilation corporelle en Grèce ancienne d'Homère à Diodore de Sicile', in L. Bodiou and V. Mehl (eds), *L'Antiquité écarlate: le sang des Anciens*, 209–36, Rennes: Presses Universitaires de Rennes.

Munteanu, D. L. (2010–11), 'The Tragic Muse and the Anti-Epic Glory of Women in Euripides' *Troades*', *CJ*, 106 (2): 129–47.

Murnaghan, S. (1999), 'The Poetics of Loss in Greek Epic', in S. Lindgren Wofford, J. Tylus and M. H. Beissingerin (eds), *Epic Traditions in the Contemporary World: The Poetics of Community*, 203–20, Berkeley, CA: University of California Press.

Murray, G. (1967), *The Rise of the Greek Epic*, Oxford: Oxford University Press.

Nappi, M. (2015), 'Women and War in the *Iliad*: Rhetorical and Ethical Implications', in J. Fabre-Serris and A. Keith (eds), *Women and War in Antiquity*, 34–51, Baltimore, MD: Johns Hopkins University Press.

Nasser-Edin, N. (2017), 'Gender Performativity in Diaspora', in J. Freedman, Z. Kivilcim and N. Özgür Baklacıoğlu (eds), *A Gendered Approach to the Syrian Refugee Crisis*, 142–54, New York: Routledge.

Nauta, R. R. (2006), 'Phyrgian Eunuchs and Roman *uirtus*: The Cult of the Mater Magna and the Trojan Origins of Rome in Virgil's *Aeneid*', in G. Urso (ed.), *Tra Oriente e Occidente: Indigeni, Greci e Romani in Asia Minore*, 79–92, Pisa: Edizioni ETS.

Ndoye, M. (2010), *Groupes sociaux et idéologie du travail dans les mondes homérique et hésiodique*, Besançon: Institut des Sciences et Techniques de l'Antiquité.

Nelis-Clément. J. (2000), *Les 'Beneficiarii': militaires et administrateurs au service de l'empire*, Bordeaux: Ausonius.

Neumann, B. (2010), 'Being Prosthetic in the First World War and Weimar Germany', *Body & Society*, 16 (3): 93–126.

Newman, S. and M. Holst (2017), *Osteological Assessment of the Human Skeletal Assemblage from Blossom Street, York*, York: York Osteoarchaeology.

Nicolay, J. (2007), *Armed Batavians*, Amsterdam: Amsterdam University Press.

Noel, A.-S. (2019), '"Prosthetic Imagination" in Greek Literature', in J. Draycott (ed.), *Prostheses in Antiquity*, 159–79, London: Routledge.

North, D. C., J. J. Wallis and B. R. Weingast (2009), *Violence and Social Orders: A Conceptual Framework for Interpreting Recorded Human History*, Cambridge: Cambridge University Press.

Nunome, J. (1979), *The Origin of Sericulture and Ancient Silks*, Tokyo: Yuzankaku.

Nutton, V. (1969), 'Medicine and the Roman Army: A Further Reconsideration', *Medical History*, 13 (3): 260–70.

O'Brien, J. (1990), 'Homer's Savage Hera', *CJ*, 86: 105–25.

O'Sullivan, T. (2011), *Walking in Roman Culture*, Cambridge: Cambridge University Press.

Ober, J. (1994), 'Classical Greek Times,' in M. Howard, G. Andreopoulos and M. Shulman (eds), *The Laws of War: Constraints on Warfare in the Western World*, 12–26, New Haven, CT: Yale University Press.

Oliensis, E. (2006), 'Sons and Lovers: Sexuality and Gender in Virgil's Poetry' in C. Martindale (ed.), *The Cambridge Companion to Virgil*, 294–311, Cambridge: Cambridge University Press.

Östenberg, I. (2018), 'Defeated by the Forest, the Pass, the Wind: Nature as an Enemy of Rome', in J. H. Clark and B. Turner (eds), *Brill's Companion to Military Defeat in Ancient Mediterranean Society*, 240–61, Leiden and Boston, MA: Brill.

Otis, B. (1964), *Virgil: A Study in Civilized Poetry*, Oxford: Clarendon Press.

Ottaway, P. (2015), 'Commercial Archaeology and the Study of Roman York 1990–2013', in M. Fulford and N. Holbrook (eds), *The Towns of Roman Britain: The Contribution of Commercial Archaeology since 1990*, 44–58, London: Roman Society.

Ousley, S., R. Jantz and D. Freid (2009), 'Understanding Race and Human Variation: Why Forensic Anthropologists are Good at Identifying Race', *AJPA*, 139: 68–76.

Oxford Essential Dictionary of the U.S. Military, The (2002), New York: Oxford University Press.

Parashar, S. (2013), 'What Wars and "War Bodies" Know about International Relations', *Cambridge Review of International Affairs*, 26 (4): 615–30.

Payen, P. (1991), 'Franchir, transgresser, résister: autour de Tomyris et Cyrus chez Hérodote', *Métis*, 6: 253–81.
Pearce, J. (2015a), 'Commemorating the Dead in Roman Britain: Monuments and Their Setting', in J. M. Álvarez, T. Nogales and I. Rodà (eds), *Proceedings of the 18th International Congress of Classical Archaeology: Centre and Periphery in the Ancient World, Merida, May 2013*, vol. 2, 1209–12, Mérida: Museo Nacional de Arte Romano.
Pearce, J. (2015b), 'Urban Exits: Commercial Archaeology and the Study of Death Rituals and the Dead in the Towns of Roman Britain', in M. Fulford and N. Holbrook (eds), *The Towns of Roman Britain: The Contribution of Commercial Archaeology since 1990*, 138–66, London: Roman Society.
Peddie, J. (1994), *The Roman War Machine*, London: Sutton.
Pegler, A. (2000), 'Social Organisations within the Roman Army', *TRAC*, 1999: 37–43.
Pelliccia, H. (1995), *Mind, Body and Speech in Homer and Pindar*, Göttingen: Vandenhoeck and Ruprecht.
Pepin-Neff, C. and A. Cohen (2021), 'President Trump's Transgender Moral Panic', *Policy Studies*, 42 (5–6): 646–61.
Peretz, D. (2005), 'Military Burial and the Identification of the Roman Fallen Soldiers', *Klio*, 87: 123–38.
Periša, D. (2009), 'Je li delmatsko područje presjekao rimski limes?', *Archaeologia Adriatica*, 2 (2): 507–17.
Perkell, C. (2008), 'Reading the Laments of *Iliad* 24', in A. Suter (ed.), *Lament: Studies in the Ancient Mediterranean and Beyond*, 93–117, Oxford: Oxford University Press.
Perring, D. (2022), *London in the Roman World*, Oxford: Oxford University Press.
Petit, T. (2004), 'Herodotus and Amathus', in V. Karageorghis and I. Taifacos (eds), *The World of Herodotus*, 9–25, Nicosia: Leventis Foundation.
Pfister, R. (1951), *Texiles de Halabiyeh (Zenobia)*, Paris: Librairie Orientalist Paul Geuthner.
Pfister, R. and R. Bellinger (1945), *Excavations at Dura-Europos. Final Report IV. Part II: The Textiles*, New Haven, CT: Yale University Press.
Pfister, R. (1934), *Textiles de Palmyre*, Paris: Les Éditions d'Art et d'Histoire.
Phang, S. E. (2008), *Roman Military Service*, New York: Cambridge University Press.
Phang, S. E., J.-J. Aubert and Z. Várhelyi (2005), 'Soldiers' Slaves, "Dirty Work", and the Social Status of Roman Soldiers', in J.-J. Aubert and Z. Várhelyi (eds), *A Tall Order. Writing the Social History of the Ancient World: Essays in Honor of William V. Harris*, 203–25, München: K.G. Saur Verlag.
Philippo, S. (2007), '"A Future for Astyanax": Alternative and Imagined Futures for Hector's Son in Classical and European Drama', *International Journal of the Classical Tradition*, 14: 321–74.
Picard, G. C. (1957), *Les trophées romains: Contributions à l'histoire de la religion et de l'art triomphal de Rome*, Paris: Diff. de Boccard.
Pitts, M. (2019), *The Roman Object Revolution: Objectscapes and Intra-cultural Connectivity in Northwest Europe*, Amsterdam: Amsterdam University Press.
Plazibat, I. (2017), 'Značenje i stalni elementi tropeja: Ilirički primjeri', MA diss., University of Zagreb.
Pliny the Elder (1942), *Natural History*, trans. H. Rackham, Cambridge, MA: Harvard University Press.
Pollitt, J. J. (1986), *Art in the Hellenistic Age*, Cambridge: Cambridge University Press.
Pope, M. (2020), 'Bodies Piled High: Lucretius, Lucan and the Unnatural Costs of War', *Classical Philology*, 115: 209–26.
Porter, D. (2010), 'The Simile at *Iliad* 16.7–11 Once Again: Multiple Meanings', *Classical World*, 103 (4): 447–54.
Porter, J. I. (ed.) (2005), *Constructions of the Classical Body*, Ann Arbor, MI: University of Michigan Press.

Portillo, S., A. Doan and A. Mog (2022), 'Maintaining Male Exclusivity: Porcelain Privilege in the Military', *Armed Forces & Society*, 48 (3): 571–88.

Postrel, V. (2020), *The Fabric of Civilization: How Textiles Made the World*, New York: Basic Books.

Potts, D. T. (2007), 'Cataphractus and Kamāndār: Some Thoughts on the Dynamic Evolution of Heavy Cavalry and Mounted Archers in Iran and Central Asia', *Bulletin of the Asia Institute*, 21: 149–58.

Pratt, L. (2007), 'The Parental Ethos of the *Iliad*', in A. Cohen and J. B. Rutter (eds), *Constructions of Childhood in Ancient Greece and Italy*, 25–40, Athens: The American School of Classical Studies at Athens.

Pritchard, D. M. (2013), *Sport, Democracy and War in Classical Athens*, Cambridge: Cambridge University Press.

Pritchett, W. K. (1974), *The Greek State at War: Part V*, Berkeley, CA: University of California Press.

Pyy, E. (2010), 'Decus Italiae virgo: Virgil's Camilla and the Formation of Romanitas', *Arctos – Acta Philologica Fennica*, 44: 181–204.

Pyy, E. (2021), *Women and War in Roman Epic*, Leiden: Brill.

Raaflaub, K. A. (2014), 'War and the City: The Brutality of War and Its Impact on the Community', in P. Meineck and D. Konstan (eds), *Combat Trauma and the Ancient* Greeks, 15–46, New York: Palgrave Macmillan.

Rabinowitz, N. S. (2011), 'Greek Tragedy: A Rape Culture?', *Eugesta*, 1: 1–21.

Rabinowitz, N. S. (2013), 'The Expansion of Tragedy as Critique', in L. Hardwick and S. Harrison (eds), *Classics in the Modern World: A Democratic Turn?* 119–30, Oxford: Oxford University Press.

Rabinowitz, N. S. (2014), 'Women and War in Tragedy', in P. Meineck and D. Konstan (eds), *Combat Trauma and the Ancient Greeks*, 185–206, New York: Palgrave Macmillan.

Rabinowitz, N. S. (2017), 'Trojan Women', in L. McClure (ed.), *A Companion to Euripides*, 197–213, Oxford: Wiley-Blackwell.

Radman-Livaja, I. (2004), *Militaria Sisciencia: nalazi rimske vojne opreme iz Siska u fundusu Arheološkoga muzeja u Zagrebu*, Zagreb: AMZ.

Radman-Livaja, I. (2010), 'The Catalogue of Finds: Gardun – The Ancient Tilurium', in I. Radman-Livaja (ed.), *Finds of the Roman Military Equipment in Croatia*, 55–111, Zagreb: AMZ.

Ramsby, T. (2010), 'Juxtaposing Dido and Camilla in the *Aeneid*', *The Classical Outlook*, 88 (1): 13–17.

Rance, P. (2020), 'Health, Wounds, and Medicine in the Later Roman Army (250–600 CE), in L. L. Brice (ed.), *New Approaches to Greek and Roman Warfare*, 173–85, Malden, MA: Wiley-Blackwell.

Ready, J. (2011), *Character, Narrator, and Simile in the* Iliad, Cambridge: Cambridge University Press.

Ready, J. and C. Tsagalis (eds) (2018), *Homer in Performance: Rhapsodes, Narrators, and Characters*, Austin, TX: University of Texas Press.

Redfern, R. C. (2006), 'A Gendered Analysis of Health from the Iron Age to the End of the Romano-British Period in Dorset, England (Mid to late 8th Century B.C. to the End of the 4th Century A.D.)', PhD thesis, University of Birmingham. Available online: http://etheses.bham.ac.uk/69/ (accessed 10 January 2024).

Redfern, R. (2010), 'A Regional Examination of Surgery and Fracture Treatment in Iron Age and Roman Britain', *International Journal of Osteoarchaeology*, 20: 443–71.

Redfern, R. C. (2016), *Injury and Trauma in Bioarchaeology: Interpreting Violence in Past Lives*, Cambridge: Cambridge University Press.

Redfern, R. C. (2018), 'Blind to Chains? The Potential of Bioarchaeology for Identifying the Enslaved of Roman Britain', *Britannia*, 49: 251–82.

Redfern, R. C. (2020), 'Changing People, Changing Settlements? A Perspective on Urbanism from Roman Britain', in S. N. DeWitte and T. K. Betsinger (eds), *The Bioarchaeology of Urbanization*, 25–47, New York: Springer-Link.

Redfern, R. C., D. R. Gröcke, A. R. Millard, V. Ridgeway, L. Johnson and J. T. Hefner (2016), 'Going South of the River: A Multidisciplinary Analysis of Ancestry, Mobility and Diet in a Population from Roman Southwark, London', *Journal of Archaeological Science*, 74: 11–22.

Redfern, R. and J. T. Hefner (2019), '"Officially Absent but Actually Present": Bioarchaeological Evidence for Population Diversity in London during the Black Death, AD 1348–50', in M. L. Mant and A. Jaagumägi Holland (eds), *Bioarchaeology of Marginalized People*, 69–114, London: Elsevier.

Redfern, R. and J. T. Hefner (2021), 'Bioarchaeological and Forensic Perspectives on Population Diversity in the Medieval World: A Case-study from London, England', in K. Coles and D. L. Kim (eds), *A Cultural History of Race in the Renaissance and Early Modern Age*, 73–90, London: Bloomsbury Academic.

Redfern, R. C., M. A. Judd and S. N. DeWitte (2017a), 'Multiple Injury and Health in Past Societies: An Analysis of Concepts and Approaches, and Insights from a Multi-Period Study', *International Journal of Osteoarchaeology*, 27: 418–29.

Redfern, R. C., M. Marshall, K. Eaton and H. Poinar (2017b), '"Written in Bone": New Discoveries about the Lives of Roman Londoners', *Britannia*, 48: 253–77.

Redfern, R. and C. A. Roberts (2019), 'Trauma', in J. E. Buikstra (ed.), *Ortner's Identification of Pathological Conditions in Human Skeletal Remains*, 211–84, New York: Elsevier.

Reinard, P. and C. Rollinger (2020), '"Meine Seele ist vom Sturm getrieben . . .": Die Debatte um antike Kriegstraumata und posttraumatische Belastungsstörungen im Lichte eines spätantiken Briefes (P.Oxy. 16/1873)', *Millennium*, 17: 163–202.

Reitz-Joosse, B., M. W. Makins and C. J. Mackie (eds) (2021), *Landscapes of War in Greek and Roman Literature*, London: Bloomsbury Academic.

Renehan, R. (1979), 'The Meaning of ΣΩΜΑ in Homer: A Study in Methodology', *California Studies in Classical Antiquity*, 12: 269–82.

Reuter, M. (2009), 'Gefallen für Rom: *Beobachtungen an den Grabinschriften*im Kampf getöteter römischer Soldaten', in Z. Visy (ed.), *Limes XIX: Proceedings of the XIXth International Congress of Roman Frontier Studies, Pécs, Hungary, 2003*, 255–64, Pécs: Pécsi Tudományegyetem.

Ribera i Lacomba, A. (2014), 'La destrucción de Valentia (75 a.c.) y la cultura material de la época de sertorio (82-75 a.c.)', in J. M. Jávega and F. S. Sellés (eds), *Las guerras civiles romanas en Hispania: una revisión histórica desde la Contestania*, 65–78, Alicante: Diputación de Alicante.

Richards, M. and J. Montgomery (2012), 'Isotope Analysis and Palaeopathology: A Short Review and Future Developments', in J. E. Buikstra and C. A. Roberts (eds), *The Global History of Paleopathology: Pioneers and Prospects*, 718–31, Oxford: Oxford University Press.

Richardson, N. (1993), *The* Iliad: *A Commentary, Vol. VI: Books 21–24*, Cambridge: Cambridge University Press.

Richlin, A. (2005), 'Cicero's Head', in J. I. Porter (ed.), *Constructions of the Classical Body*, 190–211, Ann Arbor, MI: University of Michigan Press.

Richlin, A. (2018), 'The Ones Who Paid the Butcher's Bill: Soldiers and War Captives in Roman Comedy', in J. H. Clark and B. Turner (eds), *Brill's Companion to Military Defeat in Ancient Mediterranean Society*, 213–39, Leiden and Boston, MA: Brill.

Ridgeway, V., K. Leary and B. Sudds (eds) (2014), *Roman Burials in Southwark. Excavations at 52–66 Lant Street and 56 Southwark Bridge Road, London SE1*, PCA Monograph 17, Oxford: Oxbow.

Riha, E. (1979), *Die romischen Fibeln aus Augst und Kaiseraugst*, Forschungen in Augst Band 3, Augst: Romermuseum Augst.

Rihll, T. (1993), 'War, Slavery, and Settlement in Early Greece', in J. Rich and G. Shipley (eds), *War and Society in the Greek World*, 77–107, London: Routledge.

Riley Snorton, C. (2017), *Black on Both Sides: A Racial History of Trans Identity*, Minneapolis, MN: University of Minnesota Press.

Rippon, G., L. Eliot, S. Genon and D. Joel (2021), 'How Hype and Hyperbole Distort the Neuroscience of Sex Differences', *PloS Biology*, 19 (5): 1–3.

Robert, C. ([1923] 1967), *Die griechische Heldensage,* 4th edn, vol. 3.2.1 of L. Preller (ed.), *Griechische Mythologie*, Berlin: Weidmann.

Roller, L. E. (1998), 'The Ideology of the Eunuch Priest', in M. Wyke (ed.), *Gender and the Body in the Ancient Mediterranean*, 118–35, Oxford: Blackwell.

Rollinger, C. (2018), '*Vae victae*: Frauen als Beute in der römischen Kriegsführung', in C. Walde and G. Wöhrle (eds), *Gender Studies in den Altertumswissenschaften: Gender und Krieg*, 91–126, Trier: Wissenschaftlicher Verlag Trier.

Roman Society (2021), *Race: Antiquity and Its Legacy*. Available online: https://youtu.be/UuWM0Jq6CwE (accessed 13 June 2022).

Romero Mariscal, L. (1998–9), 'Fragilidad y justicia en la literatura griega: los casos de Tomiris en Heródoto y Hécuba en Eurípides, *Praesentia*, 2–3: 235–51.

Rosenmeyer, T. G. (1960), 'Virgil and Heroism: *Aeneid* XI', *CJ*, 55 (4): 159–64.

Rossi, C., S. Savino, A. Messina and G. Reina (2015), 'Performance of Greek–Roman Artillery', *Arms Armour*, 12: 67–89.

Rost, A. and S. Wilbers-Rost (2017), 'They Fought and Died – But were Covered with Earth Only Years Later: "Mass Graves" on the Ancient Battlefield of Kalkriese', in J. Pearce and J. Weekes (eds), *Death as a Process*, 153–73, Oxford: Oxbow.

Rostovtzeff, M. I. (1943), 'The Parthian Shot', *AJA*, 57: 174–87.

Rothschild, B. (2000), *The Body Remembers: The Psychophysiology of Trauma and Trauma Treatment*, New York: Norton.

Rousseau, P. (2015), 'War, Speech, and the Bow are Not Women's Business', in J. Fabre-Serris and A. Keith (eds), *Women and War in Antiquity*, 15–33, Baltimore, MD: Johns Hopkins University Press.

Royal Navy (2022), 'Get Fit to Join: Shape Up for the Royal Navy', *Royal Navy*. Available online: https://www.royalnavy.mod.uk/what-we-do/ready-to-fight (accessed 28 March 2022).

Roymans, N. (2019), 'Caesar's Conquest and the Archaeology of Mass Violence in the Germanic Frontier Zone', in A. P. Fitzpatrick and C. Haselgrove (eds), *Julius Caesar's Battle for Gaul: New Archaeological Perspectives*, 113–34, Oxford: Oxbow.

Różycki Ł. (2021), *Battlefield Emotions in Late Antiquity: A Study of Fear and Motivation in Roman Military Treatises*, Leiden and Boston, MA: Brill.

Rudenko, S. I. (1970), *Frozen Tombs of Siberia: The Pazyryk Burials of Iron-age Horsemen*, Berkeley and Los Angeles, CA: University of California Press.

Runciman, W. G. (1998), 'Greek Hoplites, Warrior Culture, and Indirect Bias', *Journal of the Royal Anthropological Institute*, 4: 731–51.

Russell, B. (2020), 'Roman Sculptors at Work: Professional Practitioners?', in E. Stewart, E. Harris and D. Lewis (eds), *Skilled Labour and Professionalism in Ancient Greece and Rome*, 243–66, Cambridge: Cambridge University Press.

Sabin, P. (2000), 'The Face of Roman Battle', *JRS*, 90: 1–17.

Sabin, P., H. van Wees and M. Whitby (eds) (2007), *The Cambridge History of Greek and Roman Warfare, Volume 1: Greece, the Hellenistic World and the Rise of Rome*; *Volume 2: Rome from the Late Republic to the Late Empire*, Cambridge: Cambridge University Press.

Saïd, S. (2002), 'Herodotus and Tragedy', in E. J. Bakker, I. J. F. de Jong and H. van Wees (eds), *Brill's Companion to Herodotus*, 117–47, Leiden: Brill.

Salazar, C. (2000), *The Treatment of War Wounds in Graeco-Roman Antiquity*, Leiden: Brill.

Sanader, M. (2003), *Tilurium I: Istraživanja – Forschungen 1997–2001*, Zagreb: Golden Marketing.

Sanader, M. and D. Tončinić (2010), 'Gardun – The Ancient Tilurium', in I. Radman-Livaja (ed.), *Finds of the Roman Military Equipment in Croatia*, 33–54, Zagreb: AMZ.

Sanader, M. and D. Tončinić (2013), 'Das Projekt Tilurium', in M. Sanader, A. Rendić-Miočević, D. Tončinić and I. Radman-Livaja (eds), *Proceedings of the XVIIth Roman Military Equipment Conference: Weapons and Military Equipment in a Funerary Context*, 411–33, Zagreb: FF Press.
Sanader, M. and D. Tončinić (2017), 'Arheološka istraživanja 2007–2010. godine', in M. Sanader, D. Tončinić, S. Ivčević, Z. Buljević, T. Šeparović and I. Miloglav, *Tilurium IV: Arheološka istraživanja 2007–2010. godine*, 15–54, Zagreb: FF Press.
Sanader, M., D. Tončinić, S. Ivčević, Z. Buljević and T. Šeparović (2014), *Tilurium III: Istraživanja 2002–2006. godine*, Zagreb: FF Press.
Sanader, M., D. Tončinić, S. Ivčević, Z. Buljević, T. Šeparović and I. Miloglav (2017), *Tilurium IV: Arheološka istraživanja 2007–2010. godine*, Zagreb: FF Press.
Sanader, M., D. Tončinić, Z. Šimić-Kanaet, Z. Buljević, S. Ivčević, A. Pavlović and M. Vukov (2021), *Tilurium V: Arheološka istraživanja 2010–2018. godine*, Zagreb: FF Press.
Sancisi-Weerdenburg, H. (1985), 'The Death of Cyrus: Xenophon's *Cyropaedia* as a Source for Iranian History', *Acta Iranica*, 25: 459–71.
Šašel Kos, M. (2005), *Appian and Illyricum*, Ljubljana: Narodni muzej Slovenije.
Sauer, E. (2005), 'Forum Germanorum in North-west Italy: The Home Community and Life of Arguably the Earliest Known Legionary Veteran in Britain', *Oxford Journal of Archaeology*, 24: 199–214.
Scarborough, J. (1968), 'Roman Medicine and the Legions: A Reconsideration', *Medical History*, 12 (3): 254–61.
Scarry, E. (1985), *The Body in Pain*, New York: Oxford University Press.
Schaefer, A. G., R. Iyengar, S. Kadiyala, J. Kavanaugh, C. C. Engel, K. M. Williams and A. M. Kress (2016), *Assessing the Implications of Allowing Transgender Personnel to Serve Openly*, Santa Monia, CA: RAND Corporation.
Schaefer, M., S. Black and L. Scheuer (2009), *Juvenile Osteology: A Laboratory and Field Manual*, London: Academic Press.
Scheidel, W. (1996), *Measuring Sex, Age and Death in the Roman Empire: Explorations in Ancient Demography*, JRA Supplementary Series 21, Portsmouth, Rhode Island: JRA.
Scheidel, W. (2007), 'Marriages, families and survival: demographic aspects', in P. Erdkamp (ed.), *A Companion to the Roman Army*, 417–34, Oxford: Blackwell.
Schein, S. L. (1984), *The Mortal Hero: An Introduction to Homer's* Iliad, Berkeley, CA: University of California Press.
Schenk, A., H. Amoroso and P. Blanc (2012), 'Des soldats de la Legio I Adiutrix à Aventicum: A propos de deux nouvelles stèles funéraires d'Avenches', *Bulletin de l'Association Pro Aventico*, 54: 227–60.
Schierl, P. (2021), 'Konstruktionen von Männlichkeit in der römischen Literatur', *Gymnasium*, 128 (5): 397–413.
Schneider, R. M. (2007), 'Friend and Foe: The Orient in Rome', in V. S. Curtis and S. Stewart (eds), *The Age of the Parthians*, 50–86, London: I.B. Tauris.
Schönauer, S. (2000), 'Odjeća, obuća i nakit u antičkoj Dalmaciji na spomenicima iz Arheološkog muzeja u Splitu', *VAHD*, 93: 223–515.
Schultz, P. (2009), 'The North Frieze of the Temple of Athena Nike', in O. Palagia (ed.), *Art in Athens during the Peloponnesian War*, 128–67, Cambridge: Cambridge University Press.
Schvey, N. A., D. A. Klein, A. T. Pearlman, R. I. Kraff and D. S. Riggs (2020), 'Stigma, Health, and Psychosocial Functioning among Transgender Active Duty Service Members in the U.S. Military', *Stigma and Health*, 5 (2): 188–98.
Schwab, K. A. (2005), 'Celebrations of Victory: The Metopes of the Parthenon', in J. Neils (ed.), *The Parthenon: From Antiquity to the Present*, 159–97, Cambridge: Cambridge University Press.
Scodel, R. (1998a), 'Bardic Performance and Oral Tradition in Homer', *AJP*, 119 (2): 171–94.
Scodel, R. (1998b), 'The Captive's Dilemma: Sexual Acquiescence in Euripides' *Hecuba* and *Troade*', *HSCP*, 98: 137–54.

Scodel, R. (2004), 'The Story-teller and His Audience', in R. Fowler (ed.), *The Cambridge Companion to Homer*, 45–58, Cambridge: Cambridge University Press.
Scodel, R. (2009), *Listening to Homer: Tradition, Narrative, and Audience*, Ann Arbor, MI: University of Michigan Press.
Seaman, M. G. (2020), 'Early Greek Siege Warfare', in L. B. Lee (ed.), *New Approaches to Greek and Roman Warfare*, 29–38, Hoboken, NJ: Wiley-Blackwell.
Segal, C. (1971a), 'Andromache's Anagnorisis: Formulaic Artistry in *Iliad* 22.437-476', *HSCP*, 75: 33–57.
Segal, C. (1971b), *The Theme of the Mutilation of the Corpse in the* Iliad, Leiden: Brill.
Seldmayer, H. (2014), 'Le fibule del tipo Aucissa: Componente tipica dell'abbigliamento femminile in un ambito di scarsa romanizzazione', *Quaderni Friulani di Archeologia*, 24 (1), 19–32.
Serrati, J. (2013), 'The Hellenistic World at War: Stagnation or Development?', in B. Campbell and L. A. Tritle (eds), *The Oxford Handbook of Warfare in the Classical World*, 179–98, Oxford: Oxford University Press.
Seyrig, H. (1939), 'La grande statue parthe de Shami et la sculpture palmyrenienne', *Syria*, 20: 177–83.
Shahbazi, A. Sh. (1987), 'ASB i. In Pre-Islamic Iran', *Encyclopædia Iranica*, 2/7: 724–30. Available online: http://www.iranicaonline.org/articles/asb-pre-islamic-iran (accessed 30 November 2023).
Sharrock, A. (2015), 'Warrior Women in Roman Epic', in J. Fabre-Serris and A. Keith (eds), *Women and War in Antiquity*, 157–78, Baltimore, MD: Johns Hopkins University Press.
Shaw, H., J. Montgomery, R. Redfern, R. Gowland and J. Evans (2016), 'Identifying Migrants in Roman London Using Lead and Strontium Stable Isotopes', *Journal of Archaeological Science*, 66: 57–68.
Shaw, M. (2003), 'Strategy and Slaughter', *Review of International Studies*, 29: 269–77.
Shay, J. (2003), *Achilles in Vietnam: Combat Trauma and the Undoing of Character*, New York: Scribner.
Sheppard, S. (2020), *Roman Soldier versus Parthian Warrior: Carrhae to Nisibis, 53 BC–AD 217*, Oxford: Osprey.
Shotter, D. C. A. (2003), 'The Murder of Flavius Romanus at Ambleside: A Possible Context', *Transactions of Cumberland and Westmorland Antiquarian and Archaeological Society* (3rd Series), 3: 228–31.
Sidebottom, H. (2004), *Ancient Warfare: A Very Short Introduction*, Oxford: Oxford University Press.
Šimić-Kanaet, Z. (2010), *Tilurium II: keramika 1997–2006*, 2 vols, Zagreb: Golden Marketing.
Simmonds, A., N. Márquez-Grant and L. Loe (eds) (2007), *Life and Death in a Roman City: Excavation of a Roman Cemetery with a Mass Grave at 120–122 London Road, Gloucester*, Oxford Archaeology Monograph 6, Oxford: Oxford Archaeology.
Šlaus, M., A. Domić Kunić and T. Pivac (2018), 'Reconstructing the Life of a Roman Soldier Buried in Resnik Near Split, Based on the Anthropological Analysis of His Skeleton', *Collegium antropologicum*, 42 (4): 287–94.
Smith, A. (2018), 'Death in the Countryside: Rural Burial Practices' in A. Smith, M. Allen, T. Brindle, M. Fulford and L. Lodwick (eds), *Life and Death in the Countryside of Roman Britain*, 205–80, London: Roman Society.
Smith, R. J. C., F. Healy, M. J. Allen, E. L. Morris and P. J. Woodward (eds) (1997), *Excavations Along the Route of the Dorchester Bypass*, Wessex Archaeology Report 11, Salisbury: Wessex Archaeology.
Snell, B. (1953), *The Discovery of the Mind: The Greek Origins of European Thought*, trans. T. G. Rosenmeyer, Oxford: Blackwell.
Solvang, E. (2015), 'Guarding the House: Conflict, Rape, and David's Concubines', in M. Masterson, N. S. Rabinowitz and J. E. Robson (eds), *Sex in Antiquity: Gender and Sexuality in the Ancient World*, 50–66, London: Routledge.

Speidel, M. A. (2012), 'Dress for the Occasion: Clothes in the Context of the Roman Army', in M.-L. Nosch (ed.), *Wearing the Cloak: Dressing the Soldier in Roman Times*, 1–12, Oxford: Oxbow Books.
Speidel, M. A. (2015), 'The Roman Army', in C. Bruun and J. Edmondson (eds), *The Oxford Handbook of Roman Epigraphy*, 319–44, Oxford: Oxford University Press.
Speidel, M. A. (2017), 'Recruitment and Identity – Exploring the Meanings of Roman Soldiers' Origins', in D. Dana and B. Rossignol (eds), *Entrer dans l'armée romaine: bassins de recrutement desunités auxiliaires (Ier-IIe s. ap. J.-C.) HiMA*, 35–50, Paris: Société d'Édition de Revues.
Spiegel, F. (2019), 'Malfunctions of Embodiment: Man/Weapon Agency and the Greek Ideology of Masculinity', in G. M. Chesi and F. Spiegel (eds), *Classical Literature and Posthumanism*, 267–74, London: Bloomsbury.
Sporleder, R. F. (2015), 'The Bassae-frieze: 200 Years of Guesswork', in C. Klose, L. C. Bossert and W. Leveritt (eds), *Fresh Perspectives on Graeco-Roman Visual Culture, Proceedings of an International Conference at Humboldt-Universität, Berlin, 2nd–3rd September 2013*, 23–3, Berlin: Humboldt-Universität.
Stark, R. J. (2016), 'Ancient Lives in Motion: A Bioarchaeological Examination of Stable Isotopes, Nonmetric Traits, and Human Mobility in an Imperial Roman Context (1st–3rd CE)', PhD thesis, McMaster University, Hamilton, Ontario.
Stark, R. J., V. Emery, H. Schwarcz, D. Castex and T. L. Prowse (2022), 'Assessing Gallo-Roman Mobility at the Rue Jacques Brel Necropolis Site (1st to 3rd *c.* CE), France', *Journal of Archaeological Science: Reports*, 43: 103–470.
Stark, R. J., M. V. Emery, H. Schwarcz, A. Sperduti, L. Bondioli, O. E. Craig and T. Prowse (2020), 'Imperial Roman Mobility and Migration at Velia (1st to 2nd *c.* CE) in Southern Italy', *Journal of Archaeological Science: Reports*, 30: 102–217.
Startin, D. W. A. (1982), 'Excavations at the Old Vicarage, Fordington, Dorchester, Dorset', *Proceedings of the Dorset Natural History and Archaeological Society*, 103: 43–61.
Stein, A. M. (1940), *Old Routes in Western Iran*, London: MacMillan and Co.
Stewart, A. (1994), *Faces of Power: Alexander's Image and Hellenistic Politics*, Berkeley and Los Angeles, CA, and Oxford: University of California Press.
Stewart, M. E. (2016), 'The Soldier's Life: Early Byzantine Masculinity and the Manliness of War', *Byzantina Symmeikta*, 26: 11–44.
Stewart, P. (2010), 'Geographies of Provincialism in Roman Sculpture', *RIHA Journal*, 0005.
Stoevesandt, M. (2016), *Homer's* Iliad: *The Basel Commentary, Book VI*, ed. A. Bierl and J. Latacz; trans. B. W. Millis, S. Strack; ed. and trans. S. D. Olson, Berlin: De Gruyter.
Stoll, H. W. (1884–90), 'Astyanax 1', W. H. Roscher (ed.), *Ausführliches Lexikon der griechischen und römischen Mythologie*, vol. 1, 660, Leipzig: Teubner.
Stoll, O. (2019), *Vestigia Cladis – Roms Umgang mit militärischem Misserfolg: Niederlagen verdrängen, Siege betonen, Resilienz beweisen*, Berlin: Frank and Timme.
Strid, O. (2006), 'Voiceless Victims, Memorable Deaths in Herodotus', *CQ*, 56: 393–403.
Strobel, K. (2000), 'Zur Geschichte der Legiones V (Macedonica), und VII (Claudia pia fidelis), in der fruhen kaiserzeit und zur Stellung der Provinz Galatia in der augusteischen Heeresgeschichte', in Y. Le Bohec and C. Wolff (eds), *Les légions de Rome sous le Haut-Empire*, 515–28, Lyon and Paris: Diff. de Boccard.
Stuckert, C. M. (2017), *The People of Early Winchester*, Winchester Studies 9i, Oxford: Clarendon Press.
Stull, K. E., E. J. Bartelink, A. R. Klales, G. E. Berg, M. W. Kenyhercz, E. N. L'Abbé, M. C. Go, K. McCormick and C. Mariscal (2021), 'Commentary on: Bethard JD, DiGangi EA. Letter to the Editor – Moving Beyond a Lost Cause: Forensic Anthropology and Ancestry Estimates in the United States', *Journal of Forensic Sciences*, 66: 417–20.
Surtees, A. and J. Dyer (eds) (2020), *Exploring Gender Diversity in the Ancient World*, Edinburgh: Edinburgh University Press.

Sylvester, C. (2012), 'War Experiences/War Practices/War Theory', *Millennium*, 40 (3): 483–503.

Sylvester, C. (2013), *War as Experience*, London: Routledge.

Synder, J. J. and G. R. Patterson (1995), 'Individual Differences in Social Aggression: A Test of a Reinforcement Model of Socialization in the Natural Environment', *Behavior Therapy*, 26: 371–91.

Tacitus (1996), *The Annals of Imperial Rome*, trans. M. Grant, London: Penguin.

Tao, W. (2007), 'Parthia in China: a Re-examination of the Historical Records', in V. S. Curtis and S. Stewart (eds), *The Age of the Parthians*, 87–102, London: I.B. Tauris.

Tegtmeyer, C. E. and D. L. Martin (eds) (2017), *Broken Bones, Broken Bodies: Bioarchaeological and Forensic Approaches for Accumulative Trauma and Violence*, London: Lexington Books.

Temporini, H. and W. Haase (eds) (1972–), *Aufstieg und Niedergang der römischen Welt: Geschichte und Kultur Roms im Spiegel der neueren Forschung*, Berlin: W. de Gruyter.

Thimme, D. (1946), 'The Masters of the Pergamon Gigantomachy', *AJA*, 50 (3): 345–57.

Thompson, B. (2018), *Horse Games*, Ludlow: Merlin Unwin Books.

Thompson, T., J. Szigeti, R. L. Gowland and R. Witcher (2016), 'Death on the Frontier: Military Cremation Practices in the North of Roman Britain', *Journal of Archaeological Science: Reports*, 10: 828–36.

Thorne, M. (2016), 'Speaking the Unspeakable: Engaging *Nefas* in Lucan and Rwanda 1994', *Thersites*, 4: 77–119.

Tidy, J. (2019), 'War Craft: The Embodied Politics of Making War', *Security Dialogue*, 50 (3): 220–8.

Tilke, M. (1990), *Costume Patterns and Designs*, Wigston: Magna Books.

Tolia-Kelly, D. P. (2011), 'Narrating the Postcolonial Landscape: Archaeologies of Race at Hadrian's Wall', *Transactions of the Institute of British Geographers*, 36 (1): 71–88.

Tomlin, R. S. O. (2018), *Britannia Romana*, Oxford: Oxbow.

Tomlin, R. (2009), *The Roman Inscriptions of Britain*, vol. 3, Oxford: Oxbow.

Tončinić, D. (2011), *Spomenici VII. legije na području rimske provincije Dalmacije / Monuments of Legio VII in the Roman Province of Dalmatia*, Split: Arheološki muzej u Splitu.

Touchefeu, O. (1984), 'Astyanax I', *Lexicon iconographicum mythologiae classicae*, 2 (1): 929–37, Zurich: Artemis.

Tritle, L. A. (1997), 'Hector's Body: Mutilation of the Dead in Ancient Greece and Vietnam', *Ancient History Bulletin*, 11: 123–36.

Tritle, L. A. (2013), 'Men at War', in B. Campbell and L. A. Tritle (eds), *The Oxford Handbook of Warfare in the Classical World*, 279–93, Oxford: Oxford University Press.

Tsagalis, C. ([2004] 2012), *Epic Grief: Personal Laments in Homer's* Iliad, Berlin: De Gruyter.

Van Broeck, L. (2018), 'Wily Wetlands: Imperialism and Resistance in Tacitus's Batavian Revolt', in D. Felton (ed.), *Landscapes of Dread: Negative Emotion in Natural and Constructed Spaces*, 145–62, London and New York: Routledge.

van der Kolk, B. (2014), *The Body Keeps the Score: Brain, Mind, and Body in the Healing of Trauma*, New York: Viking.

van der Plas, M. (2020), 'Corpse Mutilation in the *Iliad*', *CQ*, 70: 459–72.

van Driel-Murray, C. (1997), 'Women in Forts?', *Jahresbericht der Gesellschaft Pro Vindonissa*, 1997: 55–61.

van Driel-Murray, C. and M. J. Driessen (eds) (forthcoming), *A Soldier in the Well: New Research on Well 2 in the Early Roman Fort at Velsen I, The Netherlands*, Leiden: Analecta Praehistoria Leidensia XXX.

van Lommel, K. (2013), 'The Recognition of Roman Soldiers' Mental Impairment', *Acta Classica*, 56: 155–84.

van Lommel, K. (2015), 'Heroes and Outcasts: Ambiguous Attitudes towards Impaired and Disfigured Roman Veterans', *Classical World*, 109 (1): 91–117.

van Lommel, K. (2019), 'Stoicism and War Wounds: Mucius Scaevola, Sergius Silus and Quintus Sertorius', *Journal of Ancient History and Archaeology*, 6 (2): 1–26.

van Wees, H. (1992), *Status Warriors: War, Violence and Society in Homer and History*, Amsterdam: Gieben.

van Wees, H. (1996), 'Heroes, Knights and Nutters: Warrior Mentality in Homer', in A. B. Lloyd (ed), *Battle in Antiquity*, 1–86, London: Duckworth.

Vanderpool, E. (1966), 'A Monument to the Battle of Marathon', *Hesperia*, 35 (2): 93–106.

Vermeule, E. (1979), *Aspects of Death in Early Greek Art and Poetry*, Berkeley, CA: University of California Press.

Vernant, J. P. (1991), 'A "Beautiful Death" and the Disfigured Corpse in Homer's *Iliad*', in F. I. Zeitlin (ed.), *Mortals and Immortals: Collected Essays*, 50–74, Princeton, NJ: Princeton University Press.

Virgil (2003), *The Aeneid*, trans. D. West, London: Penguin.

Vlahogiannis, N. (1998), 'Disabling Bodies', in D. Montserrat (ed.), *Changing Bodies, Changing Meanings: Studies on the Human Body in Antiquity*, 13–36, London: Routledge.

Vogelsang, W. J. (1992), *The Rise and Organisation of the Achaemenid Empire: The Eastern Iranian Evidence*, Leiden: Brill.

Voirol, A. (2000), 'Etats d'armes: Les militaria d'Avenches/Aventicum', *Bulletin de l'Association Pro Aventico*, 42: 7–92.

Vogelsang-Eastwood, G. (2000), 'The Clothing of the "Shami Prince"', *Persica*, 16: 31–47.

Vollmer, J. E. (2002), *Ruling from the Dragon Throne: Costume of the Qing Dynasty*, Berkeley, CA, and Toronto: Ten Speed Press.

Wagner, R. (1896), 'Astyanax 1', *Realencyclopädie der classischen Altertumswissenschaft*, 2.2, 1866–7, Stuttgart: J. B. Metzlerscher.

Walas, A. H. (2015), 'An Integrated Cognitive and Epigraphic Approach to Social Networks within the Community of a Roman Military Base', *TRAJ*, 2014: 17–30.

Walde, C. (2011), 'Lucan's *Bellum Civile*: A Specimen of a Roman "Literature of Trauma"', in P. Asso (ed.), *Brill's Companion to Lucan*, 283–302, Leiden and Boston, MA: Brill.

Walde, C. and G. Wöhrle (eds) (2018), *Gender Studies in den Altertumswissenschaften: Gender und Krieg*, Trier: WVT Wissenschaftlicher Verlag Trier.

Walters, B. (2020), *The Deaths of the Republic: Imagery of the Body Politic in Ciceronian Rome*, Oxford: Oxford University Press.

Walters, J. (1997), 'Invading the Roman Body: Manliness and Impenetrability in Roman Thought', in J. P. Hallett and M. B. Skinner (eds), *Roman Sexualities*, 29–43, Princeton, NJ: Princeton University Press.

Wan, F. (2009), 'Costumes Unearthed from Vault 5 of Munchaktepa in the North of Ferghana Valley of Uzbekistan', *Asian Social Science*, 5 (11): 99–108.

Ward, G. (2017), 'The Roman Battlefield: Individual Exploits in Warfare of the Roman Republic', in W. Riess and G. G. Fagan (eds), *The Topography of Violence in the Greco-Roman World*, 299–324, Ann Abor, MI: University of Michigan Press.

Warne, V. (2008), 'Artificial Leg', *Victorian Review*, 34 (1): 29–33.

Waters-Rist, A. L. and J. Palmer (2016), *Report of the Strontium and Stable Oxygen Isotope Results of Human Remains from the Roman Fort at Velsen, The Netherlands*, Leiden University. Available online: https://collectie.huisvanhilde.nl/ (accessed 1 November 2023).

Watkins, R. J. (2018), 'The Fate of Anatomical Collections in the US: Bioanthropological Investigations of Structural Violence', in C. Y. Henderson and F. Alves Cardoso (eds), *Identified Skeletal Collections: The Testing Ground of Anthropology*, 169–85, Oxford: Archaeopress Publishing Ltd.

Watkins, R. J. (2021), '"[This] system was not made for [you]:" A Case for Decolonial Scientia', *AJPA*, 175: 350–62.

Watkins, S. M. and L. E. Dunne (2015), *Functional Clothing Design: From Sportswear to Spacesuits*, London: Bloomsbury.

Watson, G. R. (1983), *The Roman Soldier*, London: Thames and Hudson.

Watson, S. (2003), *An Excavation in the Western Cemetery of Roman London: Atlantic House, City of London*, Archaeology Studies Series 7, London: Museum of London Archaeology Service.

Watt, J. C. Y. and A. E. Wardwell (1997), *When Silk Was Gold: Central Asian and Chinese Textiles*, New York: Metropolitan Museum of Art.
Webster, G. (1998), *The Roman Imperial Army of the First and Second Centuries A.D.*, Norman, OK: University of Oklahoma Press.
Webster, J. and N. Cooper (eds) (1996), *Roman Imperialism: Post-Colonial Perspectives*, Leicester: Leicester Archaeology Monographs.
Wedel, V. L. and A. Galloway (eds) (2014), *Broken Bones: Anthropological Analysis of Blunt Force Trauma*, 2nd edn, Springfield, IL: Charles C Thomas.
Wellcome Osteological Research Database (2020), *Roman Data*, London: Museum of London.
Wenham, L. P. (1968), *The Romano-British Cemetery at Trentholme Drive, York*, London: Her Majesty's Stationery Office.
Wesch-Klein, G. (2007), 'Recruits and Veterans', in P. Erdkamp (ed.), *A Companion to the Roman Army*, 435–50, Oxford: Blackwell.
Wesselingh, R. (2008), 'From *Milites Medici* to Army Medics – A Two Thousand Year Tradition of Military Medicine', *Journal of Military and Veterans' Health*, 16 (4): 28–30.
West, C. W. (1969), 'Trophies of the Persian Wars', *Classical Philology*, 64 (1): 7–19.
West, M. L. (2003), *Greek Epic Fragments from the Seventh to the Fifth Centuries BC*, Cambridge, MA: Harvard University Press.
Weststeijn, J. (2016), 'Wine, Women, and Revenge in Near Eastern Historiography: The Tales of Tomyris, Judith, Zenobia, and Jalila', *Journal of Near Eastern Studies*, 75: 91–107.
White House (2017), 'Military Service by Transgender Individuals', Presidential document, *Federal Register*, 82 (167): 41319–20.
Whitney Azoy, G. (2003), *Buzkashi: Game and Power in Afghanistan*, London: Waveland Press.
Whittaker, D. (2009), 'Ethnic Discourses on the Frontiers of Roman Africa', in T. Derks and N. Roymans (eds), *Ethnic Constructs in Antiquity*, 189–205, Amsterdam: Amsterdam University Press.
Wickert-Micknat, G. (1983), *Unfreiheit im Zeitalter der homerischen Epen*, Wiesbaden: Franz Steiner.
Widengren, G. (1956), 'Some Remarks on Riding Costume and Articles of Dress among Iranian People in Antiquity', in A. Furumark (ed), *Arctica: Essays Presented to Ake Campbell*, 228–76, Uppsala: Almqvist & Wiksell.
Wilcox, P. (1986), *Rome's Enemies: Parthians and Sassanid Persians*, Oxford: Oxbow.
Wild, J. P. (1984), 'Some Early Silk Finds in North-west Europe', *The Textile Museum Journal*, 23: 17–23.
Wilkes, J. J. (1969), *Dalmatia*, London: Routledge.
Williams, C. (1999), *Roman Homosexuality: Ideologies of Masculinity in Classical Antiquity*, Oxford: Oxford University Press.
Wilmott, T. (2010), 'Birdoswald Roman Cemetery', *Hadrian's Wall Archaeology*, 1: 43–6.
Wilson, E. (2021), 'Slaves and Sex in the *Odyssey*', in D. Kamen and C. W. Marshall (eds), *Slavery and Sexuality in Classical Antiquity*, 15–39, Madison, WI: University of Wisconsin Press.
Wise, J. E. (2019), 'Loss of Moral High-ground: The Transgender Ban, a Military Psychiatrist's Perspective and Call to Action', *Journal of Gay & Lesbian Mental Health*, 23 (2): 114–16.
Wiseman, R., B. Neil and F. Mazzilli (2021), 'Extreme Justice: Decapitations and Prone Burials at Three Late Roman Cemeteries at Knobb's Farm, Cambridgeshire', *Britannia*, 52: 119–73.
Woelcke, K. (1911), *Beiträge zur Geschichte des Tropaions*, Bonn: Universitäts-Buchdruckerei Verlag.
Wood, S. E. (1999), *Imperial Women: A Study in Public Images 40 BC–AD 68*, Leiden: Brill.
Woodward, R. and T. Winter (2007), *Sexing the Soldier: The Politics of Gender and the Contemporary British Army*, London: Routledge.
Woolf, G. (1998), *Becoming Roman: The Origins of Provincial Civilization in Gaul*, Oxford: Oxford University Press.

Wyke, M. (ed.) (1998), *Gender and the Body in the Ancient Mediterranean*, Oxford: Blackwell.
Xinyue, B. (2017), 'Imperatrix and bellatrix: Cicero's Clodia and Vergil's Camilla', in D. Campanile, F. Carlà-Uhink and M. Facella (eds), *TransAntiquity*, 164–77, London: Routledge.
Yamagata, N. (1994), *Homeric Morality*, Leiden: Brill.
Yarshater, E. (ed.) (1983), *The Cambridge History of Iran 3 (1): The Seleucid, Parthian and Sasanian Periods*, Cambridge: Cambridge University Press.
Yatsenko, S. (2014), 'Some Notes on Parthian Costume Studies', *Parthica*, 15: 117–25.
Yi-Chang, Y. (2006), 'A Study of Men's Costume of the Parthian Period in Persia', *Journal of the Korean Society of Costume*, 56 (4): 65–81.
Zaninović, M. (1996), *Od Helena do Hrvata*, Zagreb: Školska knjiga.
Zientek, L. (2021), 'The Problems with Agricultural Recovery in Lucan's Civil War Narrative', in B. Reitz-Joosse, M. W. Makins and C. J. Mackie (eds), *Landscapes of War in Greek and Roman Literature*, 91–110, London: Bloomsbury Academic.
Zink, G. and A. Kwaspen (2015), 'The Dätgen Trousers', in K. Grömer and F. Pritchard (eds), *Aspects of the Design, Production and Use of Textiles and Clothing from the Bronze Age to the Early Modern Era*, 175–84, Budapest: Archaeololingua.

INDEX

Page references in bold direct the reader to photographs, illustrations, tables or other images. Texts appear under the name of their author (e.g. the *Iliad* appears under Homer), and generally people (as opposed to mythological or mythologized figures or characters) appear as top-level headings throughout. An exception has been made for soldiers (grouped together under 'combatants' and then variously by nationality, rank, role and/or profession), and another for those authors who produced military manuals (grouped together under 'military manuals').

Aeschylus
Agamemnon 38
Eumenides 34
agriculture *see* warfare and wars, nature, relationship with
Algeria 49
Apollo 33–4
archers *see* combatants, Parthian; combatants, weapons, bow-and-arrow
Aristotle 30
Arrian 148

Barthes, Roland 83–4
battles *see* warfare and wars, battles and skirmishes
bioarchaeology *see also* injuries and wounds 8–9, 45, **46**, 57–72, 177–80
body politic (Menenius Agrippa) 12n6
Britain (Roman)
Ambleside (Cumbria) 50
Bath 53
Caerleon 51, 57
Chester 53–5, 57
Cirencester 58, **59**, 68, 71
Colchester 54–5, **59**
Hadrian's Wall (Vindolanda) 4, 51, 57–8, 196n19
Leicester 58–**9**
Lincoln 51
London 53, 58
York 57–8, **60**, 63, **64**
Winchester 58, **66**
Butler, Judith 122, 126–8

Capua *see* injuries and wounds, disabling, prostheses
centaurs 31, 146–53
Cicero 1
combatants
Amazonian 131, 147–8, **149**, 151–3
Penthesilea 134n15, 136n37
Anglesey, Marquess of 183–5
Artemisia II 142–3
armour, military dress and kit *see also* combatants, weapons **46**, **47**
belts 102–9, **104**, 131, 180
binoculars **27**, 28
brooches 106–7, **106**, 112
chlamys 141
cloaks 106–9, 114–15, 146, 193
ceremonial 181
cuirasses 109, 112, 146, 194
iconography 99
gaps in (penetration of) 34, 147–8, 156–7n11
gauntlets, lack of 182
greaves 109, 146, 180–3
helmets 99, 109–11, 141, **145**, 146, 162–3
loss and recovery of 42n38, 144, 165
nakedness 139–41, 146, 155, 165
Rees-Hall escape apparatus **17**, **20**, **21**, 28
riding dress *see* combatants, Parthian
sarissa 147
scabbards *see also* combatants, weapons, swords 103–5, 155
shields 109, 130, 144–6, 153–5, 179, 182
stirrups, lack of 80
textiles
cotton 82, **82**, **86**, 89
felt 80–2, **82**, 85, 89
hides and furs 80, 87–9, 112, 128
nylon 89
silk 82–6, **82**, 89, **89**, 93–5
wool (wool basket) 85, 89, 124–5
tunics 103–4, 153
Babylonian 31–2
body schema of (bodily boundaries, extensions) 91–3, **92**, **93**, 99, 128, 177, 182–5, 190
cavalry (mounted) *see* combatants, non-human; combatants, Parthian

(*combatants cont.*)
camps and garrisons 6–7, 49–53, 179
Caerleon 57
Camp Bastion 185
Chester (Legio XX Valeria Victrix) 53–4
Tilurium (Dalmatia) 100–9, 113–15
Salona (Dalmatia) 101, 107
York 57
camp-followers (wives and children) 72n2, 102
Carthaginian 4
Hannibal 179, 193
craftsmen (armourers, textile artisans) 78, 102, 105–7, 180–4, 191
Cleopatra VII 124
Dalmatian *see* Dalmatia
death of *see also* war and battles, representations of
age at 50–6, **55**, **56**
burial 36, 45, **47**, 50–1, 57–8, 61, 71
cost of 108
lack of 157n23
corpses, reclaiming of and respect for 30–2, 49, 144, **154**
cremated 51, 58
devoured or scavenged 32–4, 39, 41n26, 42n34, 173n49, 192
epitaphs and memorials (*stelae*) 45, 49–56, **50**, **52**, **55**, **56**, **57**, 107–9, **110**, 141, 191
execution 73n9, 151
exhaustion 193–4
funerals and other rituals 33, 58, 78
glamorization of ('beautiful death') 30, 141, 155
suicide 6, 14n24, 38
defeated
barbari 111–14, **111**
humiliation of *see* injuries and wounds; warfare and wars, second-phase
prisoners of war 30, 112–13, 175–9, 194n1
tropaea (*tropaia*, trophies) 110–15, **111**, 144, 151
deserting 31, 49, 156n6, 175
disloyal 31, 36
drunk 31, 37–8
Egyptian 155
elite *see also* combatants, leaders of 3–5, 79
enslaved *see* warfare and wars, second-phase, enslavement
Gaulish 105, 144, 179, 193
gender identity of
cross-dressing 125, 131
femaleness and femininity *see also* Virgil, *Aeneid,* Camilla 10–11, 107, 120, 123–30, 150, 181, 190
maleness and masculinity *see also* combatants, Roman, *virtus* 11, 37, 48, 107, 119–33, 145–6, 169n5, 182–4, 190
transgender, intersex and non-binary bodies 121–2, 128
gladiators *see* gladiators
Greek
Achaean *see also* Homer, *Iliad* 32–4, 161, 164–7
Alexander III 140, **140**, 147–8, 151, 196n22
Athenian 141–4, 151–5
Eumenes 142, 156n9
heroism of (semi-divine, *heros*) 139–42, 14–18, 150, 155
hoplites 9, 30, 36, 40n12, 145–7, **145**, 190
phalanx 74n14, 146–7
Pyrrhus 145
Spartan 32, 35, 142–5, 154–5, 156n2, 156n10, 177–8
Leonidas 35–6
Machanidas 147
Pausanias 36
strategoi 144, 147
heritage (Black, diverse, white) **46**, **47**, 61–2, **63**, **64**, 71–2, 73n11, 74n17
human shield 186n3, 192
infantry (rank and file, 'herd') 4, 9, **25**, 30, 40n12, 103
Jewish 6–7
Joan of Arc 125
leaders of
memorialization of 51–3, 99
mutilation of, symbolic *see also* injuries and wounds, mutilation 29–39, 43n52, 165, 190
medical staff *see* injuries and wounds, treatment
mercenary 36, 146
mutilation of *see* injuries and wounds
non-human (military animals)
elephants 13n23
horses *see also* combatants, Parthian 14n28,
dead 7, 49
rider, relationship with 14n28, **90**, 91, **93**, 147, 180–1
riding dress *see* combatants, Parthian
unsuitable terrain for 194
mules and donkeys 7
oxen and wagon teams 7
Numidian 6
Parthian
Shami Prince, the 9, 77–8, **78**, 81–90, **81**, **83**, **84**, **85**, **86**, **87**, **89**, **96**
surena 79
warrior-archers *see also* combatants, weapons, bow-and-arrow 77–80, 131

(*combatants cont.*)
Persian 31–2, 35, 144, 154–5
Cyrus (the Great) 37–9, 43n62, 190
Darius 31–2
Mardonius 36
Xerxes 35, 154
Zopyrus 31–2
physique of *see also* combatants, gender identity of 2, 9, 48, 120–2, 146
punishment of *see also* injuries and wounds 3, 30–2, 49, 58, 105, 151, 175
ownership of (soldiers as owned by the state) 5, 140–1
ransoming of 30–1
Roman
army (*exercitus*) 1–2, 7, 102
Augustus (Octavian) 13n23, 101, 112, 123
auxiliaries 53–5, 101
centurions 51, 53–4, **55**
Cestius Gallus 6
Germanicus 194
Julius Caesar 2, 49, 125
legion 48, 51–8, 101–14
Marcus Sergius Silus 178–9, 182–3
Scipio (P. Cornelius Scipio Africanus Maior) 7
Tiberius (emperor) 101, 113
virtus see also combatants, gender identity of 48–9, 54, 72, 119, 123–7, 130
Vitellius 105
Seary, Ryan 185
Spargapises 37–8
tattoos and bodily decoration 3, 10
training and routine of ('fitness', discipline) *see also* military manuals 2, 5, 18–19, **21**, **28**, 30–4, 48, 126
Tomyris, Queen of the Massegetae 29, 37–9, 190
Trojan *see also* Homer, *Iliad*; Virgil, *Aeneid* 33, 123–5, 128–9, 132
unwilling *see* warfare and wars, second-phase
veteran 11, 12n4, 18–19, 49, 53, 102, 176, 184–9
Valerius (veteran legionary) 45
weapons and ammunition *see also* combatants, armour, military dress and kit; warfare and wars, war machinery **46**, **47**
axes 125, 130
bows-and-arrows (quivers) 80, 91, 125–31, 147, 181
daggers and knives (*pugio*) 1–3, 91, 182–3
javelins **25**, 127–30
lances 80
looted 6, 36
slingshots 127–30
spears 34, 127–30, 132, 146–50, 155, 181
stones 146, 152
swords **25**, 103–5, 129–30, 142, 146–7, 155, 179
torpedoes **24**, 28
Cyprus 35, **88**

Dacia 109
Diodorus Siculus 150–2

Euripides
Andromache 160
Hecuba 42n41, 160
Ion 170n20
Trojan Women 150, 160–2, 166, 170n16, 174n61

Gaul *see* Roman Empire, Gaul
Gaca, Kathy 11, 30, 159–60
giants 146–8, 154
gladiators 3, 61, 73n16, 135n35, 181, 187n13
Greece *see also* combatants, Greek
Athens *see also* warfare and wars, Peloponnesian War 142–3, 146–7, 151, 175
Corinth 155
Sparta *see also* combatants, Greek, Spartan; warfare and wars, battles, Thermopylae 5, 32, 35–6, 142

Hadrian's Wall *see* Britain (Roman), Hadrian's Wall
Hegesistratus of Elis (diviner) 178, 183
Herodian 116n15
Herodotus 31–2, 35–8, 146–7, 153–4, 170n14, 177–8
Hesiod 147, 155
Hippocratic Corpus (Hippocrates) 124, 176
Homer
Iliad
Achaean army *see* combatants, Greek, Achaeans
Achilles 29, 33, 39, 144, 153, 163
Agamemnon 32, 35, 38
Amazonian army *see* combatants, Amazonian
Andromache 161–8
Astyanax 169n3
artistic representations of 150
childhood 165
future of, imagined 159–64
murder of, foreshadowing 166–8
Hector 146, 153
death of 162–8
postmortem mistreatment of 29, 33–4, 153
Hecuba *see also* Euripides 33–4, 39
lament (*homilia*) 166–7
Menalaeus 32, 35
Priam 33–4, 165–6
Trojan army *see* combatants, Trojan
Odyssey 31, 141, 193

illness and disease *see also* injuries and wounds
cataract 48
conjunctivitis 57
frostbite 175
infection (gangrene) 175–6
malaria 57
plague (smallpox) 57
trench foot 175
Illyricum
Dalmatia 99–100, **100**
Andetrium 113
Salona (Solin) *see* combatants, camps, Salona 101, 107
Tilurium *see* combatants, camps, Tilurium
Pannonia 101, 109
injuries and wounds *see also* illness and disease
amputation *see also* injuries and wounds, disabling 31, 58, 134n7, 176, 183–5, 186n3, 192
ankylosis 61, 64, 71
concealed, invisible 2, 184–5
cranial and dental *see also* injuries and wounds, mutilation, facial 61, 68, **69**, **70**, 73n9, 148–50
disabling 4, 11, 71, 121, 142, 191
armour, adapted 181
crutches 180, 184
cyborg *see* combatants, body schema of
eyes, loss of 175, 186n3, 192
hands and fingers, loss of 175–9, 182
legs, feet and toes, loss of 175–6, 180–1, 185
medical discharge (*missio causaria*) 186n2
prostheses 178–84, 187n10, 191
Capua Limb, the 181–3
dislocation 61, 64
fractures 45, **46**, 61, 64, 68–9, **69**, 71, 181
manual (injuries to the hands) *see also* injuries and wounds, sidedness 148, 175–9, 182, 186n3
metaphorical significance of *see also* combatants, leaders, mutilation of 2, 30–3, 38
mutilation ('despoiling') *see also* warfare and wars, second-phase 7, 30, 49
amputation *see* injuries and wounds, amputation
branding 31, 151
decapitation 7, 29, 34–9, 73n9, 153, 157n21, 190
facial (defacement) 1, 7, 31–2, 150–1, 157n21, 188n16, 191
genital (castration) 31–2, 123, 148, 160
postmortem (lack of burial, dragging, display) 29, 33–4, 153, 174n65, 190
self-inflicted 31, 175, 178
symbolic, shameful *see also* combatants, leaders, mutilation of, symbolic 31–2, 153
thighs, wounds to 43n52
'un-Greek-ness' of 36
pain 9, 139, 147–52
psychological 3, 176, 185, 189, 194n2
recidivism (repeated trauma or injury) 45, 58–63, **59–60**, **62**, **65**, **66**, **67**, 71–2, 182
scarring 31, 49, 71
Schmorl's nodes 181
sidedness 45, 68, **70**
torture 49, 151–2, 175
treatment of
Eutyches (surgeon) 176
hospitals 18
medical techniques 183–4
rehabilitation 11, 176, 186n1

Josephus, *Jewish War* 6–7
Juvenal 105

Lampon of Aegina 36
Little Iliad 171n28, 171n30
Livy 4–6, 124, 193
Lucan 186n3, 192, 195n13–14

Macedonia (Vergina) 180–1
McSorley, Kevin 5–7, 30
miltary manuals
Asclepiodotus, *Tactics* 4
Frontinus, *Stratagems* 4
Polyaenus, *Strategems* 44n67
Vegetius, *Epitome of Military Science* 2–3, 17–19, **21**, **25**, 48, 189, 192

National Museum of the Royal Navy 17–18

Ovid, *Metamorphoses* 14n27, 135n31, 135n35

Parthia *see* combatants, Parthian
Phrynichus, *The Capture of Miletus* 150–1
Plato
Gorgias 31
Republic 42n45, 124
Pliny the Elder, *Natural History* 178–9, 182
Plutarch 153–4
Polybius 3–4, 144, 151

rape *see* warfare and wars, second-phase, rape
Roman Empire *see also* Britain (Roman); combatants, Roman 7, 101, 120, 134n19
provinces *see also* Dacia; Illyricum 6, 107, 190
Gaul *see also* combatants, Gaulish 49, 105, 193
Judea 6

Sallust, *Histories* 2
Savaria 51
Skopje 49
slavery *see* warfare and wars, second-phase, enslavement
Sophocles, *Antigone* 36, 41n26
Spinoza 14n28

Tacitus
Agricola 2
Annals 49, 194
Histories 105, 194
textiles *see* combatants, armour, military dress and kit, textiles
Thucydides, *History of the Peloponnesian War* 152, 170n14
Tyrtaeus 32

Valencia 49
Vegetius *see* military manuals, Vegetius
Veneto 182
Vernant, Jean-Pierre 32
Vindolanda *see* Britain, Roman, Hadrian's Wall
Virgil, *Aeneid* 119–21
Aeneas 126
Camilla 123–33
clothing and armour 123, 129–31
death 131
fighting style and weapons (*bellatrix*) 123–31
gender 11, 123–8, 132–3
training 127–8
Diana 127, 129–30, 135n25, 135n34
Trojan army *see* combatants, Trojan
Turnus 119, 131
Vitruvius, *On Architecture* 9

warfare and wars
Afghan wars 185
American Civil War 184
Batonian war (*bellum Batonianum*) 101, 112–13
battles
Aquae Sextiae 153–4
Babylon, capture of 31
Cannae 6–7
Carrhae 79–80
Chaironeia 175
Coronea 155
Granicus 148
Hormozdagān 97n5
Jutland **26**, **28**
Lake Trasimene 193
Marathon 52, 154
Massilia (naval) 192
Oinoe 152
Pelusium 155
Plataea 29, 36
Rhegium 155
River Metaurus 13n23
Tanagra 144
Teutoberg Forest (Varian disaster) 49, 194
Thermopylae 35, 154
Waterloo 183
Carthaginian (Punic) wars *see also* combatants, Carthaginian; warfare and wars, battles, Cannae; warfare and wars, battles, Lake Trasimene 4–7, 182
First World War **25**, **26**, **28**
Graeco-Persian Wars *see also* warfare and wars, battles, Thermopylae 144
Jewish War 6
Napoleonic Wars 183–4
nature, relationship with 152–6, 195n16
Pannonion war 101
Peloponnesian War 139, 151, 142
representations of
artistic 17–19, **17**, **20**, **21**, **22**, **23**, **24**, **25**, **26**, **27**, **28**, **140**, 144–56, **154**, 172n32, 189
poetic *see also* Homer; *Little Iliad*; Lucan; Ovid; Tyrtaeus; Virgil 4, 7, 172n32, 189, 193
war memorials and funerary artwork *see* combatants, death of, epitaphs and memorials
second-phase *see also* combatants, defeated; injuries and wounds, mutilation 11, 191
displacement and refugees 7
enslavement 11, 30–1, 61, 72, 150, 159–68
murder, of civilians 30, 49, 150, 159–68
rape 30, 150, 159–64, 167–8, 169n6, 169n7, 169n11
Second Samnite War 180–1
Second World War **24**
siege 8, 18, **143**, 151–2, 179
Trojan War *see* Homer, *Iliad*
war machinery *see also* combatants, armour and kit; combatants, weapons
catapults 157n18
chariots 34, 153
siege engines 179
ships and submarines 17–19, **17**, **20**, **21**, **24**, **25**, **26**, **28**, 189, 192

Xenophon 35–6, 142, 155

www.ingramcontent.com/pod-product-compliance
Ingram Content Group UK Ltd.
Pitfield, Milton Keynes, MK11 3LW, UK
UKHW020807290926
15470UKWH00003B/999

9 781350 240858